Planning, Environment, Cities

This series is primarily aimed at students and practitioners of planning and such related professions as estate management, housing and architecture as well as those in politics, public and social administration, geography and urban studies. It comprises both general texts and books designed to make a more particular contribution, in both cases characterized by: an international approach; extensive use of case studies; and emphasis on contemporary relevance and the application of theory to advance planning practice.

* Andrew Thornley was series co-editor with Yvonne Rydin up to his retirement from the role in January 2017.

More information about this series at
https://link.springer.com/bookseries/14300

Helen Pineo

Healthy Urbanism

Designing and Planning Equitable, Sustainable and Inclusive Places

Helen Pineo
IEDE
University College London
London, UK

ISSN 2946-9589 ISSN 2946-9597 (electronic)
Planning, Environment, Cities
ISBN 978-981-16-9646-6 ISBN 978-981-16-9647-3 (eBook)
https://doi.org/10.1007/978-981-16-9647-3

This Palgrave Macmillan imprint is published by the registered company Springer Nature Singapore Pte Ltd.
The registered company address is: 152 Beach Road, #21-01/04 Gateway East, Singapore 189721, Singapore

Acknowledgements

This book is the product of experience and knowledge gained over the past 15 years in my career, spanning several jobs in public and private sector organisations, most recently in higher education. I'd like to start by thanking the people who have created opportunities for me to research healthy urbanism, particularly Peter Bonfield and Deborah Pullen, when we worked together at the Building Research Establishment, and Mike Davies and Yvonne Rydin at University College London. Beyond funding, they have each supported me with advice and encouragement to do research and share it with wide audiences.

The empirical research that informed this book involved many people working in the built environment and public health internationally. I would like to acknowledge the research participants' valuable contributions of knowledge and time. Collaborators on this research also deserve a great deal of thanks, especially Gemma Moore who was an indispensable partner in conceptualising, conducting and disseminating several qualitative studies used in this book, including the integration of THRIVES in postgraduate and professional training programmes. Thanks also to the team of multi-disciplinary researchers who contributed considerable knowledge, skills and enthusiasm, including Karla Barrantes Chaves, Isobel Braithwaite, Rosalie Callway, Elizabeth Cooper, Vafa Dianati, Kay Forster, Yuhong Wang and Ke Zhou.

Funding for the research underpinning this book is primarily from a research consultancy grant from Guy's & St Thomas' Foundation. The project followed a transdisciplinary approach and thanks are owed to Kieron Boyle, Gail Macdonald and Emily Oliver (alongside their partners) for working closely with me and Gemma in pursuit of generating new understanding to improve urban health. Another key source of funding was the Wellcome Trust–funded project Complex Urban Systems for Sustainability and Health (grant number 209387/Z/17/Z). Thanks to Mike Davies for involving me in the project and to the many co-investigators and collaborators involved in our research on transdisciplinarity and evidence use, particularly Melanie Crane, Joanna Hale, Qiyong Liu, Susan Michie, Gemma Moore, Yanlin Niu, David Osrin, Paul Wilkinson, Catherine Willan, Ke Zhou, Nici Zimmermann and José Siri at the Wellcome Trust.

I am indebted to many collaborators and professionals who have informed this book through giving tours, editing case studies and sharing experiences from their own cities about healthy urbanism. Fellows on the Salzburg Global Seminar session on Building Healthy, Equitable Communities in partnership with the Robert Wood Johnson Foundation were inspirational and shaped my thinking about systemic barriers to health. A group of the fellows provided invaluable feedback and support for THRIVES and this book, including Waleska Caiaffa, Michael Chang, Nupur Chaudhury, Evelyne de Leeuw, Frederik Leenders, Gemma McKinnon, Júnia Naves Nogueira, Tolullah Oni, Sharon Roerty, Melani Smith, Laura Taylor-Green, Rachel Toms and Kelly Worden. I'm also grateful for collaboration, tours and editing from Heather Burpee, Rod Duncan, Brad Kahn, Nat Kendall-Taylor, Jin Lim, Dan McKenna, Tony Mulhall, Paul Southon, Julia Thrift and Marja Williams. Marcus Grant, editor-in-chief of *Cities & Health*, has encouraged my work through much appreciated feedback and encouragement.

Thanks are owed to the Palgrave Macmillan book series editors, anonymous peer reviewer and editors for their encouragement and advice. I am grateful to my family, particularly Hazel and Felix for their patience on many 'site visits' during our holidays, and to Rob for reminding me to 'believe'. Finally, it must be acknowledged that many of the activities described earlier occurred during the difficult COVID-19 pandemic and I am all the more grateful for the time and support generously given to progress this work.

Contents

List of Figures

List of Tables

List of Boxes

Chapter 1
Introducing Healthy Urbanism

1.1 Introduction

In the middle of the twentieth century, the Western Harbour in Malmö, Sweden, was a port and the world's largest shipyard. Now the area is a global exemplar of sustainable urbanism. In the mid-1990s, the city began planning to redevelop this harbourside area from industrial uses to a mixed-use community based on sustainability principles. The first phase of detailed planning and construction was a Millennium Village demonstration project named Bo01, covering 22 hectares of land with over 2000 residents (Rosberg, n.d.). This well-known case study of sustainable development showcases the health and wellbeing benefits of compact and mixed-use development, meaning that homes, restaurants, offices and other uses are interspersed and in close proximity to each other.

A healthy urban development considers residents' needs from a holistic perspective of social, economic and environmental parameters. At Bo01, this is visible through diverse public spaces of different scales that support multiple uses (Fig. 1.1). Pedestrians and cyclists are clearly prioritised over motor vehicles through the design of small lanes to access housing and the provision of pedestrian and cycling paths. Waterside parks support residents of all ages to rest, exercise and play. Housing of diverse architectural styles and materials spark interest in what can be found around the next corner. The location, scale and orientation of buildings shelter residents from wind and noise. Water and green landscaping permeate the site, providing sustainable drainage and wellbeing benefits.

The achievement of sustainable and healthy urban design at Bo01 was supported by strong leadership from the city's planning department and collaboration with private sector developers. There were 20 developers involved, working with over 30 different architectural firms (Rosberg, n.d.). The city planned for environmental sustainability through energy and waste systems, soil remediation, green and blue infrastructure and mobility systems. In a time when decentralised energy systems

H. Pineo, *Healthy Urbanism*, Planning, Environment, Cities,
https://doi.org/10.1007/978-981-16-9647-3_1

Fig. 1.1 Housing diversity, shared street space and blue infrastructure examples at Bo01, Malmö, Sweden. (Source: Author, 2013)

were not typical on urban regeneration projects, Malmö achieved 100% energy production from wind, solar panels, water source heat pumps (used for heating and cooling) and biogas. Housing was built to low energy and passive house standards, which were later applied throughout Malmö (Medved, 2018). The collaboration between public and private sector partners continued to further phases of regeneration on the Western Harbour, including Varvsstaden and Masthusen, and supported the achievement of further health and sustainability principles. Both of these phases underwent certification through the British system for sustainability assessment at the neighbourhood scale, BREEAM Communities, and Masthusen was the first international project to be certified (BRE, 2016).

The vision and leadership demonstrated by Malmö city planners and politicians in the redevelopment of the Western Harbour is remarkable. Their success was aided by a collaborative relationship with private sector partners in architecture, design and engineering. Despite its achievements, the project has been criticised for failing to create a sense of community identity through participatory development approaches and lack of sufficient affordable housing, leading to a socially exclusive area (Medved, 2018). Thus, whilst environmental sustainability objectives may have been achieved, issues of equity and inclusion may have fallen short. There will always be criticisms of such large-scale urban developments because these projects inevitably involve compromises. The actors involved will encounter barriers to achieving healthy and sustainable urbanism, including development economics and lack of political will, and negotiated solutions must be found that may not please all parties. In the early 2020s, nearly 30 years after planning began on this waterside regeneration project, it is clear that sustainable and healthy urban development is not a given for any project; it is a goal that has to be actively pursued with creative thinking, determination and community involvement. Today's most pressing global challenges, such as the climate crisis and widening social inequalities, require built environment professionals to make the environmental achievements of the Western Harbour the new norm in urban development, and to act upon the lessons from these early exemplars to ensure achievement of holistic sustainability objectives, addressing social and economic needs.

1.1.1 Towards Healthy Urbanism

This book is structured around a new framework for designing and planning healthy places called the THRIVES framework, which stands for 'Towards Healthy uRbanism: InclusiVe, Equitable, Sustainable'. The framework was introduced by Pineo (2020) following a process of participatory development, as described by Pineo et al. (2020). Figure 1.2 depicts the conceptual framework visually. On the top are three scales of health impact (planetary, ecosystem and local) that have spatial and temporal dimensions. The three core principles (sustainable, equitable and inclusive) intersect the diagram with visual representations of interconnected environmental scales (from the planet at the centre to buildings on the outer edge). On the right, there are five urban scales (from regional to buildings) where decisions affecting health are made by built environment and other stakeholders. On the left, there are a selection of evidence-based design and planning goals that broadly align to the scales of health impact and decision-making, with recognition that these scales are fluid and goals can be achieved through action at multiple scales.

The THRIVES framework makes three key arguments about the challenges and solutions for healthy urbanism that are elaborated throughout this book. First, the health impacts caused by new development often occur away from the site, both in space and time. This means that design teams, planners and developers need to think beyond the 'boundaries' of development. Second, there are structural barriers that prevent people from living a healthy life. People are often constrained by their socio-economic circumstances or other characteristics (e.g. race, age and gender) that significantly reduce their access to healthy homes, neighbourhoods and workplaces. Resources should be targeted towards communities with the greatest environmental burdens and least ability to affect change. The distribution and use of these resources should be guided by inclusive processes to create places that will support everybody's health. Finally, environmental degradation is affecting health now through increased periods of extreme weather, air pollution and many other

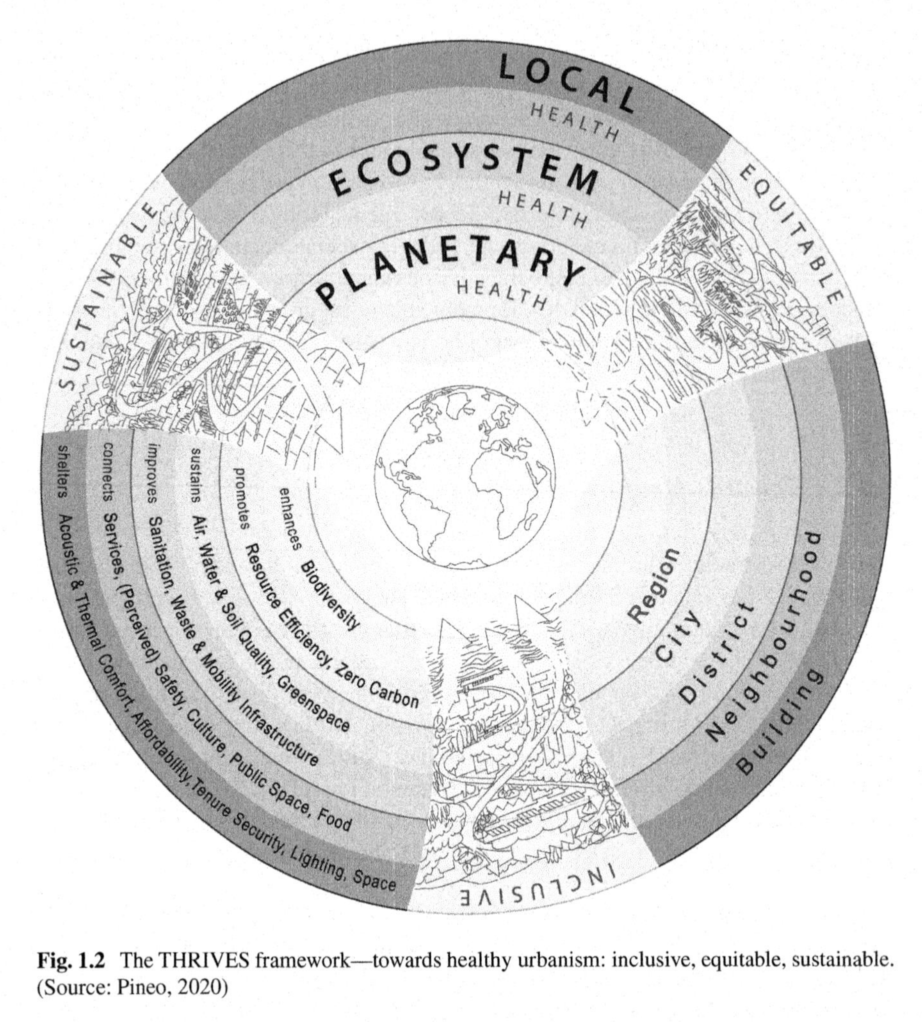

Fig. 1.2 The THRIVES framework—towards healthy urbanism: inclusive, equitable, sustainable. (Source: Pineo, 2020)

impacts. The COVID-19 pandemic highlights the overlapping nature of social and environmental health risks, not least because ecological destruction increases the risk that zoonotic diseases will jump from animals to people. There are well-established sustainable design and construction principles that can be applied at multiple urban scales to support human health, yet sustainability and health are often seen as separate problems and solutions despite these synergies. This book will give further detail about the evidence supporting these three arguments alongside many practical policy and design case study solutions.

1.1.2 Aims and Scope of the Book

This book aims to define and describe healthy urbanism as an approach to design and planning that unites human health and wellbeing with the sustainability of environmental systems. The concept brings together existing evidence and theories related to health, sustainability and urban planning as expressed in the THRIVES framework. The audience for this book includes professionals, students and researchers in built environment and health disciplines. Planners, architects, engineers, surveyors and others who plan and manage urban development will find health concepts explained in straightforward terms and linked to policy and design solutions. Public health specialists, epidemiologists, psychologists and others will benefit from understanding the health impacts of urban environments and how they can collaboratively promote health through different stages of urban planning, design, construction and management. Readers may want to focus on particular sections in the book based on their priorities and background knowledge. A brief summary of the chapters follows and the index also allows readers to navigate to specific topics quickly.

This first chapter introduces key concepts and definitions in urban health to establish a purpose and foundation for the THRIVES model, including an introduction to ecological health models (Rayner & Lang, 2012). Chapter 2 reviews a selection of approaches to healthy and sustainable urbanism, including historical and contemporary initiatives by international organisations. Chapter 3 describes the THRIVES framework in detail with an overview of its underpinning theory and concepts, including systems thinking (Meadows, 2008), ecosocial epidemiology (Krieger, 1994, 2001) and 'just sustainabilities' (Agyeman, 2013; Agyeman et al., 2003). An introduction to epidemiology is also provided to give a general understanding of the evidence base linking health and the built environment that is used in the subsequent chapters.

The following four chapters move sequentially through the framework's three scales of health impact: planetary, ecosystem and local health (the latter is covered at two scales: neighbourhoods and buildings). Chapter 4 describes design and planning measures for enhancing biodiversity and promoting resource efficiency and zero carbon systems. Chapter 5 covers the ecosystem health goals of sustaining air, water, soil and green infrastructure and improving sanitation, waste and mobility infrastructure. Chapter 6 covers local health at the neighbourhood scale, including connecting people to services, culture, public space and food, whilst increasing perceived and actual safety. Chapter 7 focuses on building scale issues (that can also be addressed at other scales), including acoustic and thermal comfort, affordable housing, tenure security and adequate lighting and space. The chapters emphasise that the design and planning goals are interconnected across scales. Each of these chapters provides a set of monitoring indicators that can be used to drive and evaluate impact.

Implementing the principles of healthy urbanism in urban policy-making and design is the focus of Chap. 8. This chapter describes policy processes and development models with insights for overcoming frequently described barriers to healthy urbanism, principally those related to development economics. Chapter 9 looks towards the future of healthy urbanism, describing healthy built environment strategies for disaster preparedness and prevention that will be necessary for future pandemics and extreme weather events. The chapter discusses the importance of framing the topic of health for non-health audiences and it considers the adequacy of relying on incremental change given the scale and urgency of urban health challenges. The book concludes with a discussion of the potential impact of proposed technological innovations and the need for transdisciplinary working to create healthy places.

1.2 Models of Health and Wellbeing

To achieve healthy urbanism, built environment professionals need to have a shared understanding of health and wellbeing, yet this has been elusive for a number of reasons. The factors that determine health and wellbeing are wide ranging and include healthcare systems, individual characteristics (e.g. age, sex and genes) and societally influenced aspects (e.g. socio-economic status, pollution and housing). Health and wellbeing are thus heavily influenced by factors outside of the control of healthcare professionals; however, there are difficulties with estimating the exact contributions of modifiable environmental factors (e.g. housing, air pollution and transport) towards ill health. A World Health Organization (WHO) report attributed '23% of global deaths and 26% of deaths among children' to these modifiable sources (Prüss-Üstün et al., 2016, p. viii). There are competing views about what health and wellbeing mean and how they can be achieved. This section establishes foundational definitions and conceptual models for health and wellbeing.

1.2.1 Defining the State of Health

The most widely adopted definition of health was set out in the preamble of the WHO constitution (1946) and it states, 'Health is a state of complete physical, mental and social well-being and not merely the absence of disease or infirmity' (p. 1). Although this definition was considered to be universal and aspirational at the time, it has been contested for over half a century. The arguments against this definition are summarised by Huber et al. (2011). First, the WHO definition is problematic because individuals may feel healthy even if they have one or more chronic conditions (e.g. diabetes) that would make them objectively unhealthy. A growing number of people have chronic conditions globally, particularly due to the demographic shift of an ageing population, and therefore many people would be viewed as having poor health regardless of their subjective wellbeing and quality of life. The second criticism is that this focus on 'complete' health contributes to the medicalisation of society. Under the biomedical model described later in this chapter, there is a tendency to continually expand the scope of the healthcare sector, including increased screening even when few diagnoses are made and increased production and prescription of drugs for new medical 'conditions' that may have otherwise not caused symptoms or premature mortality. Finally, a 'complete' state of health is seen as problematic because it is not possible to achieve or measure.

Taking into account these debates and existing models of health, Bircher and Kuruvilla (2014) offer a new definition whereby health is 'a state of wellbeing emergent from conducive interactions between individuals' potentials, life's demands, and social and environmental determinants' (p. 363). This perspective, which they term the 'Meikirch Model of Health', accounts for the accumulations of experiences and exposures throughout one's life that affect health, including social and environmental factors. Furthermore, it addresses differences in people's ability to respond to multiple physiological, psychosocial or environmental demands that can lead to poor physical or mental health. This model provides a useful conceptualisation of health and its determinants for this book.

A crucial component of the Meikirch Model of Health is that it conceives of health 'as a complex adaptive system containing ongoing interactions between individuals' potentials, the demands of life, and social and environmental determinants' (ibid., p. 376). This means that health is an 'emergent property' that arises from the interactions between parts of a wider system. It is emergent because it does not necessarily exist on its own, but rather it is created by the interactions between parts of the system (Luke & Stamatakis, 2012). The many interconnections between the urban environment and health are recognised as complex systems that require systems thinking approaches to both understand and manage healthy urban environments (Gelormino et al., 2015; Rydin et al., 2012; Northridge et al., 2003), described further in Chaps. 3 and 8. In Bircher and Kuruvilla's view, the broad conceptualisation of health outlined in the Meikirch Model supports multiple sectors and disciplines to better understand their own roles in promoting health.

1.2.2 *Defining Wellbeing*

This book often refers to health as a catch-all phrase to encompass physical and mental health, alongside wellbeing, because the environmental characteristics that support health significantly overlap with those that support wellbeing. The two states of health and wellbeing are interdependent, but these terms are not interchangeable. Wellbeing is recognised as a 'complex and multi-faceted concept' (Pollard & Lee, 2003, p. 60) that continues to be debated, particularly with regard to its definition and measurement (Dodge et al., 2012; Marsh et al., 2020). Approaches to understanding wellbeing can be divided into *hedonic* and *eudaimonic*. The hedonic tradition focuses on 'happiness, positive affect, low negative affect, and satisfaction with life' (Dodge et al., 2012, p. 223). Critiques of the hedonic tradition highlight that pursuing happiness may actually lead people towards behaviours that are negative for health and happiness over the long term, such as drinking alcohol or gambling. In contrast, the eudaimonic tradition stems from Aristotle's idea of eudaimonia (or wellbeing) and it is focused on 'positive psychological functioning and human development' (ibid.). Marsh et al. (2020) developed a method to measure wellbeing that aims to move beyond these two perspectives. Their approach mirrors the Organisation for Economic Co-operation and Development (OECD) and the WHO, by including aspects of both hedonic and eudaimonic conceptualisations of wellbeing, and like the WHO, Marsh and colleagues equate wellbeing with positive mental health. They outline five constructs that are part of wellbeing: competence, self-acceptance, autonomy (which have an individual focus), empathy and prosocial behaviour (which have an interpersonal focus). However, they stop short of providing a definition for wellbeing.

This book adopts a wellbeing definition developed by Dodge et al. (2012) that attempts to address challenges with the dynamic state of wellbeing. In their view, wellbeing is 'the balance point between an individual's resource pool and the challenges faced' (p. 230). They offer a visual of this definition whereby wellbeing is above the fulcrum of a see-saw, with resources on the left and challenges on the right. Resources and challenges can both include psychological, social and physical aspects. In other words, Dodge and colleagues explain that 'stable wellbeing is when individuals have the psychological, social and physical resources they need to meet a particular psychological, social and/or physical challenge. When individuals have more challenges than resources, the see-saw dips, along with their wellbeing, and vice-versa' (ibid.). This perspective aligns very well with Bircher and Kuruvilla's (2014) definition of health in which people have varying degrees of 'potentials' (akin to resources) to deal with 'life's demands', and in turn their definition builds on Amartya Sen's (1999) idea of capabilities. Sen's approach has been influential in the OECD to develop alternative measures of economic progress that are more focused on quality of life and wellbeing, and also better reflect sustainable development principles (Raworth, 2017; Stiglitz et al., 2017). In summary, this book adopts definitions for health and wellbeing that recognise the important influence of forces beyond an individual (e.g. socio-economic and environmental) that determine whether and how they can respond to life's challenges. This perspective allows urban environment professionals to connect their work to the improvement of health and wellbeing.

1.2.3 Competing Models of Public Health

Just as the definition of health is debated, there are similarly divergent views on the mechanisms that influence health. The biomedical model of health focuses on human (molecular) biology as the source of diseases and cures. George Engel's (1977) critique of this perspective argues that a sufficient model for understanding and treating disease also needs to account for 'the patient, the social context in which he lives, and the complementary system devised by society to deal with the disruptive effects of illness' (p. 196). He went on to propose a biopsychosocial model which broadened scholarly debates, yet did not end the reductionist biomedical perspective which persists today (Farre & Rapley, 2017).

The ecological model of health takes a much broader view of the factors that influence health and wellbeing. It encompasses the learning from early environmental health practitioners and epidemiologists, such as Edwin Chadwick (1842), John Snow (1854) and Sir Benjamin Ward Richardson (1876), whose influential reports linked poor urban environmental conditions with disease. At a time when the medical profession considered disease to be spread through miasma or 'bad air', their work showed that the conditions of British industrialised cities, including air pollution, overcrowding, open sewage, and contaminated drinking water, were responsible for causing or spreading many diseases. Importantly, they also demonstrated that the poor or working-class population were more commonly afflicted by these environmental and associated health conditions, introducing a social justice component to health (see Chap. 2). Although these landmark shifts in knowledge occurred in the mid- to late nineteenth century, the biomedical model prevailed until the 1970s, and still dominates in many disciplines and settings.

The ecological model gained traction among public health professionals following a Pan American Health Organization conference in Ottawa, Canada, in 1973. From this conference the Lalonde Report (1974) set out 'A New Perspective on the Health of Canadians', and it was grounded in the recognition of social, economic and environmental determinants of health. Nearly a decade later, another international conference organised by the WHO produced the Ottawa Charter (WHO, 1986) which sparked a new approach to public health and health promotion, and gave rise to new conceptualisations of health and urban environments, which were strongly informed by British sanitarians (Hancock, 2011). This new approach espoused an ecological (Hancock, 1985) or social determinants model of health (Dahlgren & Whitehead, 1991, 2006) that accounted for individuals' characteristics and wider factors. The social determinants of health are 'the circumstances in which people are born, grow up, live, work and age, and the systems put in place to deal with illness' (WHO, 2012). Göran Dahlgren and Margaret Whitehead's rainbow-shaped model (Fig. 1.3) is still widely used in public health training to represent the important health influence of lifestyles, social networks, working and living conditions and wider socio-economic, cultural and environmental factors. Where the biomedical model is focused almost exclusively on the centre of this diagram (people's age, sex and constitutional factors), the ecological approach takes a much broader view of the factors that affect health, encompassing diet and exercise (so-called lifestyle factors), community networks and macro-level economic conditions.

A key point of differentiation in the biomedical and ecological health models is their respective foci on curing disease and preventing poor health and wellbeing. Public health professionals are concerned with avoiding circumstances that contribute to disease and promoting conditions that lead to good health and wellbeing. In addition to the two models that have been described (biomedical and ecological), Lang and Rayner (2012) identify three further public health models that can be related to urban environments: sanitary-environmental, social-behavioural and

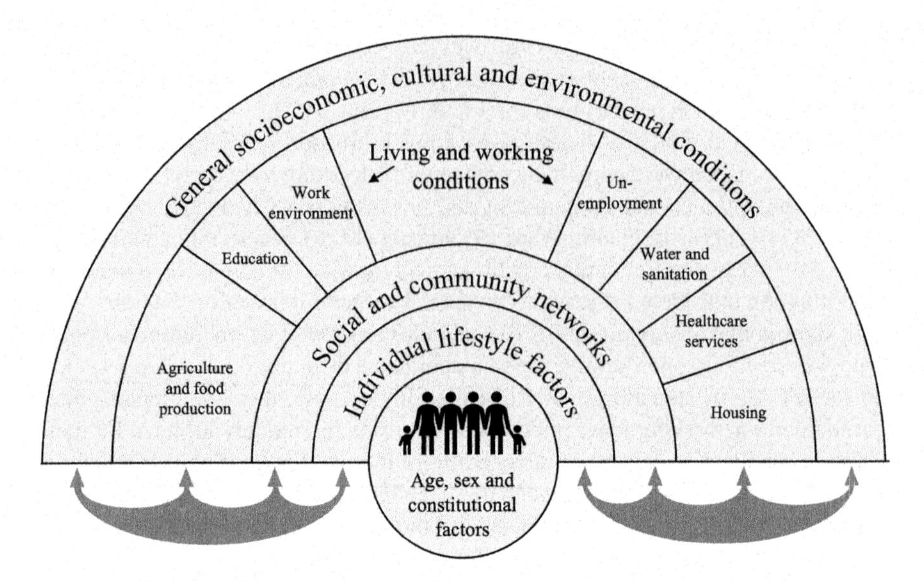

Fig. 1.3 Reproduction of Dahlgren and Whitehead's 'main determinants of health' framework. (Source: Dahlgren and Whitehead, 1991)

Table 1.1 Lang and Rayner's (2012) five public health models with additional examples of how these are expressed by built environment professions

	Sanitary-environmental	Biomedical	Social-behavioural	Techno-economic	Ecological public health
Core idea	'The environment is a threat to health'	'Health improvement requires understanding of biological causation'	'Health is a function of knowledge and behaviour patterns'	'Economic and knowledge growth is prime elevator of health'	'Health depends on successful co-existence of the natural world and social relationships'
Example expression by built environment professions	Modern waste, water and sanitation infrastructure	Medical architecture	Restrictive licensing for alcohol, tobacco and fast-food outlets	Enabling economic growth through urban planning	Holistic models of sustainable development

techno-economic. These competing models are set out in Table 1.1 with example descriptions of how they are expressed by built environment professions.

The different models of health described by Lang and Rayner are understood and expressed by built environment professionals in different ways. Likewise, not all health professionals will have an aligned view of what health means and how it can be influenced. Individuals may adopt biomedical approaches, whilst still being aware of the importance of ecological or social-behavioural models. This book argues for adoption of an ecological public health perspective, yet this does not dismiss the need for biomedical research and practice. Many of today's urgent challenges are related to health, and they will not be addressed by the health sector alone, but require action from house builders, planners, food companies, transport services and other sectors.

1.3 Global Population Trends

As centres of public life and institutions, economic engines, and cultural melting pots, cities have long been focal points for the intersection of health and the built environment. With most of the global population now living in cities, they are the point of convergence for trends related to population ageing, poverty and health. These interrelated trends are essential background to understand why cities are paradoxically considered to be causes of ill health through concentrating pollution and crime and places of good health through access to healthcare and jobs.

1.3.1 Urbanisation

The rapid increase of people living in cities is a significant trend influencing health and the quality of the urban environment. The United Nations' (2019a) statistics on world population changes show that 55% of the global population lived in urban areas in 2018. Roughly half of those urbanites lived in settlements of less than 500,000 and only 7% lived in megacities of 10 million or more. However, megacities are growing quickly with their population increasing by 3.5 times between 1990 and 2018. Although urban growth is a global phenomenon, it is more pronounced in less developed regions, such as Africa and Asia. In 2018, urban populations in less developed regions were three times higher than more developed regions (e.g. Europe and North America) equating to 3.2 billion and 1 billion residents, respectively (Fig. 1.4). In future, cities are predicted to continue growing, albeit at a slower pace, with urban populations set to reach 5.6 billion by 2050, with 83% living in less developed regions and 17% in more developed regions (accounting for 87% and 13% of the world population, respectively). These trends prompt geographers to say that the future is urban, yet there are many questions about what this shift to city living will mean for people's economic circumstances and quality of life.

The driving forces of urbanisation are debated; they vary across settings and they are related to health. It is difficult to accurately estimate urban populations because the definition of 'urban' varies considerably. The UN (2019a) attributes urban population growth to three factors: natural increases (more births than deaths in cities), migration (from rural areas or other countries) and reclassification (e.g. where land adjacent to an existing urban area is incorporated into the city). Each of these factors

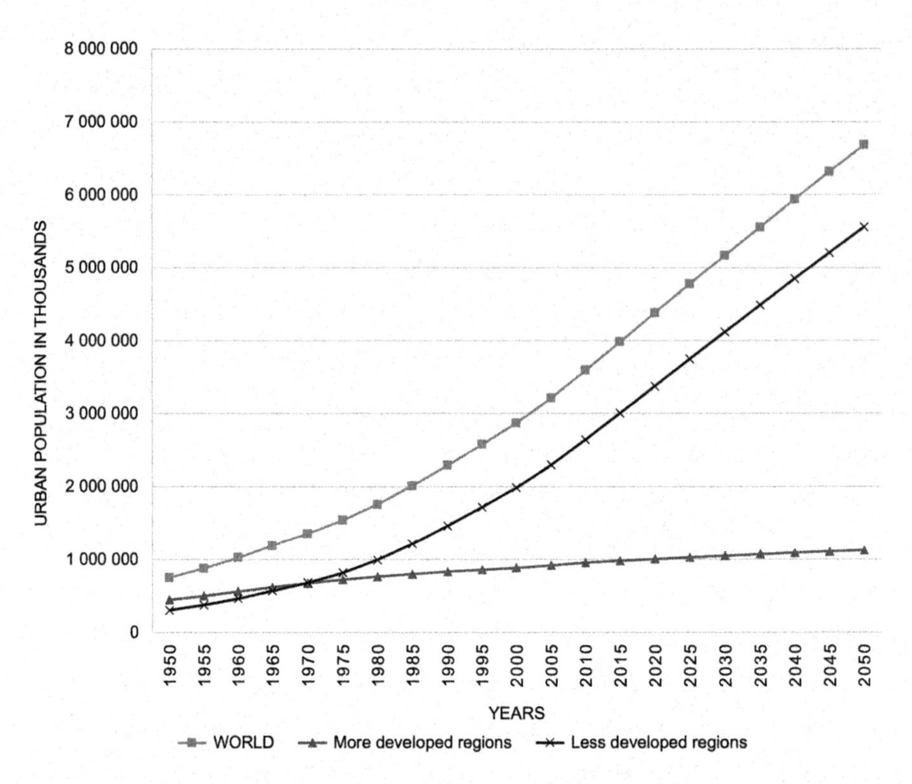

Fig. 1.4 Urban population in thousands (1950–2050) in more and less developed regions. (Source: United Nations, 2019a)

is explained, in turn, with a brief summary of how they relate to health and how they vary internationally.

Natural Population Changes

Natural increases in the urban population of developing countries are partly caused by high fertility and low mortality rates (Jedwab et al., 2017). Fertility levels vary internationally, for example, with high rates in sub-Saharan Africa, where there are roughly 4.72 live births per woman, falling to low rates of 1.66 in Europe and Northern America and 1.84 in Australia and New Zealand (UN, 2019b). Countries with higher rates of fertility are experiencing increasing urbanisation, such as in Africa and Asia. Conversely, countries with low rates of fertility, such as those in Europe and North America, have had primarily urban populations for decades. For example, the USA's urban population rose from 64% to 82% between 1950 and 2018 (ibid.).

The other component of natural population increase, low mortality, is explained by the complex process of *epidemiological transition*, in which improvements to

Table 1.2 Estimated causes of global deaths in 2017 reported by the Global Burden of Disease

Causes of death	No. (1000s)	%
Non-communicable diseases	41,071.1	73.4
Communicable, maternal, neonatal, and nutritional diseases	10,389.9	18.6
Injuries	4484.7	8.0
All causes	55,945.7	100.0

Source: Roth et al. (2018)

standards of living and healthcare systems reduce deaths from communicable diseases, malnutrition and other conditions and people therefore live longer (Omran, 1971). The Global Burden of Disease (GBD) study, run by the University of Washington's Institute for Health Metrics and Evaluation (IHME), estimates the causes of death (mortality) and disease (morbidity) using data from 195 countries. Table 1.2 shows that 73.4% of global deaths in 2017 were caused by non-communicable diseases (NCDs) (Roth et al., 2018). Only 18.6% of deaths were caused by communicable, maternal, neonatal and nutritional causes (ibid.). This shift to a greater burden of NCDs is global, yet it disproportionately affects low- and middle-income countries where, in 2016, 78% of deaths from NCDs occurred (WHO, 2018a). Although rates of death are declining globally, the GBD group highlight that this is not guaranteed to continue and that there are worrying variations in this generally positive trend. For example, in 2017, there were rising death rates caused by conflict, disasters and HIV epidemics in some countries, including Eastern Europe, Southeast Asia, Latin America and the USA (Dicker et al., 2018). At the time of writing, statistics for COVID-19 were changing rapidly, yet the IHME's (2021) projected estimate for cumulative excess deaths due to COVID-19 up to 1 December 2021 was 11,603,000 (which would not be enough to shift the overall burden of disease from primarily non-communicable to communicable diseases).

Migration

Migrants to cities may come from rural areas within the same country, or they may travel internationally. The UCL–*Lancet* Commission on Migration and Health reports that most global migration occurs within low- and middle-income countries, rather than from those countries to high-income countries as commonly discussed in the media (Abubakar et al., 2018). Despite common misconceptions, migrants tend to contribute more to their host societies than they cost, and the health risks posed from migrants to host populations are generally low (ibid.). China's urbanisation has been driven by industrialisation in cities, fuelled by migrant labour from rural areas (Yang et al., 2018). In 2010, the Sixth Census reported that the number of migrants living in urban areas was 225.96 million people (Qin & Zhang, 2014). The total number of migrants in China in 2010 (nearly 261 million) had more than doubled from the previous census in 2000, largely driven by the labour needs of industrialisation. A significant challenge for Chinese rural to urban migrants is that they do not have access to the same benefits as urban residents, such as public schools, housing and health services, because of the *Hukou* system that registers residents to a particular place (Qin & Zhang, 2014; Yang et al., 2018). This has many implications for health and city services. Considering migration from the perspective of urban design and planning, it is important to remember that this population may reside in poor-quality living environments that adversely affect their health. This is evident in refugee camps or informal settlements (Shackelford et al., 2020) and poor urban villages within cities (Yang et al., 2018). Environmental health issues can also affect migrants in high-income settings like the UK where they have an increased risk of tuberculosis compared to the UK-born population (Aldridge et al., 2016), which can be exacerbated by unstable and overcrowded housing and poor access to healthcare (Weller et al., 2019).

Reclassification of Land

Reclassification is the final factor that the UN considers to be a significant influencer of urban growth. The term reclassification refers to multiple phenomena that involve land being classified as urban for political and administrative purposes, often with the aim of promoting social and economic development (Farrell, 2017). There can be negative consequences from reclassification, including urban sprawl and removal of agricultural land (ibid.). Both of these impacts of reclassification could negatively affect health through ecosystem destruction, overconsumption of natural resources and increased poverty. In addition to the driving forces of urbanisation discussed above (natural increases, migration and reclassification), there are two other significant global population trends that relate to urban health, namely the ageing population and widening economic inequalities.

1.3.2 Population Ageing

Population ageing is one of the four global demographic 'megatrends' identified by the UN (2019c) alongside population growth, migration and urbanisation. The ageing of the world's population is an important factor affecting health in cities. Statistics from the UN show that globally in 2018, the population of people aged 65 and over outnumbered that of children under five. This was a landmark shift that had never been recorded previously and it is set to continue. By 2050, the population of people aged 65 and older is expected to be more than double that of children under 5, and older people will also outnumber those aged 15–24 years. This shift is caused by historically low levels of fertility in many countries and reducing levels of mortality. Cities are home to 43.2% of people aged 65 or older and this has important urban policy implications (OECD, 2015). City planners and design teams need to ensure that inclusive design enables older people to participate in all aspects of urban society, alongside the provision of appropriate housing and healthcare services (van Hoof et al., 2018).

Although an ageing urban population should not be framed as a problem, it does create challenges related to health. Older people are more likely to have one or more chronic diseases and to suffer from social isolation, which, in turn, is a risk factor for chronic diseases (OECD, 2015). Social isolation relates to the 'number of social relationships an individual has and describes a denuded social network', and it is considered to be separate to loneliness which involves a subjective negative perception of the quality of social relations (Yang & Victor, 2011, p. 1369).

Loneliness and isolation are important for health and quality of life, and they are particularly problematic for older people and those with limited mobility. Studies have found associations between loneliness/isolation and heart disease, depression, dementia and suicide, leading some scholars to conclude that this risk factor is comparable to smoking and obesity (ibid.). Social isolation also results in increased vulnerability to disasters and the impacts of climate change, such as overheating, when emergency services may not be aware that isolated people need help (OECD, 2015). After Hurricane Sandy hit Manhattan, USA, and surrounding areas in 2012, *The New York Times* catalogued and mapped deaths caused by the storm. Tragically, there were more than 100 deaths attributed to the storm and more than half were aged 65 or older, many of whom died from drowning, hypothermia and other causes (Keller, 2012). The built environment can be designed to strengthen social interaction and networks, thereby protecting against social isolation, unhappiness and mental illness, and this is described in Chap. 6.

1.3.3 Poverty and Inequity

The overall trend of increasing life expectancy is not equally enjoyed across society. Multiple studies have demonstrated that income inequalities reduce population health and wellbeing (Pickett & Wilkinson, 2015). Health inequities are broader than income inequalities, they are '*systematic, socially produced* (and therefore modifiable) and *unfair*' differences in health between different groups in society (Dahlgren & Whitehead, 2006, p. 2). Michael Marmot et al. (2010, 2020) conducted independent reviews of health inequalities in England. They highlighted the so-called social gradient in health, whereby people living in the most deprived neighbourhoods have shorter lives and spend more time in ill health than people in the least deprived neighbourhoods (Figs. 1.5 and 1.6). The life expectancy gap in England (2016–2018) between the most and least deprived groups was 9.5 years for males and 7.7 years for females (Marmot et al., 2020). Within a single city, the life expectancy gaps between rich and poor can be significant, and Glasgow is an oft-cited example of this problem in the UK. Seaman et al. (2015) found differences of 12.2 and 6.6 years between the most and least deprived males and females in Glasgow, respectively, which are primarily explained by socio-economic deprivation.

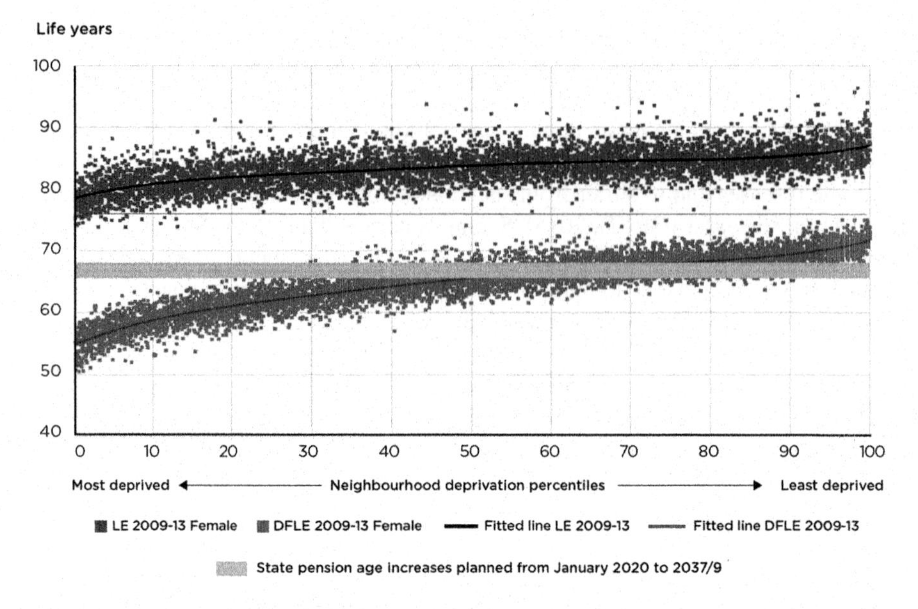

Fig. 1.5 Life expectancy at birth (females) by neighbourhood deprivation percentiles, 2009–2013, England. Reproduced with permission from Marmot et al. (2020). (Source: Office for National Statistics (ONS) and Department for Work and Pensions. Note: Each dot represents life expectancy (LE) or disability-free life expectancy (DFLE) of a neighbourhood (middle-level super output area))

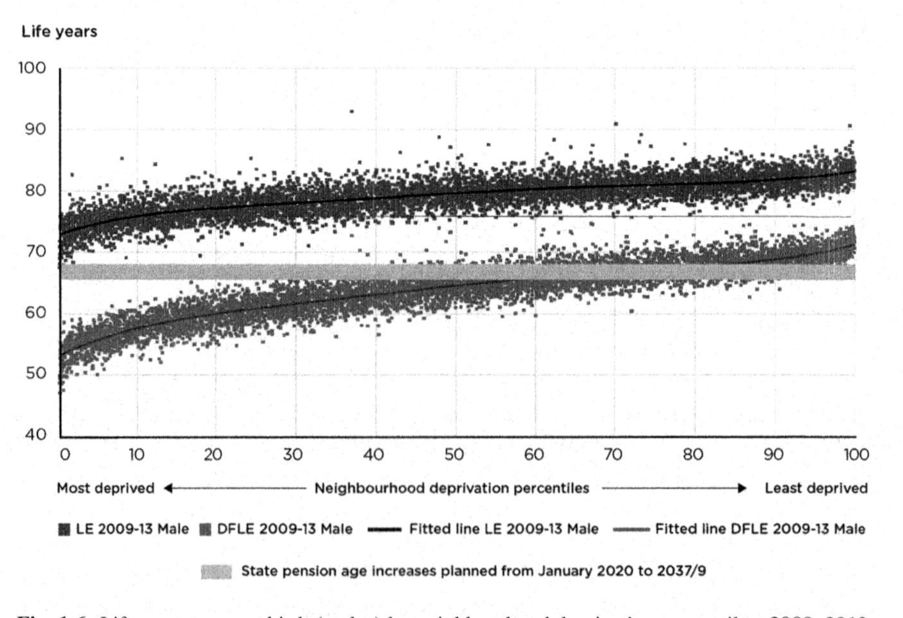

Fig. 1.6 Life expectancy at birth (males) by neighbourhood deprivation percentiles, 2009–2013, England. Reproduced with permission from Marmot et al. (2020). (Source: Office for National Statistics (ONS) and Department for Work and Pensions. Note: Each dot represents life expectancy (LE) or disability-free life expectancy (DFLE) of a neighbourhood (middle-level super output area))

There are multiple debates about the terminology for these unfair differences in health and how they should be eliminated. Braveman (2014) argues the need for a common approach to health equity, which she describes as 'the principle underlying a commitment to reduce—and, ultimately, eliminate—disparities in health and in its determinants, including social determinants' (p. 6). A growing understanding of these inequities emerged from the WHO's Commission on Social Determinants of Health that aimed to better understand the causes of avoidable differences in health and policies to reduce such differences (CSDH, 2007, 2008). The Commission reported that the neoliberal economic models that became prevalent in the 1980s had global effects—shifting governments' focus from equity to efficiency and reducing developing countries' spending on the social sector. Both of these effects decreased socially disadvantaged groups' access to healthcare and other services. The causes of health inequities are described by the Commission as:

> the unequal distribution of power, income, goods, and services, globally and nationally, the consequent unfairness in the immediate, visible circumstances of peoples [sic] lives—their access to health care, schools, and education, their conditions of work and leisure, their homes, communities, towns or cities—and their chances of leading a flourishing life. (CSDH, 2008, p. 1)

This description demonstrates the vast range of societal factors that are implicated in reducing individuals' potential to live a healthy life.

Inequities are often exacerbated in urban environments where the rich and poor live in vastly different circumstances, not only in terms of income but also related to housing, infrastructure and pollution (WHO & UN Habitat, 2010). Disadvantaged populations in cities 'typically have fewer choices open to them and are in locations and settings that are less conducive to good health, with little ability to move away from unhealthy working and living environments' (WHO, 2012, p. 4). The built environment contributes to health inequities through a number of pathways including the concentration of environmental burdens in deprived neighbourhoods such as air and noise pollution, poor-quality housing, limited access to parks and poorly maintained streets (Gelormino et al., 2015; Northridge & Freeman, 2011). Inequities are a key component of this book and the link between the built environment and health inequities is further explored in each chapter.

1.3.4 Changing Patterns of Disease

The global trend of increased life expectancy is partly attributed to the epidemiological transition, whereby economic development increases living standards and improves healthcare systems, resulting in reductions in communicable, nutritional and other diseases. An analysis by Santosa et al. (2014) found that countries have moved through the stages of epidemiological transition proposed by Omran (1971) over different timescales and with varying accordance to the classic theoretical pathways. In general, until the mid-twentieth century, life expectancies in low- and middle-income countries were much lower than high-income industrialised countries due to deaths from infectious diseases. After the Second World War, there was rapid progress in reducing infectious diseases. In 2017, NCDs accounted for 73.4% of global deaths, with communicable, maternal, neonatal and nutritional diseases at 18.6% and finally injuries at 8.0% (see Table 1.2). However, many cities suffer from the 'triple threat of disease', whereby poor living and social conditions result in high levels of NCDs, infectious diseases and injuries (WHO & UN Habitat, 2010). From the perspective of built environment professionals, it is vital to understand how the physical and natural environment contributes to changing patterns of disease.

Non-communicable Diseases

NCDs are described by the WHO (2014) as non-transmissible conditions that afflict people for long durations, therefore they are also called chronic diseases. They are caused by the 'effects of globalization on marketing and trade, rapid urbanization and population ageing' (ibid., p. vii) among other factors. In 2017, the leading causes of NCD deaths were cardiovascular diseases, neoplasms (i.e. cancers), chronic respiratory diseases, neurological diseases (Alzheimer's disease and other dementias) and diabetes (Table 1.3). Many people will live for years with one or more chronic conditions that may significantly reduce their quality of life. In the USA, roughly six out of ten adults have a chronic condition, and four out of ten have two or more conditions (Buttorff et al., 2017).

Although NCDs are the most common and expensive diseases, how countries prioritise funding for treating and preventing these conditions is not proportionate to their ill-health burden. In the USA, 90% of the country's US$3.5 trillion annual healthcare spending is for chronic disease and mental health treatment (National Center for Chronic Disease Prevention and Health Promotion, 2020). In comparison, in low- and middle-income countries, less than 2% of development assistance

Table 1.3 Number and percentage of global deaths in 2017 from non-communicable disease categories reported by the Global Burden of Disease

Causes of death	No. (1000s)	%
Cardiovascular diseases	17,790.9	31.8
Neoplasms (i.e. cancers)	9556.2	17.1
Chronic respiratory diseases	3914.2	7.0
Neurological disorders	3094.2	5.5
Diabetes and kidney diseases	2611.2	4.7
Digestive diseases	2377.7	4.2
Other non-communicable diseases	1153.3	2.1
Musculoskeletal disorders	121.3	0.2
Skin and subcutaneous diseases	100.3	0.2
Mental disorders	0.3	0.0

Source: Roth et al. (2018)

for health funding is spent on NCDs (IHME, 2018; WHO, 2018b), even though in 2016, 78% of deaths from NCDs occurred in these countries (WHO, 2018a). A set of 'best buy' policy options to decrease NCDs developed by the WHO offer cost-effective ways to address this challenge. They estimate that every US$1 spent on the best buy policies in low- and middle-income countries will yield at least US$7 in returns to society through increased employment, productivity and life expectancy (WHO, 2018b). Given the significant cost of NCDs to healthcare services and the wider economy, it is important for urban environment professionals to consider ways to reduce risk factors for these diseases through design and planning.

NCDs can be prevented by reducing their four main risk factors of physical inactivity, unhealthy diet and tobacco and alcohol consumption. Scholars have critiqued policy-makers' focus on these risk factors as amenable to individual 'lifestyle' changes, claiming that these factors are significantly determined by wider societal structures, including poverty and the built environment (Kelly & Russo, 2018). Allen and Feigl (2017) call for NCDs to be renamed and reframed because their current description over-emphasises individual responsibility and fails to reflect that some NCDs are in fact communicable.

Built environment efforts to address the risk factors of NCDs are supportive of such critiques by recognising the societal decisions that influence individual behaviour. Physical activity and diet are most strongly related to actions that can be taken by urban design and planning professionals, for example, through the concepts of 'active design' and 'active travel' (see Chaps. 6 and 7). Zoning codes have sought to reduce oversupply of alcohol (see Hippensteel et al., 2019). In addition to these individual risk factors, other environmental conditions cause NCDs, principally air and noise pollution (see Chaps. 5 and 7). Designing the built environment to prevent NCDs requires a shift in thinking away from biomedical solutions towards a much broader understanding of how people's living and working conditions affect health.

Communicable Diseases

In the century preceding COVID-19, communicable diseases received far less focus among built environment professionals in high-income countries compared to chronic diseases, due to the decline in deaths caused by infectious diseases. The origins of both public health and urban planning professions stem from the late nineteenth century when poor urban environments in cities like London and New York were linked with the spread of communicable diseases such as cholera (see Chap. 2). Today the leading causes of death in low- and middle-income countries continue to include conditions that relate to overcrowding and poor-quality water, waste and sanitation infrastructure, including lower respiratory infections, diarrhoeal diseases and tuberculosis (Neiderud, 2015; WHO, 2018c). These unhealthy environmental conditions are exacerbated in urban slums, where 29.7% of urban populations in developing regions lived in 2014 (UN, 2015). Neiderud's (2015) review of urbanisation and infectious diseases highlights that poor-quality and overcrowded housing can contribute to the growth of disease vectors, including rodents and insects, which can spread zoonotic diseases (e.g. dengue) and parasitic infections (e.g. Chagas disease). Some infectious diseases that were primarily considered as rural are now affecting urban health such as those spread through mosquitos and domesticated animals brought to urban settings.

The global emergence of the novel coronavirus, COVID-19, refocused attention in early 2020 on the role of urbanisation and urban design in infectious disease transmission (see Box 1.1). The SARS epidemic in 2003 is thought to have emerged from the SARS-like coronavirus found in bats that was transmitted to humans through handling of bats in exotic food markets (Neiderud, 2015). Similarly, COVID-19 appears to have been transmitted to humans from animals in the Chinese city of Wuhan (Heymann & Shindo, 2020) and then spread internationally through human-to-human transmission. Neiderud (2015) argues that rapid and unplanned urbanisation has increased the likelihood of animals spreading diseases to humans for a number of reasons. Growing cities are encroaching on previously undeveloped land where new housing is in close contact with ecosystems that may host zoonotic diseases, such as the Ebola virus. People may introduce these diseases into local human populations through close contact with domesticated or wild animals. Deforestation for timber or infrastructure development (among other reasons) puts people in closer contact with disease hosts, including bats and primates. Finally, when new zoonotic diseases spread in urban environments, they are particularly hard to control in overcrowded housing. Those living in informal settlements are especially vulnerable due to lack of basic water, waste and sanitation infrastructure, among other challenges (Corburn et al., 2020).

Box 1.1 The COVID-19 syndemic and the built environment

It became clear in the early months of the pandemic that the common slogan in the UK and elsewhere—'we are all in this together'—was grossly inaccurate. The suffering was not equal across society and some people were disproportionately at risk of infection and death. This led the editor of *The Lancet*, Richard Horton, to describe COVID-19 as a syndemic. To understand this perspective and its value in identifying the built environment's role, it is helpful to consider the meanings of epidemic, pandemic and syndemic.

An *epidemic* describes the 'occurrence in a community or region of cases of an illness, specific health-related behavior, or other health-related events clearly in excess of normal expectancy' (Porta, 2014). In other words, epidemics are not only about infectious diseases, such as cholera outbreaks. Opioid drug addiction in the USA is called an epidemic, as is the obesity crisis. A *pandemic* is an 'epidemic occurring over a very wide area, crossing international boundaries, and usually affecting a large number of people' (ibid.). Examples of pandemics include HIV/AIDS and SARS. In the late twentieth century, medical anthropologists developed the concept of syndemics in recognition that diseases are not isolated entities, but are influenced by a patient's pre-existing conditions and external economic, social

(continued)

Box 1.1 (continued)

and environmental forces (Singer & Clair, 2003). A *syndemic* describes 'the clustering of two or more health conditions within a particular context; interaction of those conditions via biological, social, or psychological pathways; and involvement of social, political, economic, or ecological drivers' (Mendenhall & Singer, 2019). There is debate about whether syndemic can apply to conditions that spread globally as the local context is important. Nevertheless, syndemic has been used to explain and link obesity, undernutrition and climate change.

The term syndemic is useful to understand the role of the built environment in a disease like COVID-19. Minority ethnic and low-income populations were disproportionately affected by the pandemic (see Chap. 3, Box 3.1) and this was likely linked to other social determinants of health. Analyses found links between air pollution, household overcrowding and disease severity in multiple countries (Aldridge et al., 2021; Chen & Krieger, 2020; Cole et al., 2020; Wu et al., 2020). This raised attention to systemic racism and inequities in society and specifically in the built environment. Through discriminatory urban planning and uneven investment processes, minority ethnic and low-income populations are more likely to live in neighbourhoods that contribute to risk of COVID-19 infection and many other health problems (Berkowitz et al., 2020).

Injuries

Injuries of three broad types comprise the final category used to describe global deaths and disease, including unintentional injuries, transport injuries and finally, self-harm and interpersonal violence. The built environment should be designed and adequately maintained to avoid injuries such as falls, overheating in buildings or deaths from fires (see Chaps. 6 and 7). The WHO (2018d) status report on road safety makes it clear that road injuries are preventable, yet they are the leading cause of death among people aged 5–29 years. They are the eighth leading cause of death globally, causing 1.35 million deaths and up to 50 million injuries each year. Half of these traffic deaths are pedestrians, cyclists and motorcyclists who are not well-protected by transport and mobility infrastructure. Finally, road injury death rates are three times higher in low- and middle-income countries than high-income countries, demonstrating the layers of health threats experienced in developing countries. Rapid urbanisation contributes to transport and building safety challenges.

1.4 Cities and Health

As cities continue to grow in population, they are also growing physically, with multiple ramifications for people and the planet. Cities face competing and interconnected challenges in the twenty-first century that span economic, environmental and social domains, highlighting the complexity of urban systems. Cities are simultaneously exposure environments, contributors to environmental degradation and governance institutions that can create a better future for residents. There are many health and wellbeing advantages to urban life, such as access to jobs, higher education and cultural institutions. In rapid growth settings, such as Asia and Africa, the challenge is that unplanned growth can result in unhealthy environments characterised by urban sprawl, lack of infrastructure (mobility, energy, water, waste and sanitation) and increased air pollution (Floater & Rode, 2014). Established cities also experience these problems and low-income and minority ethnic populations are more likely to live in neighbourhoods with poor infrastructure provision and pollution (Agyeman et al., 2003; CSDH, 2008). The heterogeneity of urban settlements (and variations in how they are classified) makes it difficult to compare urban health challenges and solutions internationally. This section focuses on the urban environment and how it is associated with health conditions. It demonstrates that urban growth challenges, such as urban sprawl and poor-quality infrastructure, are not exclusive to rapidly urbanising settings.

1.4.1 Urban Context as Exposure

Epidemiologists have a number of theories about whether and why cities are fundamentally good or bad for human health. Epidemiology is the study of the determinants and the distribution of disease within populations. As will be shown in Chap. 3, epidemiologists study associations between specific exposures (e.g. air pollution) and health outcomes (e.g. heart disease), and they investigate why some population groups (e.g. poor people) get ill when others do not. The relatively new field of urban health is concerned with 'the determinants of health and diseases in urban areas and with the urban context itself as the exposure of interest' (Galea & Vlahov, 2005, p. 342). Historically, cities have been viewed as overcrowded and polluted places that caused an 'urban health penalty', yet today's urban populations are often healthier than their rural counterparts resulting in an 'urban health advantage' (Vlahov et al., 2005). This advantage is caused by 'greater resources, higher density, and better infrastructure and service availability' in cities compared to rural areas, yet it is not universal and poor populations are less likely to benefit (WHO, 2016, p. 18). It can be argued that cities are the human habitat that best supports human, planetary and ecosystem health; yet achieving this ultimate goal requires managing urban development in a way that is equitable and inclusive for everybody in society.

When considering the urban context as an exposure, Galea and Vlahov (2005) propose three categories to group theories and causal pathway mechanisms: 'the physical environment, the social environment, and the availability of and access to health and social services' (p. 344). Broadly speaking, the physical environment includes parks, buildings, air, greenspace and other parts of the built and natural environment. The social environment is a very broad set of economic, social, cultural, governance and related factors that influence or are influenced by the built environment. The descriptions of non-communicable and communicable diseases, and injuries in the previous section introduced some of the connections between Galea and Vlahov's categories and health. These will be elaborated in Chap. 3 through a description of existing urban health frameworks. Chapters 4, 5, 6 and 7 will review the causal pathways that connect urban environments to specific health outcomes in detail.

1.4.2 Causes of Environmental Degradation

Urban growth, particularly when it occurs rapidly, increases the environmental 'footprint' of a city. A working group for the Fifth Assessment Report of the Intergovernmental Panel on Climate Change (Seto et al., 2014) found that cities account for over 71–76% of global carbon dioxide emissions and 67–76% of global energy use. Urban form is a key driver of both emissions and energy use, and the relevant measures to reduce these problems are density, land use mix, connectivity and accessibility. The report highlights the strong evidence that 'co-locating higher residential densities with higher employment densities, coupled with significant public transit improvements, higher land use mixes, and other supportive demand management measures can lead to greater emissions savings in the long run' (ibid., p. 928). The construction industry, providing infrastructure for growing cities, is the largest consumer of raw materials (Krausmann et al., 2017) and emits roughly 40% of energy-related carbon dioxide emissions, the primary greenhouse gas from human activities (UNEP & IEA, 2017). Infrastructure for rapidly growing cities could release 226 gigatonnes (Gt) of carbon dioxide by 2050, equal to four times the emissions released to build infrastructure in the existing developed world (Bai et al., 2018). Thus, cities are significant contributors to climate change and other environmental degradation.

Climate change and other global environmental harms (e.g. biodiversity loss and resource depletion) affect human health in multiple interrelated ways and the effects can be both proximate to and distant from sources of harm. Costello et al. (2009) called climate change 'the biggest global health threat of the twenty-first century', creating health impacts through 'changing patterns of disease, water and food insecurity, vulnerable shelter and human settlements, extreme climatic events, and population growth and migration' (p. 1693). One significant injustice of global environmental degradation is that poor populations, particularly those living in informal settlements, who have contributed the least carbon emissions (and other harms) will suffer the most from the health impacts of climate change, such as malnutrition and death or injury from extreme weather (Butler, 2016; Satterthwaite et al., 2018). A growing field of research called planetary health is investigating the interlinkages between human health, civilisation and unprecedented environmental degradation, and seeking to drive transformative action to solve these challenges in ways that redress inequities (Whitmee et al., 2015).

1.4.3 Governance Institutions

Cities are recognised as financial and political centres of power that can more quickly respond to threats and opportunities than higher tiers of government. The urgency of the climate crisis and other urban challenges underscores the need for leadership and collaborative approaches to achieve the scale and pace of change required. Moore (2015) calculated that transitioning to 'one planet living' levels of consumption would require on average 73% reductions in household energy use and 96% reductions in car ownership. To convince decision-makers to make such significant changes, it is essential to demonstrate the economic advantages of sustainable city governance. Floater and Rode's (2014) '3C Model' for sustainable urbanisation proposes that *compact* growth, *connected* infrastructure and *coordinated* governance can help city leaders deal with the challenge that large infrastructure decisions are 'largely irreversible' and they 'will lock-in economic and climate benefits—or costs—for decades to come' (p. 4). In support of their approach, Floater and Rode explain that 'Stockholm reduced transport, heating and electricity emissions by 35% between 1993 and 2010 from a low starting point' and 'the city's economic output grew by 41% over the same period—one of the highest growth rates in Europe' (p. 32).

The UN has played a fundamental role in positioning cities as a focus for achieving the Sustainable Development Goals (SDGs) by including a specific urban goal and by introducing monitoring indicators at the city scale (Barnett & Parnell, 2016). Goal 11 of the SDGs aims to 'make cities and human settlements inclusive, safe, resilient and sustainable' and the associated targets cover a broad range of topics including housing, transport, participatory planning and so on (see Chap. 2) (United Nations General Assembly, 2015). The SDGs provide a broad structure to guide urban policies that will promote human health alongside environmental sustainability, including by addressing climate change; yet the SDGs require local interpretation to guide implementation in differing economic, social, cultural and environmental circumstances (Tan et al., 2019).

1.5 Conclusion

Cities are facing many challenges that affect health, yet they are well-positioned to meet these challenges with locally appropriate solutions. The WHO Healthy Cities movement demonstrated that achieving a healthy city is a contextually specific process and an outcome that cannot be achieved once with lasting results, but must be continually pursued as circumstances change (Hancock & Duhl, 1986; Rydin et al., 2012). Existing models for healthy and sustainable urban development pathways can be manifest at different urban scales and led by different stakeholders. Through the THRIVES framework, this book introduces a new way of conceptualising healthy urbanism that brings together the interconnected objectives of sustainability, equity and inclusion to ensure that human health is protected across three scales with spatial and temporal dimensions (planetary, ecosystem and local) affecting current and future populations. The next chapter will explore historical and contemporary approaches to planning and designing healthy places, introducing current drivers for healthy urbanism.

References

Abubakar, I., Aldridge, R. W., Devakumar, D., Orcutt, M., Burns, R., Barreto, M. L., Dhavan, P., Fouad, F. M., Groce, N., Guo, Y., Hargreaves, S., Michael Knipper, J., Miranda, J., Madise, N., Kumar, B., Mosca, D., McGovern, T., Rubenstein, L., Sammonds, P., … Zhou, S. (2018). The UCL–Lancet Commission on migration and health: The health of a world on the move. *The Lancet, 392*, 2606–2654.

Agyeman, J. (2013). *Introducing just sustainabilities: Policy, planning and practice, Just sustainabilities*. Zed Books.

Agyeman, J., Bullard, R. D., & Evans, B. (Eds.). (2003). *Just sustainabilities: Development in an unequal world*. Earthscan.

Aldridge, R.W., Pineo, H., Fragaszy, E., Eyre, M.T., Kovar, J., Nguyen, V., Beale, S., Byrne, T., Aryee, A., Smith, C., Devakumar, D., Taylor, J., Katikireddi, S.V., Fong, W.L.E., Geismar, C., Patel, P., Shrotri, M., Braithwaite, I., Patni, N., Navaratnam, A.M.D., Johnson, A., Hayward, A. (2021). Household overcrowding and risk of SARS-CoV-2: analysis of the Virus Watch prospective community cohort study in England and Wales [version 1; peer review: awaiting peer review]. Wellcome Open Res 2021, 6:347 (https://doi.org/10.12688/wellcomeopenres.17308.1)

Aldridge, R. W., Zenner, D., White, P. J., Williamson, E. J., Muzyamba, M. C., Dhavan, P., Davide Mosca, H., Thomas, L., Lalor, M. K., Abubakar, I., & Hayward, A. C. (2016). Tuberculosis in migrants moving from high-incidence to low-incidence countries: A population-based cohort study of 519 955 migrants screened before entry to England, Wales, and Northern Ireland. *The Lancet, 388*, 2510–2518.

Allen, L. N., & Feigl, A. B. (2017). What's in a name? A call to reframe non-communicable diseases. *The Lancet Global Health, 5*, e129–e130.

Bai, X., Dawson, R. J., Ürge-Vorsatz, D., Delgado, G. C., Barau, A. S., Dhakal, S., Dodman, D., Leonardsen, L., Masson-Delmotte, V., Roberts, D. C., & Schultz, S. (2018). Six research priorities for cities and climate change. *Nature, 555*, 23–25.

Barnett, C., & Parnell, S. (2016). Ideas, implementation and indicators: Epistemologies of the post-2015 urban agenda. *Environment and Urbanization, 28*, 87–98.

Berkowitz, R. L., Gao, X., Michaels, E. K., & Mujahid, M. S. (2020). Structurally vulnerable neighbourhood environments and racial/ethnic COVID-19 inequities. *Cities Health, 0*, 1–4. https://doi.org/10.1080/23748834.2020.1792069

Bircher, J., & Kuruvilla, S. (2014). Defining health by addressing individual, social, and environmental determinants: New opportunities for health care and public health. *Journal of Public Health Policy Basingstoke, 35*, 363–386.

Braveman, P. (2014). What are health disparities and health equity? We need to be clear. *Public Health Report, 129*(Suppl 2), 5–8.

BRE. (2016). Masthusen, Malmö, Sweden. *BREEAM*. Retrieved March 19, 2020, from https://www.breeam.com/case-studies/communities/masthusen-malmo-sweden/

Butler, C. D. (2016). Sounding the alarm: Health in the anthropocene. *International Journal of Environmental Research and Public Health, 13*, 665.

Buttorff, C., Ruder, T., & Bauman, M. (2017). *Multiple chronic conditions in the United States*. RAND Corporation.

Chadwick, E. (1842). *Report to Her Majesty's principal secretary of state for the home department, from the poor law commissioners, on an inquiry into the sanitary condition of the labouring population of Great Britain* (House of Commons Sessional Paper). W. Clowes and Sons, London.
Chen, J., & Krieger, N. (2020). *Revealing the unequal burden of COVID-19 by income, race/ethnicity, and household crowding: US county vs. ZIP code analyses* (HCPDS Working Paper, Volume 19, Number 1). Harvard T.H. Chan School of Public Health, Boston, MA.
Cole, M. A., Ozgen, C., & Strobl, E. (2020). Air pollution exposure and Covid-19 in Dutch municipalities. *Environmental and Resource Economics, 76*, 581–610.
CSDH. (2007). *A conceptual framework for action on the social determinants of health.* Discussion paper for the Commission on Social Determinants of Health. World Health Organization, Geneva.
CSDH. (2008). *Closing the gap in a generation: Health equity through action on the social determinants of health.* Final Report of the Commission on Social Determinants of Health. World Health Organization, Geneva.
Corburn, J., Vlahov, D., Mberu, B., Riley, L., Caiaffa, W. T., Rashid, S. F., Ko, A., Patel, S., Jukur, S., Martínez-Herrera, E., Jayasinghe, S., Agarwal, S., Nguendo-Yongsi, B., Weru, J., Ouma, S., Edmundo, K., Oni, T., & Ayad, H. (2020). Slum health: Arresting COVID-19 and improving well-being in urban informal settlements. *Journal of Urban Health, 97*, 348–357.
Costello, A., Abbas, M., Allen, A., Ball, S., Bell, S., Bellamy, R., Friel, S., Groce, N., Johnson, A., Kett, M., Lee, M., Levy, C., Maslin, M., McCoy, D., McGuire, B., Montgomery, H., Napier, D., Pagel, C., Patel, J., … Patterson, C. (2009). Managing the health effects of climate change. *The Lancet, 373*, 1693–1733.
Dahlgren, G., & Whitehead, M. (1991). *Policies and strategies to promote social equity in health.* Institute for Futures Studies.
Dahlgren, G., & Whitehead, M. (2006). *European strategies for tackling social inequities in health: Levelling up Part 2.* WHO Collaborating Centre for Policy Research on Social Determinants University of Liverpool of Health, Liverpool.
Dicker, D., Nguyen, G., Abate, D., Abate, K. H., Abay, S. M., Abbafati, C. M., Abbasi, N., Abbastabar, H., Abd-Allah, F., Abdela, J., Abdelalim, A., Abdel-Rahman, O., Abdi, A., Abdollahpour, I., Abdulkader, R. S., Abdurahman, A. A., Abebe, H. T., Abebe, M., … Murray, C. J. L. (2018). Global, regional, and national age-sex-specific mortality and life expectancy, 1950–2017: A systematic analysis for the Global Burden of Disease Study 2017. *The Lancet, 392*, 1684–1735.
Dodge, R., Daly, A. P., Huyton, J., & Sanders, L. D. (2012). The challenge of defining wellbeing. *International Journal of Wellbeing, 2*, 222–235.
Engel, G. L. (1977). The need for a new medical model: A challenge for biomedicine. *Science, 196*, 129–136.
Farre, A., & Rapley, T. (2017). The new old (and old new) medical model: Four decades navigating the biomedical and psychosocial understandings of health and illness. *Healthcare, 5*, 1–9.
Farrell, K. (2017). The rapid urban growth triad: A new conceptual framework for examining the urban transition in developing countries. *Sustainability, 9*, 1407.
Floater, G., & Rode, P. (2014). *Cities and the new climate economy: The transformative role of global urban growth, cities and the new climate economy.* London School of Economics and Political Science, LSE Cities.
Galea, S., & Vlahov, D. (2005). Urban health: Evidence, challenges, and directions. *Annual Review of Public Health, 26*, 341–365.
Gelormino, E., Melis, G., Marietta, C., & Costa, G. (2015). From built environment to health inequalities: An explanatory framework based on evidence. *Preventive Medical Reports, 2*, 737–745.
Hancock, T. (1985). The mandala of health: A model of the human ecosystem. *Family & Community Health, 8*, 1–10.

Hancock, T. (2011). The Ottawa charter at 25. *Canadian Journal of Public Health*. Can. Santee Publique, *102*, 404–406.

Hancock, T., & Duhl, L. J. (1986). *Healthy cities: Promoting health in the urban context* (No. WHO Healthy Cities Paper #1). World Health Organization Regional Office for Europe.

Heymann, D. L., & Shindo, N. (2020). COVID-19: What is next for public health? *The Lancet, 395*, 542–545.

Hippensteel, C. L., Sadler, R. C., Milam, A. J., Nelson, V., & Debra Furr-Holden, C. (2019). Using zoning as a public health tool to reduce oversaturation of alcohol outlets: An examination of the effects of the new "300 Foot Rule" on packaged goods stores in a Mid-Atlantic City. *Prevention Science, 20*, 833–843.

Huber, M., Knottnerus, J. A., Green, L., Horst, H. van der, Jadad, A. R., Kromhout, D., Leonard, B., Lorig, K., Loureiro, M. I., van der Meer, Jos W. M, Schnabel, P., Smith, R., van Weel, C., & Smid, H. (2011). How should we define health? *BMJ* Online London, *343*, d4163.

IHME. (2018). *Financing global health 2017: Funding universal health coverage and the unfinished HIV/AIDS agenda*. IHME, Seattle, WA.

IHME. (2021). COVID-19 results briefing: Global. IHME, Seattle, WA.

Jedwab, R., Christiaensen, L., & Gindelsky, M. (2017). Demography, urbanization and development: Rural push, urban pull and … urban push? *Journal of Urban Economics, 98*, 6–16.

Keller, J. (2012). Mapping Hurricane Sandy's Deadly Toll. *N. Y. Times*.

Kelly, M. P., & Russo, F. (2018). Causal narratives in public health: The difference between mechanisms of aetiology and mechanisms of prevention in non-communicable diseases. *Sociology of Health & Illness, 40*, 82–99.

Krausmann, F., Wiedenhofer, D., Lauk, C., Haas, W., Tanikawa, H., Fishman, T., Miatto, A., Schandl, H., & Haberl, H. (2017). Global socioeconomic material stocks rise 23-fold over the 20th century and require half of annual resource use. *Proceedings of the National Academy of Sciences, 114*(8), 1880–1885.

Krieger, N. (1994). Epidemiology and the web of causation: Has anyone seen the spider? *Social Science & Medicine, 39*, 887–903.

Krieger, N. (2001). Theories for social epidemiology in the 21st century: An ecosocial perspective. *International Journal of Epidemiology, 30*, 668–677.

Lang, T., & Rayner, G. (2012). Ecological public health: The 21st century's big idea? An essay by Tim Lang and Geof Rayner. *BMJ, 345*, e5466.

Luke, D. A., & Stamatakis, K. A. (2012). Systems science methods in public health: Dynamics, networks, and agents. *Annual Review of Public Health, 33*, 357–376.

Marmot, M., Allen, J., Boyce, T., Goldblatt, P., & Morrison, J. (2020). *Health equity in England: The Marmot Review 10 years on*. Institute of Health Equity, London.

Marmot, M., Allen, J., Goldblatt, P., Boyce, T., McNeish, D., Grady, M., & Geddes, I. (2010). Fair society, healthy lives: The Marmot review; strategic review of health inequalities in England post-2010. *Marmot Review*, London.

Marsh, H. W., Huppert, F. A., Donald, J. N., Horwood, M. S., & Sahdra, B. K. (2020). The well-being profile (WB-Pro): Creating a theoretically based multidimensional measure of well-being to advance theory, research, policy, and practice. *Psychological Assessment, 32*, 294–313.

Meadows, D. H. (2008). *Thinking in systems: A primer*. Chelsea Green Pub.

Medved, P. (2018). Exploring the 'Just City principles' within two European sustainable neighbourhoods. *Journal of Urban Design, 23*, 414–431.

Mendenhall, E., & Singer, M. (2019). The global syndemic of obesity, undernutrition, and climate change. *The Lancet, 393*, 741.

Moore, J. (2015). Ecological footprints and lifestyle archetypes: Exploring dimensions of consumption and the transformation needed to achieve urban sustainability. *Sustainability, 7*, 4747–4763.

National Center for Chronic Disease Prevention and Health Promotion. (2020). Health and economic costs of chronic disease. Retrieved February 04, 2020, from https://www.cdc.gov/chronicdisease/about/costs/index.htm

Neiderud, C.-J. (2015). How urbanization affects the epidemiology of emerging infectious diseases. *Infection Ecology & Epidemiology, 5*, 27060.
Northridge, D. M. E., Sclar, D. E. D., & Biswas, M. P. (2003). Sorting out the connections between the built environment and health: A conceptual framework for navigating pathways and planning healthy cities. *Journal of Urban Health, 80*, 556–568.
Northridge, M. E., & Freeman, L. (2011). Urban planning and health equity. *Journal of Urban Health, 88*, 582–597.
OECD. (2015). *Ageing in cities*. OECD Publishing.
Omran, A. R. (1971). The epidemiologic transition: A theory of the epidemiology of population change. *Milbank Memorial Fund Quarterly, 49*, 509–538.
Pickett, K. E., & Wilkinson, R. G. (2015). Income inequality and health: A causal review. *Social Science & Medicine, 128*, 316–326.
Pineo, H. (2020). Towards healthy urbanism: Inclusive, equitable and sustainable (THRIVES)—an urban design and planning framework from theory to praxis. *Cities Health, 0*, 1–19. https://doi.org/10.1080/23748834.2020.1769527
Pineo, H., Moore, G., & Braithwaite, I. (2020). Incorporating practitioner knowledge to test and improve a new conceptual framework for healthy urban design and planning. *Cities Health, 0*, 1–16. https://doi.org/10.1080/23748834.2020.1773035
Pollard, E. L., & Lee, P. D. (2003). Child well-being: A systematic review of the literature. *Social Indicators Research, 61*, 59–78.
Porta, M. (2014). *A dictionary of epidemiology*. Oxford University Press.
Prüss-Üstün, A., Wolf, J., Corvalan, C., Bos, R., & Neira, M. (2016). *Preventing disease through healthy environments: A global assessment of the burden of disease from environmental risks*. World Health Organization.
Qin, B., & Zhang, Y. (2014). Note on urbanization in China: Urban definitions and census data. *China Economic Review, 30*, 495–502. https://doi.org/10.1016/j.chieco.2014.07.008
Raworth, K. (2017). *Doughnut economics: Seven ways to think like a 21st-century economist*. Random House Business Books.
Rayner, G., & Lang, T. (2012). *Ecological public health: Reshaping the conditions for good health*. Routledge.
Richardson, B. W. (1876). *Hygeia, a city of health*. Macmillan.
Rosberg, G. (n.d.). Western Harbour—A new sustainable city district in Malmö.
Roth, G. A., Abate, D., Abate, K. H., Abay, S. M., Abbafati, C., Abbasi, N., Abbastabar, H., Abd-Allah, L., Abdela, J., Abdelalim, A., Abdollahpour, I., Abdulkader, R. S., Abebe, H. T., Abebe, M., Abebe, Z., Abejie, A. N., Abera, S. F., Abil, O. Z., Abraha, H. N., & Murray, C. J. L. (2018). Global, regional, and national age-sex-specific mortality for 282 causes of death in 195 countries and territories, 1980–2017: A systematic analysis for the Global Burden of Disease Study 2017. *The Lancet, 392*, 1736–1788.
Rydin, Y., Bleahu, A., Davies, M., Dávila, J. D., Friel, S., De Grandis, G., Groce, N., Hallal, P. C., Hamilton, I., Howden-Chapman, P., Ka-Man Lai, C. J., Lim, J. M., Osrin, D., Ridley, I., Scott, I., Taylor, M., Wilkinson, P., & Wilson, J. (2012). Shaping cities for health: Complexity and the planning of urban environments in the 21st century. *The Lancet, 379*, 2079–2108.
Santosa, A., Wall, S., Fottrell, E., Högberg, U., & Byass, P. (2014). The development and experience of epidemiological transition theory over four decades: A systematic review. *Global Health Action, 7*, 23574.
Satterthwaite, D., Archer, D., Colenbrander, S., Dodman, D., Hardoy, J., & Patel, S. (2018). *Responding to climate change in cities and in their informal settlements and economies*. International Institute for Environment and Development.
Seaman, R., Mitchell, R., Dundas, R., Leyland, A. H., & Popham, F. (2015). How much of the difference in life expectancy between Scottish cities does deprivation explain? *BMC Public Health Lond., 15*, 1057.
Sen, A. (1999). *Development as freedom*. Oxford University Press.

Seto, K. C., Dhakal, S., Bigio, A., Blanco, H., Delgado, G. C., Dewar, D., … Zwickel, T. (Eds.). (2014). Climate change 2014: Mitigation of climate change. In *Contribution of Working Group III to the Fifth Assessment Report of the Intergovernmental Panel on Climate Change*. Cambridge University Press, Cambridge, United Kingdom and New York, NY, USA.

Shackelford, B. B., Cronk, R., Behnke, N., Cooper, B., Tu, R., D'Souza, M., Bartram, J., Schweitzer, R., & Jaff, D. (2020). Environmental health in forced displacement: A systematic scoping review of the emergency phase. *Science of the Total Environment, 714*, 136553.

Singer, M., & Clair, S. (2003). Syndemics and public health: Reconceptualizing disease in bio-social context. *Medical Anthropology Quarterly, 17*, 423–441.

Snow, J. (1854). The Cholera near golden square, and at Deptford, letter to the editor. *Medical times and Gazette, 9*, 321–322.

Stiglitz, J. E., Sen, A., & Fitoussi, J. -P. (2017). *Report by the commission on the measurement of economic performance and social progress*. French National Institute of Statistics and Economic Studies.

Tan, D. T., Siri, J. G., Gong, Y., Ong, B., Lim, S. C., MacGillivray, B. H., & Marsden, T. (2019). Systems approaches for localising the SDGs: Co-production of place-based case studies. *Globalization and Health, 15*, 85.

UNEP, International Energy Agency. (2017). *Towards a zero-emission, efficient, and resilient buildings and construction sector*. Global Status Report 2017. UNEP.

United Nations. (2015). *The millennium development goals report*. United Nations, New York.

United Nations, Department of Economic and Social Affairs, Population Division. (2019a). World urbanization prospects: The 2018 revision, File 3: Urban Population at Mid-Year by Region, Subregion, Country and Area, 1950–2050 (thousands).

United Nations, Department of Economic and Social Affairs, Population Division. (2019b). World population prospects 2019: Volume II: Demographic profiles.

United Nations, Department of Economic and Social Affairs, Population Division. (2019c). World population prospects 2019: Highlights (ST/ESA/SER.A/423).

United Nations General Assembly. (2015). Resolution adopted by the general assembly on 25 September 2015: Transforming our world: The 2030 Agenda for Sustainable Development.

van Hoof, J., Kazak, J. K., Perek-Białas, J. M., & Peek, S. T. M. (2018). The challenges of urban ageing: Making cities age-friendly in Europe. *International Journal of Environmental Research and Public Health, 15*, 2473.

Vlahov, D., Galea, S., & Freudenberg, N. (2005). The urban health "Advantage". *Journal of Urban Health, 82*, 1–4.

Weller, S. J., Crosby, L. J., Turnbull, E. R., Burns, R., Miller, A., Jones, L., & Aldridge, R. W. (2019). The negative health effects of hostile environment policies on migrants: A cross-sectional service evaluation of humanitarian healthcare provision in the UK. *Wellcome Open Res., 4*, 109.

Whitmee, S., Haines, A., Beyrer, C., Boltz, F., Capon, A. G., de Souza Dias, B. F., Ezeh, A., Frumkin, H., Gong, P., Head, P., Horton, R., Mace, G. M., Marten, R., Myers, S. S., Nishtar, S., Osofsky, S. A., Pattanayak, S. K., Pongsiri, M. J., Romanelli, C., … Yach, D. (2015). Safeguarding human health in the Anthropocene epoch: Report of The Rockefeller Foundation–Lancet Commission on planetary health. *The Lancet, 386*, 1973–2028.

WHO. (1946). Preamble to the Constitution of the World Health Organization as adopted by the International Health Conference, 19–22 June 1946; and entered into force on 7 April 1948. WHO, New York.

WHO. (1986). *Ottawa charter for health promotion*. World Health Organization, Geneva.

WHO. (2012). *Addressing the social determinants of health: The urban dimension and the role of local government*. WHO, Copenhagen, Denmark.

WHO. (2014). *Global status report on noncommunicable diseases 2014: Attaining the nine global noncommunicable diseases targets; a shared responsibility*. WHO, Geneva.

WHO. (2016). *Global report on urban health: Equitable, healthier cities for sustainable development*. WHO, Geneva.

WHO. (2018a). *Noncommunicable diseases country profiles 2018*. WHO, Geneva, Switzerland.

WHO. (2018b). *Savings lives, spending less: A strategic response to noncommunicable diseases*. WHO, Geneva, Switzerland.
WHO. (2018c). The top 10 causes of death. Retrieved April 13, 2020, from https://www.who.int/news-room/fact-sheets/detail/the-top-10-causes-of-death
WHO. (2018d). *Global status report on road safety*. WHO, Geneva.
WHO, United Nations Human Settlements Programme (Eds.). (2010). *Hidden cities: Unmasking and overcoming health inequities in urban settings*. WHO; UN-HABITAT, Kobe, Japan.
Wu, X., Nethery, R. C., Sabath, M. B., Braun, D., & Dominici, F. (2020). Air pollution and COVID-19 mortality in the United States: Strengths and limitations of an ecological regression analysis. *Science Advances, 6*, eabd4049.
Yang, J., Siri, J. G., Remais, J. V., Cheng, Q., Zhang, H., Chan, K. K. Y., Sun, Z., Zhao, Y., Cong, N., Li, X., Zhang, W., Bai, Y., Bi, J., Cai, W., Chan, E. Y. Y., Chen, W., Fan, W., Hua, F., He, J., … Gong, P. (2018). The Tsinghua—Lancet commission on healthy cities in China: Unlocking the power of cities for a healthy China. *The Lancet, 391*, 2140–2184.
Yang, K., & Victor, C. (2011). Age and loneliness in 25 European nations. *Ageing and Society, 31*, 1368–1388.

Chapter 2
Shifting Priorities for Healthy Places

2.1 Introduction

Examining the emergence of urban settlements in ancient cultures reveals many similar dynamics to today's urbanisation trends. People created and moved to cities to increase their quality of life through better economic and environmental circumstances. Ancient cities were important social, cultural and political centres with planned public spaces to facilitate gatherings and governance. Just as in today's cities, there were social divisions between ruling and labouring classes that impacted health. Urban designers and planners over the centuries have revered ancient approaches, extending traditional models to serve political, nationalistic and ideological positions (Smith & Hein, 2017). Barton (2017) provides a detailed overview of the influence that ancient and medieval urban forms have had on modern planning and architecture, particularly with regard to health. He praises ancient understandings of human ecology and the links between cities, land and climate. The beginning of this chapter will review the health impacts of ancient cities and infrastructure, noting that some ancient models of urban design and infrastructure created health inequities and depended upon slavery. The chapter then moves to the modern beginnings of public health and planning through a review of the British sanitarians and social reformers. Twentieth-century approaches to designing healthy and sustainable places are considered, including high-rise housing, neighbourhood-centred design and so-called 15-minute cities. The chapter then describes the international movements that underpin healthy urban design and planning, including the WHO Healthy Cities programme and the New Urban Agenda.

H. Pineo, *Healthy Urbanism*, Planning, Environment, Cities,
https://doi.org/10.1007/978-981-16-9647-3_2

2.2 Health in Ancient Cities and Infrastructure

Historical accounts of the earliest cities show that urban form was influenced by social objectives, including health (de Leeuw, 2017). One example is shown through Etruscan cities, developed in the sixth century BC in modern Italy and later conquered by the Romans. The urbanisation of Etruria increased interaction across socio-economic strata and highlighted the marked power imbalance between aristocrats and the public, who occupied segregated spaces in Etruscan cities (Cerchiai, 2017). The resulting social conflict led to changes in urban form towards a grid street pattern that was meant to signify 'the superiority of the collective norm over any particularism' (ibid., p. 636), which included abandoning or demolishing wealthier neighbourhoods within some Etruscan cities in the fifth century. Whether the Etruscans linked city planning, socio-economic status and health is unknown, however, scholars have suggested that they were aware of environment pollution and its negative health impacts. Harrison et al. (2010) suggest that 'city-wide purification' rituals and even the abandonment of many settlements could have been partly related to heavy metal pollution in soil and water from mining and metalworking.

The remains of sewers and aqueducts from ancient civilisations are evidence that health and hygiene were supported through substantial infrastructure, yet not all urban residents had equal access. City builders of the 3rd millennium BC in modern Crete, Egypt and Pakistan created water and sanitation infrastructure, including wells, pipes, toilets, baths, drains and sewer systems. Archaeological evidence from the city of Harappa in the Indus Civilisation in northwest India and Pakistan shows that sanitation and water infrastructure were provided for elite groups, which may have caused higher risk of disease and death among lower-status residents (Schug et al., 2013). Water and health were linked very early in human history (Angelakis et al., 2020). The role of nature and the environment (including clean water), in addition to individual constitutional factors, were recognised in Hippocratic medicine, an early influence on modern medicine (ibid.). Angelakis et al. noted the necessity of natural springs or aquifers in selecting the location of ancient hospitals, *Asclepieia*, in Greece. As in many societies, the water was used in healing rituals, therapies and other treatments for physical and mental health illnesses.

The Roman Baths, *Aquae Sulis*, located in Bath, England, provide a well-preserved specimen of ancient thermal baths. Eleri Cousins (2020) has analysed the spa's history, beginning with its supposed creation by King Bladud of the Britons because it cured his leprosy. Cousins documents the bath's medieval, Georgian and Victorian healing sanctuary periods, building her argument that there is very little evidence that the Romans created the baths for curative purposes. While water was undoubtedly used for healing in antiquity, Cousins argues that in Bath, the *Aquae Sulis* functioned as a place of pilgrimage, religious worship, social gathering and bathing. In Roman culture, bathing was an important social and individual cleansing

Box 2.1 Public baths in support of health and hygiene

Public baths are a form of water engineering that has enabled health and hygiene in many cultures. While a minimum level of hygiene is accepted as essential to avoid disease, there is also evidence that bathing in water can reduce anxiety, pain and stress—effects that result from specific physiological changes induced by increased pressure and certain water temperatures. Public baths are also a social good with positive wellbeing impacts from social interaction.

Japanese hot springs, or *onsen*, can be found in rural and urban parts of the country and they are a fundamental component of wellbeing and social life. Visiting an *onsen* involves a ritual of individual cleansing before entering the communal baths, where small groups of family, friends or colleagues relax and chat in pools of all shapes and sizes. Some *onsen* play soothing music and have variable water temperatures and other features, while others are simple pools in natural settings. *Onsen* are considered to be an important component of Japan's traditional culture and heritage, thus their social function goes beyond being places to gather.

Ancient and foreign public bathing and water therapy practices have been reinterpreted around the world for health, social and economic purposes. An Irish entrepreneur, Richard Barter, developed a hydro-therapy spa business in the mid-1800s that was exported to Britain and other parts of Europe. Foley (2014) describes the assemblage of traditions, medical practices and new technology that informed such facilities and their use by different groups within society. There was a sense that medical practices at that time were insufficient to cure patients' ills, and many sought alternative treatments. Baths in the 1800s were built in cities and at asylums with clear 'medico-moral purposes', whereby good health, cleanliness, hygiene and morals were all wrapped together in a social belief. Other baths, both urban and rural, were associated with luxury and social interaction, or health retreat identities.

act (see Box 2.1). Modern visitors to the remains of the *Aquae Sulis* marvel at the Roman engineering that harnessed the hot springs to feed the Great Bath at the centre of the bathing complex. Although Roman bathing culture is notably described as being available for all classes in society, slaves were used to operate the baths. In the hot rooms adjacent to the baths, *caldarium*, the floor rested on top of pillars between which child slaves placed fuel to heat the air using the Roman hypocaust system (Fig. 2.1). For a modern visitor, it is clear that slaves would have suffered hardship and toxic air to produce the health-promoting environment enjoyed by some in society.

Reflecting on ancient practices of city building, infrastructure and public baths, there are several conclusions for modern practitioners. Investing considerable effort and expense into designing places that promote the health of the public is not new. People have understood the health benefits and harms of urban design and infrastructure for millennia, even if the specific causal pathways from exposure to health

Fig. 2.1 Hot room, *caldarium*, in the Roman *Aque Sulis* in Bath, England, depicting the hypercaustic heating system. The floor rested on top of the pillars. (Source: Author, 2019)

outcome were unclear. In designing healthy places, built environment professionals should consider the range of pathways that may result in health and wellbeing benefits and how these differ across society. Furthermore, health impacts should be considered broadly including those arising from construction practices, international supply chains and building operation, noting that a 'healthy building' for occupants may create negative impacts for other groups in society.

2.3 Social Reform, Sanitarians and the Rise of Public Health and Planning

Modern understanding of the links between urban environments and health, and the development of policy solutions to improve health, rests on the work of the British sanitarians and social reformers. Edwin Chadwick's (1842) national investigation of the living conditions of the British working class detailed the extensive environmental pollution and poor-quality housing that afflicted their health. In highlighting the discrepancies across social classes, his work was contentious, but ultimately led to the 1848 Public Health Act that addressed issues related to sanitation, water, waste and health boards. These progressive policies were not unique to the UK, but they were also emerging in the USA and Canada, based on similar investigations into sanitary conditions. Chadwick's contribution was a starting point for policy related to social and environmental justice that continues to be advocated for today.

The transition from miasma to germ theories in the late nineteenth century was influenced by the investigations into proposed remedies to environmental exposures; however, it did not depend entirely on an accurate theory of disease transmission. 'Miasma' is a Greek term that was linked to religious and moral explanations of disease pre-dating Hippocrates, who focused on impurities, vapours and foul airs (Curtis, 2007). Cholera was thought to spread through miasma. Epidemiology is famously based on the physician John Snow's study of cholera deaths in two outbreaks in London (1848–1849 and 1853–1854). In one of his analyses, he mapped the deaths in each household in a central London area and he was able to pinpoint the Broad Street pump as the source of disease. He then successfully advocated for the pump handle to be removed. The waterborne transmission of disease that Snow uncovered was not consistent with the prevalent miasma theory explanation for cholera, but the problem was resolved regardless.

Still prior to germ theory, Sir Benjamin Ward Richardson's (1876) plan for healthy cities 'Hygeia: A City of Health' describes a detailed proposal for a utopian city with specifications for building heights, sewage infrastructure, street layout and an underground train. Of particular interest are the details of residential streets and neighbourhoods, which can be identified in the buildings and neighbourhoods that were built in the following decades (Fig. 2.2) and relate to contemporary ideas of healthy design:

> The streets from north to south which cross the main thoroughfares at right angles, and the minor streets which run parallel, are all wide, and owing to the lowness of the houses, are thoroughly ventilated, and in the day are filled with sunlight. They are planted on each side of the pathways with trees, and in many places with shrubs and evergreens. All the interspaces between the backs of the houses are gardens. The churches, hospitals, theatres, banks, lecture-rooms, and other public buildings, as well as some private buildings such as warehouses and stables, stand alone, forming parts of streets, and occupying the position of several houses. They are surrounded with garden space, and add not only to the beauty but to the healthiness of the city. The large houses of the wealthy are situated in a similar manner. (p. 21)

Fig. 2.2 Tree-lined residential streets in North London built in the early 1900s. Cars now have a dominant position in the street and in some front gardens. (Source: Author, 2021)

Richardson's ideals, like other reformers, were hotly contested because he proposed that it was not, as some suggested, the moral depravity of poorer groups in society that determined their health, instead it was a problem of unhealthy living environments.

The social reform struggles of the 1800s are an important part of the history of healthy urbanism. Harold Platt (2007) tells the story of Ancoats, a slum in Manchester, England, where social and environmental reforms were manifested through grassroots activism to overcome corrupt local government in the 1880s. Mancunian Charles Rowley formed the Healthy Homes Society in 1884 to advocate for environmental policy to improve Ancoats. Various scientific studies of health and environmental conditions in the area demonstrated that mortality rates were far higher in Ancoats than other parts of the city. The Thresh Report blamed the health disparities on social and physical conditions but was particularly harsh of the city's failure to properly remove sewage from the outdoor privy system of pail closets. Although wealthy neighbourhoods had indoor flush toilets, city leaders held the view 'that the masses were too stupid and deprived to keep modern plumbing in running order' (Platt, 2007, p. 763). The Thresh Report findings were politically damaging after a previous scandal had exposed the responsible public utility of dumping animal and human faeces into the river upstream of Ancoats. Platt argues that in 1889, social protests and the emergence of germ theory were both instrumental in the eventual creation of city policies, building and health codes that forced landlords to improve dwellings, required the installation of indoor plumbing for all classes, and spurred government-led housing projects.

The British sanitarians and social reformers played a critical role in linking the environment and health, which became the foundation of the public health and urban planning professions. These progressive ideas were adopted by the English town planner Ebenezer Howard in his *Garden Cities of To-morrow* (1902) vision to overcome the pollution and poor housing of industrialised nineteenth-century cities. Howard's plans for Garden Cities aimed to provide high-quality housing, greenspaces, fresh air and amenities for everybody. The plans not only were physical but also outlined communal ownership of land and other assets so that income gained from these assets could be re-invested into the community over time. Letchworth in Hertfordshire, England, was the first Garden City, followed by Welwyn. The model has been adopted elsewhere in the UK and internationally. Moving into the twentieth century, other planners and architects began developing physical solutions for social and health problems, many of which have been criticised, as described further.

2.4 Twentieth-Century Ideas for Healthy Communities

The social and environmental justice ideals that emerged from the British sanitarians informed the purpose of the newly forming public health and urban planning professions in the late nineteenth century. However, the contemporaneous shift towards germ theory moved the focus from socio-ecological models of health towards biomedical models (as introduced in Chap. 1). Jason Corburn (2015) argues that urban designers and planners operated in this paradigm by adopting the model of laboratory science to legitimise their inherently social and political work. Techniques such as land use zoning, strategic planning and cost-benefit analyses were developed in the early twentieth century to drive the efficient use of land. Corburn gives the examples of Howard's Garden Cities movement and Clarence Perry's (1929) Neighbourhood Unit as two proposals that 'offered a physical ideal that tended to ignore the often contested, gendered, variegated and value-laden characteristics of cities', exemplifying his argument that planners sought universally applicable approaches to lend credibility to their work (p. 39). The Neighbourhood Unit concept was adopted in American slum clearance initiatives following the Second World War, often resulting in displacement and weakening of social networks that supported health.

2.4.1 High-Rise Housing for Health

Following both the First and Second World Wars, there were extensive house building efforts in both Europe and the USA that sought to improve residents' health and wellbeing. Le Corbusier's modernist high-rise housing was meant to take Howard's Garden City model into a vertical format to promote residents' health and quality of life. Alexi Marmot's (1981) examination of Le Corbusier's legacy highlights his desire to improve the housing of all urban dwellers, not just those who were least affluent, which was unusual among his contemporaries in modernist architecture. His *Unité d'Habitation* model was developed between the wars and finally funded by the French Ministry of Reconstruction and Urban Planning after the Second World War for construction in Marseilles at the *Cité radieuse* (radiant city). The building is now akin to a pilgrimage for architecture students and influenced much post-war public housing in Europe and beyond. The model envisages a 17-storey building situated in a parkland and raised on pillars. The healthy design features included sound-proofing between dwellings, large interiors, ample daylight and ventilation, views of nature and integrated services such as shops, a gym, a kindergarten and other amenities.

Marmot argued that the problem with many emulations of Le Corbusier's mass housing model is that they failed to properly maintain the buildings and often did not put in place the required services to support residents, leading to disrepair and social problems. This is still a problem in mid-century and more recent high-rise housing. Some modernist architecture of that era remains popular among residents who appreciate its ample interior space, minimalist aesthetic and the views from upper story flats, as exemplified by the Barbican Centre in London, a highly successful nod to Le Corbusier's vision. A well-known example of failed high-rise public housing is the Pruitt-Igoe project in St Louis, Missouri. Katharine Bristol's (1991) account of the post-war urban redevelopment project challenges the popular myth that the expression of modernist architecture was the cause of the failed housing, instead pointing to institutional and structural sources, including racism. Chad Freidrichs' documentary film, *The Pruitt-Igoe Myth*, illustrates Bristol's argument with historical footage alongside contemporary interviews with people who lived in Pruitt-Igoe. Their stories demonstrate that the buildings were doomed to fail from the outset, not by their design but through lack of proper management.

The United States Housing Act of 1949 funded the Pruitt-Igoe development in St Louis, Missouri, USA, to solve the city's problem of racially segregated slums, reinforced by the countrywide 'White flight' trend where middle-class White families moved to the suburbs to fulfil the 'American dream'. The city's plan to upgrade the poor-quality housing inhabited by a predominately Black community was to sell land to private developers to fund public housing which included market rate units. The developers received public money to build Pruitt-Igoe, yet there was no funding for its ongoing management, which was meant to come through tenants' fees. The residents in Freidrichs' film speak of the initial excitement at living in the 'the poor man's penthouse', yet very quickly the punitive policies of the social welfare system and the rapid breakdown of the buildings' state of repair caused problems. Market rate units were not popular and the tenants' fees were not enough to cover maintenance costs. Blaming the architectural style was a convenient way to side-step government responsibility for the failed housing. In addition to exposing the social injustice and racism that heavily contributed to the failure of Pruitt-Igoe, this case serves as an important reminder that the ongoing management and maintenance of buildings is equally essential to their functionality as their design.

2.4.2 *Urban Sprawl to '15-Minute Cities'*

The development of suburbia, underwritten by highways infrastructure investment and lower land values, and settled by the new post-war middle-class, was a huge contributor to the urban sprawl that characterises American cities. Sprawling development is not healthy for people or the planet (Frumkin, 2004), nor is it uniquely American. For example, the negative health impacts of rapid car-centric growth in the outskirts of Australian cities have been documented (Christian et al., 2017; Giles-Corti et al., 2013). The low residential density and lack of mixed land uses of many suburban landscapes reinforce the need for car ownership. Many suburbs do not have adequate public transport or pavements (sidewalks) and walking along their strip mall-lined commercial streets feels unsafe, or at least unpleasant.

The environmental and social impacts of poorly planned urbanisation and suburban areas drew attention to the need for new models of sustainable development in the 1970s and 1980s. These new models integrated health considerations, sought to reverse the sprawling development patterns of post-war America and aimed to mitigate against the causes of climate change. Simon Joss (2015) synthesised three seminal approaches to sustainable cities, including Richard Register's (1987) urban ecology, Jeffrey R. Kenworthy's eco-city dimensions (2006) and Steffan Lehmann's (2010) green urbanism. Whilst there are subtle differences between these approaches, their commonalities include a range of social, economic and environmental aspirations, such as compact mixed-use communities, prioritisation of public transport over cars, ample greenspace, affordable housing, resource conservation and inclusive planning processes. The urban ecology approach has strong inclusivity principles, drawing attention to social justice and the needs of women, ethnic minorities and disabled people.

Concurrently, frameworks and common approaches for sustainable urban design and planning at the scale of urban infill, extension and new communities were produced internationally. The principles and Charter of the Congress for New Urbanism, founded in 1996 integrate walkability and accessible public space goals to promote wellbeing (Talen, 2013). Sustainable building standards broadened to cover large-scale development, such as LEED Neighbourhood Development and BREEAM Communities in the early 2010s. The EcoDistricts (2018) framework and protocol (originating in Portland, Oregon, USA) outlines several measures to promote health and wellbeing, including goals for physical activity, health care, food, environmental pollution and safety. The extent to which health considerations were integrated into these various sustainable design approaches varies, yet through the aligned goals of environmentalism and social cohesion, they are likely to support health.

Creating neighbourhoods as broadly self-sufficient urban units was advocated by New Urbanism, Transit Oriented Development, Urban Villages and other approaches as a solution to the low-density sprawling development of the 1960s, most recently badged as so-called 15- or 20-minute cities/neighbourhoods. This urban planning approach made media headlines during the COVID-19 pandemic as people were in 'lockdown' and restricted from travelling outside of their local area. The goal of these compact and complete neighbourhood environments is to reduce the need to travel long distances for daily activities, especially by car, thereby reducing pollution and improving wellbeing and social networks. There is substantial debate about whether this model is desirable and practical.

Pozoukidou and Chatziyiannaki (2021) explored the history and current manifestation of this neighbourhood focus noting its origins in Clarence Perry's Neighbourhood Unit and the work of Christopher Alexander. They analysed examples of the 15- or 20-min city in Paris (France), Portland (USA) and Melbourne (Australia) against criteria related to inclusion, health and safety. Pozoukidou and Chatziyiannaki concluded that there are commonalities with these strategies and other neighbourhood-focused design approaches, mainly the use of accessibility, walkability, density, land use mix and design diversity as core objectives. However, there is also a shift in some strategies towards proximity over accessibility which would require spatially clustering everything that a resident needs within a short travel distance from their home. This goal raises practical challenges for land use, employment models and the potential unintended consequence of displacing low-income residents as property prices are likely to rise. In conclusion, they argue that 15- or 20-min city plans are 'neither radical nor a -fit for all- idea', but they can promote wellbeing if they are informed by community engagement and implemented with provision of affordable and rental housing (ibid., p. 22).

Another component to debates about the merits of urban planning models is the extent to which the built environment can directly influence behaviour, known as environmental determinism. For instance, New Urbanism has been criticised as over-claiming for the attribution of social effects to the physical environment (Carmona, 2021). Petter Næss (2016) summarises the main arguments against recognising causal effects of the urban environment, which he sees as stemming from 'philosophical positions and disciplinary traditions that completely reject the notion of causality in studies of humans and the social world, or deny that the physical world can have any causal influence on social life' (p. 54). In response to critiques of environmental determinism, the concepts of environmental 'possibilism' (recognises environmental constraints but puts more emphasis on social conditions and choice) and 'probabilism' (sees the environment as creating opportunities that may discourage or encourage certain behaviours) were proposed. In Næss' view, the latter is a more accurate reflection of the complex set of potential outcomes arising from interactions between people and the environment. Using a critical realist philosophy of science perspective, Næss moves beyond determinism, possibilism and probabilism:

> The ways in which human actions depend on built environment characteristics should not be conceived of as either voluntarism or determinism, but instead in terms of tendencies or dispositions, where the 'structural imperatives' levied by the built environment are adapted, augmented or counteracted by a number of other structural and individual conditions. Furthermore, the relationships between the built environment and the actions of individuals are not unidirectional but a matter of two-way influences. (p. 61)

He goes on to state that the causal effects of buildings or other urban infrastructure are not automatically triggered, 'but they can (usually in interaction with other causal powers) enable, amplify, facilitate, restrain, suppress or prevent the occurrence of events and situations' (ibid.). Næss' explanation is useful for this book's focus on improving the health impacts of the urban environment because it accounts for the complexity of this system, in which people's actions and behaviours (e.g. sleep, physical activity and diet) are driven by political, cultural, social and physical factors (among others). His conceptualisation of causality in the built environment highlights the requirement for multidisciplinary teams to unpick the diverse factors that may influence 'tendencies or dispositions', considering how these may change over time and space, to increase understanding and inform action. The issue of determinism is discussed again in Chap. 3, regarding public health policy and its effects on human health.

This chapter has thus far provided evidence that health and wellbeing have been integrated into urban design since the first human settlements. Environmental pollution, climate change and inequity are not exclusively modern issues, and there are lessons to be learned from the successes and failures of prior civilisations. City leaders and architects have historically blamed the 'unhealthy' behaviours of low-income or ethnic minority groups rather than take responsibility for not providing adequate living environments for everybody in society. These attitudes are evident in contemporary cities and they need to be challenged. By learning from historical successes and failures and new knowledge about the urban environment's effect on health, built environment professionals can use inclusive processes to develop and trial new approaches to healthy urbanism.

2.5 Mobilising Non-health Sectors

Contemporary healthy urban development agendas are tied to the global challenges of increasing inequities, climate change and chronic disease. The COVID-19 pandemic heightened built environment professionals' awareness of the interconnections among these challenges. But even a decade prior to the pandemic, there was a renewed interest among built environment professional institutions and organisations about their role in supporting health. This shift was partly driven by the public health goal of preventing illness by changing the so-called upstream determinants of health. As Chap. 1 explained, people are living longer and obesity and physical inactivity have risen sharply, resulting in global increases in non-communicable diseases (NCDs) and costly pressures on healthcare systems. The WHO and governments around the world have recognised that preventing NCDs reduces human suffering and is far more cost-effective than treating these conditions, which can last for many years. Health leaders and governments have therefore sought to influence 'upstream' societal factors (e.g. affordable housing and jobs) that, in turn, shape 'downstream' health risks (e.g. smoking and unhealthy diet), also known as 'distal' and 'proximal' factors in public health terms. Whilst this distinction is relatively easy to grasp, Krieger (2008) critiques these terms for obscuring the true causal relations between different exposures and health outcomes, instead proposing a focus on levels, pathways and power. Krieger's work and other ways to conceptualise the health impact of the urban environment are described in Chap. 3. This section explores the ways through which non-health sectors are mobilised through introductions to the WHO Healthy Cities Movement, the New Urban Agenda and government initiatives to support healthy urban design and planning.

The complexities of urban health mean that there is not a single focus or agenda for 'healthy' building and place-making. There are multiple agendas competing for attention among built environment disciplines that are communicated with differing terms, rationales and underlying evidence. International bodies such as the WHO and United Nations have considerably shaped ideas about how health and sustainability can be achieved through urban development. Guidance from these bodies has informed national and urban policies to promote health. In recent years, voluntary standards have emerged that demonstrate the health credentials of new buildings and communities (e.g. WELL and Fitwel) and are linked to the financial value of a development (see Chap. 8). Prior to and during the COVID-19 pandemic, there has been interest in the office real estate sector in increasing the productivity and retention of employees through office design that supports health and wellbeing. It is important to understand the motivations of these agendas and how they relate to or compete with each other so that those who apply specific principles and frameworks understand the potential health benefits and disbenefits for different populations groups.

2.5.1 *The WHO Healthy Cities Movement*

As new models of sustainable development emerged in the 1970s and 1980s, there were also shifting understandings of the factors that determine a person's health, whereby the biomedical model that had dominated much of the twentieth century was challenged with the renewed focus on socio-ecological models. When the WHO Healthy Cities programme was created in 1986, there was a growing understanding of the role that actors across local government play in improving population health. The programmed originated in Europe and rapidly grew internationally (Hancock, 1993). Local government members of the network committed to developing a city profile based on a series of indicators that were used to inform evidence-based policy development across local government departments (Breuer, 1998).

The emphasis of the Healthy Cities movement is the ongoing process of improving health, rather than achieving a final end state. Trevor Hancock and Leonard Duhl, two leading figures in the movement, provided the WHO's definition for a healthy city in a publication outlining the programme aims and process:

> A healthy city is one that is continually creating and improving those physical and social environments and expanding those community resources which enable people to mutually support each other in performing all the functions of life and in developing their maximum potential. (Hancock & Duhl, 1986, p. 24)

The policy domains that were required to take action were broad as they reflected the ecological model of health provided by Hancock's (1985) work for the City of Toronto in Canada. The parameters of health that underpinned the Healthy Cities programme included: the physical environment, stable and sustainable ecosystems, community cohesion, public participation, basic needs being met (food, water, shelter, work, etc.), social interaction, strong local economy, heritage, urban form that supports the preceding factors, health services and good health status (Hancock & Duhl, 1986, p. 33).

Research on governance for health shows the importance of the WHO Healthy Cities movement in increasing cross-sectoral healthy city policies (de Leeuw et al., 2014; de Leeuw & Skovgaard, 2005). For example, it has influenced urban planners to collaborate with public health partners and integrate health into urban policy and development. Progress against the network's goals is variable internationally and there is a continued need to emphasise the key messages to local government officers and politicians. Evelyne de Leeuw and Jean Simos' (2017) book provides a detailed account of how the movement has shaped urban health governance internationally.

Following the launch of the WHO Healthy Cities programme, the United Nations Rio 'Earth Summit' in 1992 was described by Hugh Barton (2005) as a pivotal moment for healthy urban planning. He viewed the conference and the years following as instigating a return to considering settlements as human habitats rather than 'simply as physical or aesthetic constructs, or manifestations of economic forces' (p. 339). The Local Agenda 21 movement resulting from Rio called for increased collaboration across local government policy domains to promote sustainable development and the involvement of communities in governance processes (Rydin, 2010). Concurrently, the WHO Healthy Cities programme and like-minded organisations were considering how health could be integrated into strategic and environmental impact assessments to develop holistic appraisals of projects, programmes and policies. In Barton's view, these changes were nothing less than a 'paradigm shift, a new collective mind set, and a revitalized vision of what is appropriate and possible in settlement planning' (2005, p. 340). Approaching the turn of the century, sustainable community and city strategies were receiving significant attention from policy-makers, with consideration of economic, social and environmental goals.

2.5.2 *Health in the New Urban Agenda*

The result of efforts from the WHO, United Nations and other international bodies to promote consideration of health in all government policies has been profound. Although there is still considerable progress to be made, the UN 2030 Agenda for Sustainable Development has integrated health as a foundational requirement for producing and measuring progress against the Sustainable Development Goals (SDGs). The Agenda's vision makes clear that human health depends on functioning environmental systems and equitable access to social and economic resources, such as education and work. The 17 SDGs cut across many policy domains (e.g. food, transport and energy), demonstrating the need for coordinated action across government tiers and multiple sectors. The detail for each of the SDGs is set out in 169 targets, for which monitoring indicators are applied to measure progress at global, national and local scales.

The 2030 Agenda was seen as a significant shift in international sustainable development priorities. Parnell (2016) outlines five major changes compared to the prior UN Habitat processes, whereby the SDGs: (1) are universal rather than aimed at the Global South, (2) give greater emphasis to ecological limits and climate change, (3) allow for local, national and global indicators, (4) acknowledge the requirement to connect development priorities with finance and (5) note the importance of sub-national government and cities as pathways to sustainable development, through the inclusion of goal 11 (see Table 2.1). The increased attention to cities and environmental performance stimulated by the SDGs process marked an opportunity to specifically align the urban health and sustainable development agendas. Dora et al. (2015) emphasised the importance of environmental determinants of health and their integration in the SDGs by proposing specific monitoring indicators that account for health equity and sustainability goals. Table 2.1 summarises many of their proposed indicators, highlighting that health is cross-cutting through the SDGs. The importance of indicators to drive and monitor progress towards healthy urbanism is emphasised throughout the book, specifically by providing evidence-based design strategies coupled with suitable indicators.

Table 2.1 Selected Sustainable Development Goals (UN, 2015) and monitoring indicators for health equity proposed by Dora et al. (2015)

Selected SDGs	Selected indicators to monitor the SDG health equity impacts, summarised from Dora et al. (2015)
Goal 2. End hunger, achieve food security and improved nutrition and promote sustainable agriculture	(1) Prevalence of stunting in children younger than 5 years; (2) Prevalence of adults (≥18 years), young children (≤5 years), older children and adolescents (5–18 years) who are overweight (or obese); (3) % of calories from saturated and unsaturated fat; (4) consumption of red meat (kg/per capita per day); (5) % of adult population (≥ 18 years) who eat less than five servings of fruit and vegetables, on average, per day; (6) household dietary diversity (HDDS) score.
Goal 3. Ensure healthy lives and promote well-being for all at all ages	(1) % of the urban population exposed to small or fine urban particulates (PM_{10} or $PM_{2.5}$) in concentrations exceeding WHO Air Quality Guidelines; (2) Estimated burden of disease from urban ambient air pollution.
Goal 6. Ensure availability and sustainable management of water and sanitation for all	(1) % of population using basic drinking water; (2) % of population using basic sanitation; (3) % of population with hand-washing facilities at home; (4&5) % of health care facilities and schools with basic drinking water, basic sanitation and hygiene.
Goal 7. Ensure access to affordable, reliable, sustainable and modern energy for all	(1) % of households using modern fuels or technologies, as defined by WHO guidelines, for all cooking, heating and lighting activities; (2) mortality and morbidity attributed to household air pollution; (3) % of health facilities with access to clean and reliable electricity; (4) global mortality and morbidity attributed to outdoor ambient air pollution at levels above WHO guidelines.
Goal 11. Make cities and human settlements inclusive, safe, resilient and sustainable	(1) % of people living in urban slums; (2) % of urban households living in durable structures; (3) % of urban households with access to 'modern' energy sources for heating, cooking and lighting; (4) % of trips or passenger kilometres travelled by public transport, cycling and walking; (5) number of traffic injury deaths, including among vulnerable road users per 1000 km of non-motorised travel.

2.5.3 *Urban Health Governance*

The challenge raised through the integrated and holistic model of healthy and sustainable development proposed by the 2030 Agenda is that actors are required to work across government silos and across public and private sectors. In some places, such as China, national strategies aim to achieve this level of integration (see Box 2.2). However, countries do not often mandate the required amount of cooperation from the national level. Outside of health departments and the healthcare sector, there is little incentive to spend limited resources on policy initiatives that will promote health unless a clear argument can be made for benefits that align to the respective organisation's mission or remit. To address this, advocates of healthy urban development have started using the concept of *co-benefits* to describe policies that will support multiple challenges simultaneously. For example, investing in urban green infrastructure can reduce the urban heat island effect and flood risk, whilst supporting biodiversity and human wellbeing. This approach of seeking win–win solutions builds on the governance principles for other complex challenges like climate change and increasing income inequalities, which all require cooperative action from multiple sectors and disciplines.

Complex problems require governance based on systems thinking, multi-stakeholder collaboration, co-production and multisectoral action. Scholars consider urban health to be a complex adaptive system (Corburn & Cohen, 2012; Gatzweiler et al., 2017; Glouberman et al., 2006), requiring new governance models that involve greater attention to local knowledge to inform experimental policy-making that is coupled with close monitoring. This systems-based conceptualisation of governance for health has been advocated by the WHO. Governance for health means 'the attempts of governments or other actors to steer communities, countries or groups of countries in the pursuit of health as integral to well-being through both whole-of-government and whole-of-society approaches' (Kickbusch & Gleicher, 2012, p. vii). The idea behind whole-of-government and whole-of-society approaches recognises that complex urban health issues necessitate multiple government departments (and tiers), the public and private sectors and civil society to collaborate in understanding the issues and creating solutions. A key component of the whole-of-government concept is health-in-all-policies, meaning that all public sector policy portfolios should consider health. Kickbusch and Gleicher (2012) describe health-in-all-policies as 'an innovation in governance' that responds to a persistent government focus on 'care and cure rather than on health promotion and disease prevention' (pp. 39–40). The Healthy China 2030 strategy described in Box 2.2 is an example of a national strategy that adopts a whole-of-government approach and calls for health-in-all-policies (Yang et al., 2018).

The principles advocated by the WHO and scholars can be identified in urban policy documents and professional practice, particularly through increased interest in systems thinking and co-production/co-design methods. Adopting a systems thinking approach is necessary for multiple challenges affecting urban development, including the climate crisis and demographic change. The WHO and UN-Habitat communicate the complementarities of equitable, healthy and sustainable urban and territorial planning in a sourcebook publication with reference to international case studies (UN-Habitat and WHO, 2020). Through tools such as asset-based planning, health impact assessment and advocacy frameworks, urban planners are encouraged to work with health professionals to drive holistic policy and decision-making. There has been variable progress on the principles of healthy urban planning due to poor understanding of the wider determinants of health, the perceived additional upfront costs and other factors (described in Chap. 8). In England, the National Health Service (NHS) sought to promote healthy urban

Box 2.2 Healthy China 2030 strategy

The Healthy China 2030 strategy (Central Committee of Chinese Communist Party and State Council, 2016) has been a significant driver for intersectoral action for health in China. The strategy puts forward that 'the construction of healthy cities and towns should be regarded as an important starting point to promote the construction of healthy China' (ibid.), laying the foundations for integrated healthy design in buildings, new communities and cities (Wang et al., 2020). It has been described as a 'dramatic departure from traditional strategies' for health service improvement towards a more holistic governance model covering multiple environmental and social determinants of health (Yang et al., 2018, p. 3). The strategy calls for increased environmental monitoring and public participation, among other measures. In 2019, the State Council established the Promotion Committee of Healthy China, which focuses on monitoring and evaluation of Healthy China 2030 via indicators (National Health Commission, 2019).

Pineo et al.'s (2021a) research on the use of evidence in urban sustainability and health governance in two Chinese cities identified the key role of the Healthy China 2030 strategy in driving local action. It was frequently mentioned by study participants as defining and putting weight behind their work because it demonstrated political commitment. Lessons from other national policy agendas in China identify a gap between aims and local implementation. The sustainable urbanisation agenda has been promoted in each of China's strategic Five-Year Plans since the 1992 Rio Summit (Chung et al., 2018), with a national ecological civilisation agenda since 2013, putting significant weight on social and environmental over economic goals. However, de Jong et al. (2016) uncovered a lack of understanding by the central state of the importance of local networks between local government officials and land developers, thus undermining local progress. It will therefore be important to closely monitor progress on the environmental components of the Healthy China 2030 plan, alongside other initiatives to support local implementation.

development through its Healthy New Towns programme (Box 2.3), which captured lessons about the process for wider dissemination among the public and private sectors. This project occurred in the context of renewed collaboration between planning and public health in England spurred on by legislative changes (McKinnon et al., 2020). The Health and Social Care Act 2012 changed the structure and function of public health within government, moving public health teams from the NHS into local government, where they could more closely work with planners and other services that influence health, although cuts to public sector spending have dampened some of the potential benefits of this restructure (see Carmichael et al., 2019).

Box 2.3 NHS England Healthy New Towns programme
The National Health Service (NHS) was established in 1948 to provide free healthcare for all citizens. In 2015, NHS England launched a Healthy New Towns programme aimed at understanding how to develop healthy new places to improve the social determinants of health. Ten demonstrator sites were selected to take part in the programme. The sites were large-scale proposed developments (ranging between 900 and 15,000 homes) at different stages of the design and planning process. Through participating in the Healthy New Towns programme, the demonstrator sites were encouraged to share learning with each other and contribute to other dissemination activities including the 'Putting Health Into Place' series of guidance documents (NHS England, 2019). Many lessons were learned through the programme, including the following key points for built environment practitioners: early engagement with existing and future communities is important to understand and respond to their needs; collaboration with the NHS supports better integration of health and care facilities in new places; and drawing upon the latest guidance and standards to design healthy places and benchmark proposals against such frameworks ensures quality is achieved. Chapter 6 includes a case study of one of the demonstrator sites, Barton Park in Oxford, describing how the developers engaged with the local community and funded local services.

2.6 Conclusion

Contemporary approaches to planning for health and wellbeing have been diversely conceptualised as planning for liveability, quality of life, community wellbeing and other wide-reaching objectives. Rather than considering this a specialist area, Barton (2017) positions wellbeing as *the key objective* of the planning system, but one that requires specific understanding and knowledge. Despite this integrative vision, many planning scholars and practitioners recognise that other planning objectives, such as economic growth, may take priority in decision-making and cause indirect negative health effects (Kent & Thompson, 2019). This chapter has reviewed historical approaches to planning healthy cities, noting the challenges and enduring lessons from past examples. The UN SDGs clearly articulate the interconnected nature of health, equity and sustainability, yet there is a need to better define how this can be achieved for all built environment stakeholders. The next chapter describes gaps in existing understandings of healthy urban development and describes the evidence and theory underpinning the THRIVES framework, calling for a reconceptualisation of healthy urbanism that prioritises equity, inclusion and sustainability.

References

Angelakis, A. N., Antoniou, G. P., Yapijakis, C., & Tchobanoglous, G. (2020). History of hygiene focusing on the crucial role of water in the Hellenic Asclepieia (i.e., Ancient Hospitals). *Water, 12*, 754.

Barton, H. (2005). A health map for urban planners: Towards a conceptual model for healthy, sustainable settlements. *Built Environment, 31*, 339–355.

Barton, H. (2017). *City of well-being: A radical guide to planning*. Routledge; Taylor & Francis Group.

Breuer, D. (1998). *City health profiles: A review of progress*. World Health Organization, Regional Office for Europe.

Bristol, K. G. (1991). The Pruitt-Igoe myth. *Journal of Architectural Education, 44*, 163–171.

Carmichael, L., Townshend, T. G., Fischer, T. B., Lock, K., Petrokofsky, C., Sheppard, A., Sweeting, D., & Ogilvie, F. (2019). Urban planning as an enabler of urban health: Challenges and good practice in England following the 2012 planning and public health reforms. *Land Use Policy, 84*, 154–162.

Carmona, M. (2021). *Public places urban spaces: The dimensions of urban design* (3rd ed.). Routledge.

Central Committee of Chinese Communist Party, State Council. (2016). *The plan for healthy China 2030. Government of the People's Republic of China*, Beijing.

Cerchiai, L. (2017). Urban civilization. In A. Naso (Ed.), *Etruscology*. De Gruyter, Berlin.

Chadwick, E. (1842). *Report to Her Majesty's principal secretary of state for the home department, from the poor law commissioners, on an inquiry into the sanitary condition of the labouring population of Great Britain* (House of Commons Sessional Paper). W. Clowes and Sons, London.

Christian, H., Knuiman, M., Divitini, M., Foster, S., Hooper, P., Boruff, B., Bull, F., & Giles-Corti, B. (2017). A longitudinal analysis of the influence of the neighborhood environment on recreational walking within the neighborhood: Results from RESIDE. *Environmental Health Perspectives, 125*, 077009.

Chung, C. K. L., Zhang, F., & Wu, F. (2018). Negotiating green space with landed interests: The urban political ecology of Greenway in the Pearl River Delta, China. *Antipode, 50*, 891–909.

Corburn, J. (2015). Urban inequities, population health and spatial planning. In H. Barton, S. Thompson, M. Grant, & S. Burgess (Eds.), *The Routledge handbook of planning for health and well-being: Shaping a sustainable and healthy future* (pp. 37–47). Taylor and Francis.

Corburn, J., & Cohen, A. K. (2012). Why we need urban health equity indicators: Integrating science, policy, and community. *PLoS Medicine, 9*, e1001285.

Cousins, E. H. (2020). *The sanctuary at bath in the Roman Empire, Cambridge classical studies*. Cambridge University Press.

Curtis, V. A. (2007). Dirt, disgust and disease: A natural history of hygiene. *Journal of Epidemiology and Community Health, 61*, 660–664.

de Jong, M., Yu, C., Joss, S., Wennersten, R., Yu, L., Zhang, X., & Ma, X. (2016). Eco city development in China: Addressing the policy implementation challenge. *Journal of Cleaner Production*, Special Volume: Transitions to Sustainable Consumption and Production in Cities, *134*, 31–41.

de Leeuw, E. (2017). Cities and health from the neolithic to the anthropocene. In E. De Leeuw & J. Simos (Eds.), *Healthy cities: The theory, policy, and practice of value-based urban planning* (pp. 3–30). Springer.

de Leeuw, E., & Simos, J. (Eds.). (2017). *Healthy cities: The theory, policy, and practice of value-based urban planning*. Springer.

de Leeuw, E., & Skovgaard, T. (2005). Utility-driven evidence for healthy cities: Problems with evidence generation and application. *Social Science & Medicine, 61*, 1331–1341.

de Leeuw, E., Tsouros, A. D., Dyakova, M., & Green, G. (Eds.) (2014). *Healthy cities, promoting health and equity, evidence for local policy and practice: Summary evaluation of Phase V of the WHO European Healthy Cities Network*. World Health Organization Regional Office for Europe, Copenhagen, Denmark.

Dora, C., Haines, A., Balbus, J., Fletcher, E., Adair-Rohani, H., Alabaster, G., Hossain, R., de Onis, M., Branca, F., & Neira, M. (2015). Indicators linking health and sustainability in the post-2015 development agenda. *The Lancet, 385*, 380–391.

EcoDistricts. (2018). *EcoDistricts protocol: The standard for urban and community development*, version 1.3. EcoDistricts, Portland, Oregon.

Foley, R. (2014). The Roman–Irish Bath: Medical/health history as therapeutic assemblage. *Social Science & Medicine, 106*, 10–19.

Frumkin, H. (2004). *Urban sprawl and public health: Designing, planning, and building for healthy communities*. Island Press.

Gatzweiler, F. W., Zhu, Y.-G., Roux, A. V. D., Capon, A., Donnelly, C., Salem, G., Ayad, H. M., Speizer, I., Nath, I., Boufford, J. I., Hanaki, K., Rietveld, L. C., Ritchie, P., Jayasinghe, S., Parnell, S., & Zhang, Y. (2017). *Advancing health and wellbeing in the changing urban environment: Implementing a systems approach, urban health and wellbeing*. Springer.

Giles-Corti, B., Bull, F., Knuiman, M., McCormack, G., Van Niel, K., Timperio, A., Christian, H., Foster, S., Divitini, M., Middleton, N., & Boruff, B. (2013). The influence of urban design on neighbourhood walking following residential relocation: Longitudinal results from the RESIDE study. *Social Science & Medicine, 77*, 20–30.

Glouberman, S., Gemar, M., Campsie, P., Miller, G., Armstrong, J., Newman, C., Siotis, A., & Groff, P. (2006). A framework for improving health in cities: A discussion paper. *Journal of Urban Health, 83*, 325–338.

Hancock, T. (1985). The mandala of health: A model of the human ecosystem. *Family & Community Health, 8*, 1–10.

Hancock, T. (1993). The evolution, impact and significance of the health cities/healthy communities movement. *Journal of Public Health Policy, 14*, 5–18.

Hancock, T., & Duhl, L. J. (1986). *Healthy cities: Promoting health in the urban context* (No. WHO Healthy Cities Paper #1). World Health Organization Regional Office for Europe.

Harrison, A. P., Cattani, I., & Turfa, J. M. (2010). Metallurgy, environmental pollution and the decline of Etruscan civilisation. *Environmental Science and Pollution Research, 17*, 165–180.

Joss, S. (2015). *Sustainable cities: Governing for urban innovation, Planning, environment, cities*. Palgrave Macmillan.

Kent, J., & Thompson, S. (2019). Planning Australia's healthy built environments. In *Routledge research in planning and urban design*. Routledge.

Kenworthy, J. R. (2006). The eco-city: Ten key transport and planning dimensions for sustainable city development. *Environment and Urbanization, 18*, 67–85.

Kickbusch, I., & Gleicher, D. (2012). *Governance for health in the 21st century*. World Health Organization, Regional Office for Europe, Copenhagen.

Krieger, N. (2008). Proximal, distal, and the politics of causation: What's level got to do with it? *American Journal of Public Health, 98*, 221–230.

Lehmann, S. (2010). *The principles of green urbanism: Transforming the city for sustainability*. Earthscan.

Marmot, A. F. (1981). The legacy of Le Corbusier and high-rise housing. *Built Environment, 7*, 82–95.

McKinnon, G., Pineo, H., Chang, M., Taylor-Green, L., Johns, A., & Toms, R. (2020). Strengthening the links between planning and health in England. *BMJ, 369*. https://doi.org/10.1136/bmj.m795

National Health Commission. (2019). Healthy China action (2019–2030). Retrieved October 25, 2020, from http://www.gov.cn/xinwen/2019-07/15/content_5409694.htm

Næss, P. (2016). Built environment, causality and urban planning. *Planning Theory & Practice, 17*, 52–71.

NHS England. (2019). *Putting health into place: Executive summary*. NHS England, London.

Parnell, S. (2016). Defining a global urban development agenda. *World Development, 78*, 529–540.

Perry, C. A. (1929). City planning for neighborhood life. *Social Forces, 8*, 98–100.

Pineo, H., Zhou, K., Niu, Y., Hale, J., Willan, C., Crane, M., Zimmermann, N., Michie, S., Liu, Q., & Davies, M. (2021a). Evidence-informed urban health and sustainability governance in two Chinese cities. *Buildings and Cities, 2*, 550–567. https://doi.org/10.5334/bc.90

Platt, H. L. (2007). From Hygeia to the garden city: Bodies, houses, and the rediscovery of the slum in Manchester, 1875–1910. *Journal of Urban History, 33*, 756–772.

Pozoukidou, G., & Chatziyiannaki, Z. (2021). 15-Minute city: Decomposing the new urban planning Eutopia. *Sustainability, 13*, 928.

Register, R. (1987). *Ecocity Berkeley: Building cities for a healthy future*. North Atlantic Books.

Richardson, B. W. (1876). *Hygeia, a city of health*. Macmillan.

Rydin, Y. (2010). *Governing for sustainable urban development* (1st ed.). Earthscan.

Schug, G. R., Blevins, K. E., Cox, B., Gray, K., & Mushrif-Tripathy, V. (2013). Infection, disease, and biosocial processes at the end of the Indus civilization. *PLoS One San Franc., 8*, e84814.

Smith, M. E., & Hein, C. (2017). The ancient past in the urban present. In C. Hein (Ed.), *The Routledge handbook of planning history* (1st ed., pp. 109–120). Routledge.

Talen, E. (2013). *Charter of the new urbanism: Congress for the new urbanism* (2nd ed.). McGraw-Hill Education.

United Nations. (2015). *The millennium development goals report*. United Nations, New York.

Wang, Q., Deng, Y., Li, G., Meng, C., Xie, L., Liu, M., & Zeng, L. (2020). The current situation and trends of healthy building development in China. *Chinese Science Bulletin, 65*, 246–255.

WHO. (2020). *Personal interventions and risk communication on air pollution*. WHO, Geneva, Switzerland.

Yang, J., Siri, J. G., Remais, J. V., Cheng, Q., Zhang, H., Chan, K. K. Y., Sun, Z., Zhao, Y., Cong, N., Li, X., Zhang, W., Bai, Y., Bi, J., Cai, W., Chan, E. Y. Y., Chen, W., Fan, W., Hua, F., He, J., ... Gong, P. (2018). The Tsinghua—Lancet commission on healthy cities in China: Unlocking the power of cities for a healthy China. *The Lancet, 391*, 2140–2184.

Chapter 3
A Framework for Healthy Urbanism

3.1 Introduction

The dominance of the biomedical model of health has made it difficult for the WHO and other organisations to communicate the important health impacts of social, economic and environmental factors. In many ways, the biomedical focus on healthcare and pharmaceutical treatments is easier for governments to manage than the diffuse activities which are required to improve health through all city policies and sectors. A key first step in addressing the distributed set of factors that affect health is in conceptualising this complex urban health system.

Kate Raworth (2017) makes a strong argument for the communicative power of visual models in her *Doughnut Economics* book. Complex processes can be simplified and understood by a wide audience through a simple picture. For Raworth, this picture is essentially a doughnut with human thriving being in the doughy portion between a minimum foundation of social and environmental needs at the centre and an ecological ceiling that society cannot overshoot at the outer boundary of the doughnut. For public health practitioners, the Dahlgren and Whitehead (1991) rainbow-shaped social determinants of health diagram has been effective at communicating the complex factors affecting health, encompassing individual behaviours, community networks, ecosystems and macro-level economic conditions. As discussed in Chap. 1, a broad ecological model of health informed key international consensus documents, such as the Ottawa Charter (1986), leading to the creation of the WHO Healthy Cities movement. These conceptual diagrams effectively explain complex phenomena to relatively broad audiences. A more specific and targeted model is needed to communicate the health impact of the urban built environment.

H. Pineo, *Healthy Urbanism*, Planning, Environment, Cities,
https://doi.org/10.1007/978-981-16-9647-3_3

3.2 Conceptualising Health Impacts of Urban Development

The research community has been developing models to inform both policy and research about the causal relations between urban environment exposures and health outcomes. Similar models have been produced by industry and professional bodies with the aim of informing built environment audiences. Pineo (2020) reviewed 15 frameworks of built environment and health relations as part of the development of THRIVES. The international publications within which these frameworks are contained form valuable professional guidance for planners, architects and other built environment professionals.[1] This review identified four common visual models of the urban environment and health: (1) logic or causal chain model, (2) nested circles, (3) multi-nodal and (4) hybrids (Fig. 3.1). As Raworth (2017) described in developing her doughnut model, these types of visuals have an important communicative value, but they are not equally clear to all audiences and each has its own strengths and weaknesses.

3.2.1 Logic or Causal Chain Models

A logic or causal chain model represents environmental exposures (typically on the left side of a figure) with directional arrows to various effects (as shown in Fig. 3.1), including behaviours and health outcomes. The model in Fig. 3.2, produced by Amy Schulz, Mary Northridge and colleagues, makes explicit the relations between macro-, meso- and micro-scale built environment exposures (or risk factors) and specific health outcomes (Northridge et al., 2003; Schulz & Northridge, 2004). Many similar diagrams have been produced covering broad urban health factors (Berke & Vernez-Moudon, 2014; Giles-Corti et al., 2016; Rydin et al., 2012), or focusing on a sub-set such as transport (Nieuwenhuijsen, 2016).

The benefit of logic or causal chain models is the clear depiction of aspects of the built environment that are causally associated with specific health outcomes, alongside the inclusion of broad contextual factors such as inequity that affect the causal pathway. However, it is challenging to visually show all of the interconnections between cause and effect and some relations may exist but are not yet evidence-based. Pineo et al.'s (2018b) causal pathways framework attempts to show the strength of evidence supporting a causal relationship between different exposures and effects using dotted and solid lines. Their framework contains many more causal lines than Fig. 3.2, meaning that a policy-maker could trace cardiovascular disease back to air pollution, for example. However, the many intersecting lines in the visual appear complicated and could be unapproachable for some audiences. In

[1] An online database of healthy urban planning and design guidance documents is available at http://www.healthyurbanism.net/guidance-for-healthy-urbanism/. The database can be filtered by country, urban decision-making scales and other factors.

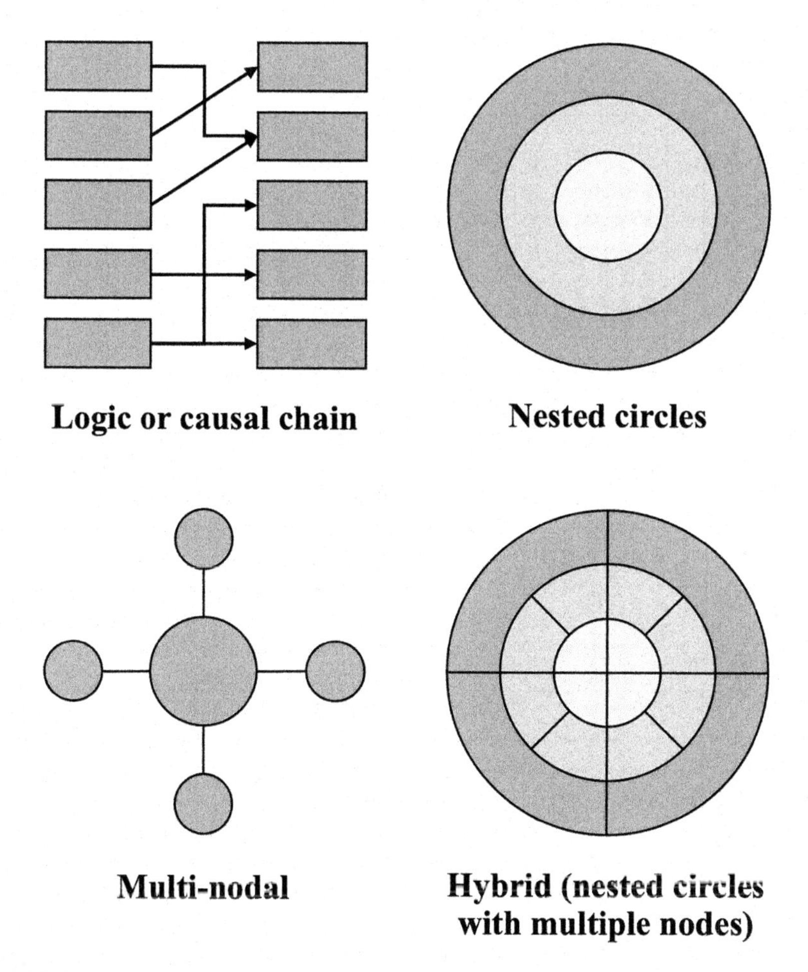

Fig. 3.1 Types of visual models typically used to represent built environment impact on health and wellbeing. (Source: Author)

applying the model and associated BRE Healthy Cities Index with city government audiences in Dubai and London, they found that the causal chain model helped to raise awareness about the specific links between the environment and health, among other benefits.

3.2.2 Nested Circles and Multi-nodal Models

A nested circles model, such as Dahlgren and Whitehead's social determinants of health model (Fig. 1.3), shows how individual health is affected by multiple scales of interacting factors. Individuals' personal characteristics are mostly fixed, yet the larger layers of behaviours, living and working environments and societal factors

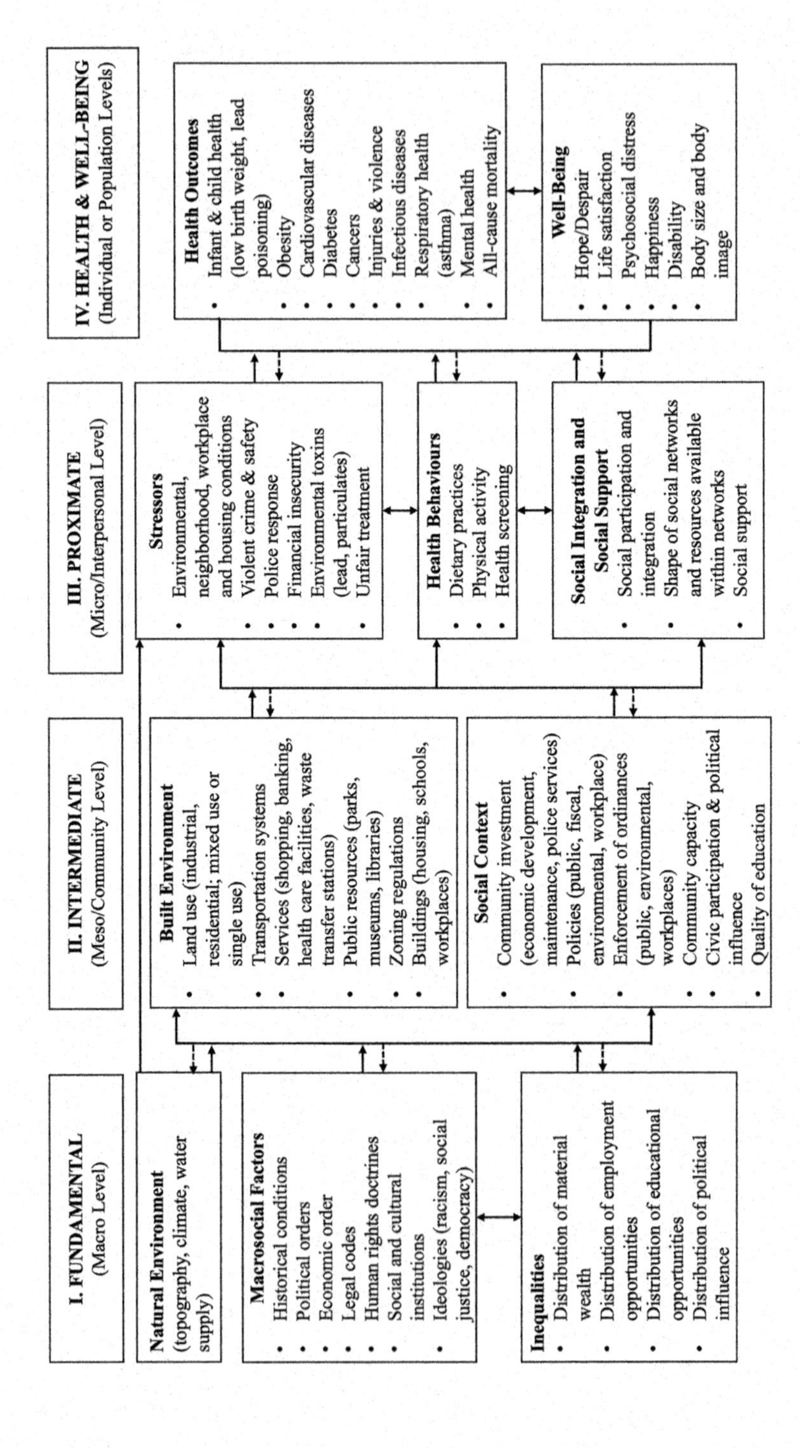

Fig. 3.2 Conceptual model of social determinants of health and environmental health promotion. (Source: Reproduced from Schulz and Northridge, 2004)

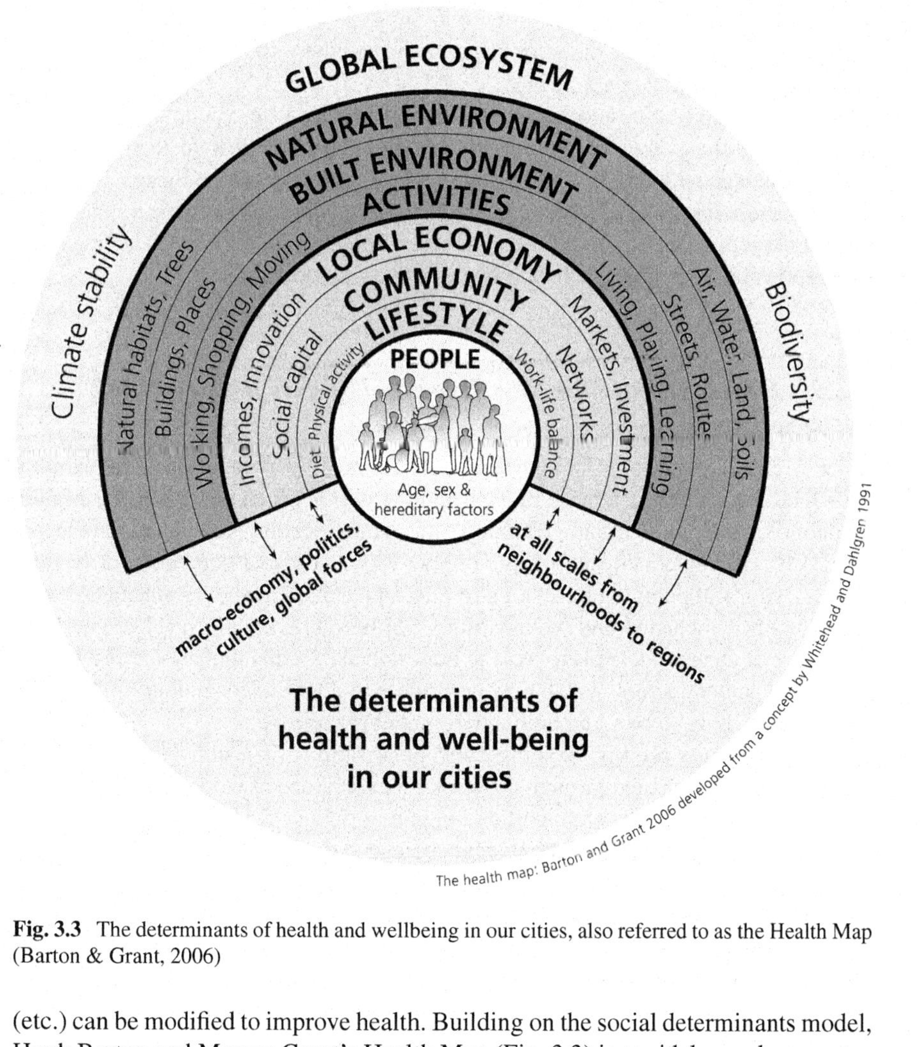

Fig. 3.3 The determinants of health and wellbeing in our cities, also referred to as the Health Map (Barton & Grant, 2006)

(etc.) can be modified to improve health. Building on the social determinants model, Hugh Barton and Marcus Grant's Health Map (Fig. 3.3) is a widely used conceptualisation of the complex factors affecting health that can be modified through settlement planning and development. Barton (2005) explained that systems theory and Kevin Lynch's (1981) theory of the ecosystem influenced the model's form of concentric circles and multiple nodes. Each factor and layer in the diagram can be considered to interact with other factors and layers, yet the visual depiction does not include arrows between each interconnection. Through the Health Map, Barton sought to redress a problem he perceived in the sustainable development agenda at that time. He felt that social dimensions had been side-lined, and although heavily contested, he argued that planning had a key role to play in 'lifestyle, social capital, equity and access' (ibid., p. 345). Barton and Grant have been very active in supporting healthy urban planning practice, particularly through the work of the WHO

Healthy Cities movement and United Nations (see UN Habitat and WHO, 2020) and their model forms the basis for key texts for neighbourhood and urban planning (Barton, 2017; Barton et al., 2021).

A disadvantage of nested circle/multi-nodal models where people are shown at the centre is the perceived emphasis on individual and lifestyle factors (Pineo, 2020). When considering that these models are competing with audiences' conceptual understanding of health being geared heavily towards a biomedical model, the central depiction of people and lifestyle does little to challenge the false idea that the responsibility to improve health lies mostly with individuals.

3.2.3 Development of a New Healthy Urbanism Framework

Pineo's (2020) review of existing frameworks for healthy urban design and planning highlighted gaps between these models and the topics which are highlighted in the epidemiological and urban planning literature, leading to development of THRIVES. A key conclusion of this review was that visual depictions of healthy urban design and planning have not strongly represented the interconnected principles of equity, inclusion and sustainability. Furthermore, the portrayal of people at the centre of such models overplays the importance of individual choice and behaviour over societal factors, which may contribute to poor understanding of the structural barriers to health among built environment practitioners.

The impetus to reframe existing conceptualisations of healthy urban environments emerged over time through the author's experience of working with different stakeholders in urban development and planning policy. This experiential learning alongside wider field advancements and societal shifts (see next section) informed the author's view that a new way of defining and considering the emerging practice of healthy urbanism was required. Initial sketches of the framework, interviews with practitioners and conversations with colleagues occurred iteratively with a more formal process of literature review leading to a preliminary model. Existing theories and epidemiological evidence underpin THRIVES, as reported in Pineo (2020), including systems thinking (Meadows, 2008), ecological health models (Rayner & Lang, 2012), ecosocial epidemiology (Krieger, 1994, 2001) and 'just sustainabilities' (Agyeman, 2013). The review of existing frameworks (described earlier) helped to identify gaps and methods to visually display the complex urban health and environment system. The design and planning goals, scales of health impact and scales of urban decision-making in THRIVES were informed by the review of frameworks and previous research on urban health indicators (Pineo et al., 2018a). The preliminary model of THRIVES was presented at conferences and tested through a process of extended peer review (Pineo et al., 2020b), leading to the version published in Pineo (2020), which remains open to adaptations and improvements over time.

3.3 Reframing Healthy Urbanism in Response to New Knowledge

Knowledge of the social, economic, environmental and political mechanisms that mediate the urban environment's impact on health have changed considerably in recent years. Society's readiness to tackle key challenges, such as the climate crisis and environmental injustices, has been influenced by the growing scientific evidence base regarding health impacts and extreme weather events experienced globally over the past decade. The global COVID-19 pandemic exposed and exacerbated existing inequalities and built environment professionals have become increasingly aware of the important impact their work can make towards improving health and wellbeing. This section explores three broad shifts that informed the development of THRIVES, including (1) the need for increased attention to the structural factors that determine individual health, moving away from the previous focus on individual 'lifestyle' choices, (2) the urgency of environmental breakdown, including the climate crisis and biodiversity loss, and (3) the negative health impacts of urban development experienced by groups that are under-represented in decision-making.

3.3.1 Structural Barriers to Health

Built environment professionals deal with macro-, meso- and micro-scale policy and design decisions that affect people's living environment. In that sense, they are involved in designing environments that affect both structural and individual determinants of health. For example, at the macro and meso scales, urban planners create policies outlining requirements for the location, quantity and type of new housing needed in a city in relation to other services, such as public transport facilities, schools and employment. These policy decisions contribute to structural determinants of health, such as whether low-income groups can afford to live in housing near transport hubs, which, in turn, affects their ability to access work and education. A micro-scale decision about the location and design of stairs within a building can increase physical activity in building occupants, which is an individual determinant of health.

Recent debates in public health argue that both structural and individual determinants of health must be addressed in policy and that neither should result in deterministic thinking about population-level risks and individual health outcomes, giving due weight to individual agency (Lundberg, 2020). Much health guidance for built environment professionals has focused on individual choices and healthy 'lifestyles' (e.g. exercise and diet), under-representing how structural factors play a significant role in population health. Built environment professionals are regularly involved in decisions that will significantly constrain or enable individual agency and capability to live a healthy life, thus a model of healthy urban environments needs to emphasise the scale and impact of these built environment decisions for population health.

Social epidemiology theory can help explain why structural factors are so important for health and how they can be barriers to individual health. Specifically, Nancy Krieger's ecosocial theory (1994, 2001) outlines how layers of disadvantage can affect health throughout a person's life and across generations, from exposure to pollutants during pregnancy through to poverty and race discrimination. Her theory uses five core ecological concepts that closely relate to built environment factors: scale, level of organisation, dynamic states, mathematical modelling and understanding unique phenomena in relation to general processes. Through combining biological, ecological and social concepts to understand drivers and patterns of social inequalities in health, Krieger proposes that people 'literally incorporate, biologically, the world around us, a world in which we simultaneously are but one biological species among many—and one whose labour and ideas literally have transformed the face of this earth' (2001, p. 668). Krieger's ecosocial theory has informed healthy urban planning approaches, notably Corburn's (2013) adaptive urban health justice framework.

There are four constructs in ecosocial theory that explain how factors influenced by built environment practitioners affect health, including embodiment, pathways of embodiment, cumulative interplay, and accountability and agency (see definitions in Table 3.1). Krieger (2001) used these four constructs to explore the risk of hypertension in African American populations. Economic and social deprivation in African Americans has been caused by residential and occupational segregation. African Americans live in neighbourhoods with reduced access to healthy foods, higher pollution and toxins, and other health risks (known as spatial inequalities).

Table 3.1 Constructs in Nancy Krieger's (2001) ecosocial theory, reproduced from p. 672

Construct	Description
Embodiment	'A concept referring to how we literally incorporate, biologically, the material and social world in which we live, from conception to death'
Pathways of embodiment	'Structured simultaneously by: (a) societal arrangements of power and property and contingent patterns of production, consumption, and reproduction, and (b) constraints and possibilities of our biology, as shaped by our species' evolutionary history, our ecological context, and individual histories, that is, trajectories of biological and social development'
Cumulative interplay between exposure, susceptibility and resistance	'Expressed in pathways of embodiment, with each factor and its distribution conceptualised at multiple levels (individual, neighbourhood, regional or political jurisdictions, national, inter- or supra-national) and in multiple domains (e.g. home, work, school, other public settings), in relation to relevant ecological niches, and manifested in processes at multiple scales of time and space'
Accountability and agency	'Expressed in pathways of and knowledge about embodiment, in relation to institutions (government, business and public sector), households and individuals, and also to accountability and agency of epidemiologists and other scientists for theories used and ignored to explain social inequalities in health'

They are also subject to socially inflicted trauma and reduced access to healthcare. Krieger tied all of these factors from the 'material and social world in which we live' and 'societal arrangements of power and property' into the ecosocial construct of embodiment and the multiple pathways and interplays between exposure, susceptibility and resistance to these factors as an explanation for increased risk of hypertension in African American communities (p. 673). Under an ecosocial approach, any posited '"racial" differences in biology' are recast as 'mutable and embodied biological expressions of racism' (ibid.), thus requiring attention to accountability and integration of biological and social lenses in epidemiology.

Corburn (2013) reinterprets Krieger's constructs for healthy urban planning. He argues that planners and public health officials should gain greater understanding of the history and context of the people and places where they seek to improve health. Corburn highlights Krieger's attention to interplays between multiple exposures over time and place, noting that 'urban policies, institutions and practices shape and influence these factors, such as policies promoting racial segregation, taxation and government spending, infrastructure, transport and environmental policies and inclusive or exclusive decision-making processes' (p. 16). With regard to Krieger's fourth construct of accountability and agency, Corburn argues for the use of better urban health monitoring systems that identify intra-city inequities, legitimate multiple sources of knowledge and attribute responsibility for positive and negative health outcomes (described further in Chap. 8).

Krieger has provided evidence about the health impacts of structural racism in the USA and potential solutions with Zinzi Bailey and colleagues (2017), including recent analyses related to COVID-19 (see Box 3.1). One solution of particular relevance to built environment practitioners is the success of the Purpose Built Communities project. This started in Atlanta, Georgia, but is now a nationwide network. Community leaders worked closely with residents to plan and implement the regeneration of public housing in the East Lake neighbourhood. They focused on high-quality construction, walkable streets, education programmes, job creation and more, resulting in measurable successes that informed the expansion and replication of their model in other states to break the cycle of disinvestment and decline in low-income and minority ethnic communities. Ecosocial theory provides a lens to understand how built environment decisions create structural barriers to health and how health can be improved through inclusive and equitable urban development processes.

3.3.2 *Urgency of Environmental Breakdown*

Scientists have been investigating the health impacts of environmental pollution since the mid-nineteenth century. Decades of environmentally unsustainable practices, such as energy and food production processes that deplete natural resources and emit pollutants, have caused significant changes to Earth's systems that affect health. These processes are interconnected and they occur across multiple scales

Box 3.1 Structural determinants of health and COVID-19
It became apparent very early in the COVID-19 pandemic that low-income and minority ethnic groups disproportionately suffered increased risk of infection and death in countries, including the USA and the UK. Environmental, social and economic factors are all likely to contribute to this inequitable impact (Berkowitz et al., 2020). Low-income and minority ethnic populations are likely to live in lower-quality housing with poorer ventilation, lack of private outdoor space and overcrowding, contributing to their risk of infection and potential mental health challenges associated with lockdown measures. In the USA, Chen and Krieger (2020) found that COVID-19 deaths were higher in counties with higher proportions of people of colour, household overcrowding and poverty. In the UK, Aldridge et al. (2020) found that COVID-19 deaths were higher in all Black, Asian and Minority Ethnic (BAME) groups than White British and White Irish, with the highest deaths in Indian and Black Caribbean groups. Potential explanations for these differences include socio-economic factors that are structural determinants of health. Aldridge et al. describe how minority ethnic groups are more likely to work in low paid, zero hours contract or non-salaried jobs, reducing their ability to adhere to social distancing restrictions. People in these groups are also more likely to live in overcrowded housing or suffer from pre-existing conditions that increase their risk of infection or severity of COVID-19. The authors highlight that ethnicity is a social construct that is poorly correlated with biology, and therefore biological factors are highly unlikely to explain the observed differences across ethnicities. Air pollution is another environmental health risk that is typically higher in low-income and minority ethnic neighbourhoods. Analyses in Europe, China and the USA found associations between short- and long-term exposure to several air pollutants and increased risk of COVID-19 infection or severity (Cole et al., 2020; Conticini et al., 2020; Wu et al., 2020; Zhu et al., 2020). Applying the concepts of ecosocial theory demonstrates how the novel coronavirus does not affect everybody equally in society. People living in neighbourhoods with high pollution and low-quality housing are also likely to be those groups with low income and other disadvantages that increase their risk of infection and the severity of the disease, whilst also reducing their ability to cope with lockdown measures. This complex layering of disadvantages is not unique to COVID-19 but representative of the interconnected structural barriers that affect health.

from local to global. There is a growing body of evidence demonstrating that environmental degradation is a huge threat to human health causing death and disease now with the expectation that this will increase over time. Analysis from *The Lancet* Countdown for climate change and health, underscores the urgency of health risks quantifying the excess deaths and morbidity from heatwaves, extreme weather (e.g.

wildfires), food scarcity, infectious disease transmission and other impacts through its annual reports (Watts et al., 2021) and these are described in more detail in Chap. 4.

Environmental justice is the movement advocating to change the unfair and disproportionate health (and other) impacts of environmental degradation experienced by low-income and racial minority populations. From the 1980s, the work of Robert Bullard (see Bullard, 2005, 2007) and others in the USA exposed the environmental discrimination against African Americans and other marginalised groups that occurred through waste, housing, transport and other land use management decisions. At the beginning of the twenty-first century, it was already well-document that people living in lower-income countries who have contributed the least environmental damage are at higher risk of health impacts than high-income and high-polluting countries (Agyeman et al., 2003). Marginalised populations are most vulnerable to the health risks of environmental breakdown and least able to change their circumstances by moving to a safer neighbourhood or improving the safety and comfort of their homes. Built environment professionals need to rapidly gain new knowledge about these existing disparities and how they can be avoided through design and planning. A clear example relates to the quality and location of social housing which should be designed for climate change adaptation. In places like the UK, this would mean designing for thermal comfort to avoid overheating, reducing the risk of flooding, and considering other harmful exposures such as noise and air pollution.

In addition to greater knowledge of how environmental degradation affects health, and the unfair impacts on low-income and minority populations, public perceptions on this topic are shifting—demonstrating clear support for policy action. Although there has been increasing polarisation of views and climate denialism in some countries including Australia, the USA and the UK, overall there is evidence of increasing global public concern about climate change (Capstick et al., 2015). A global poll of 1.2 million people by the UN Development Programme and University of Oxford found that '64% of people said that climate change was an emergency' and '59% said that the world should do everything necessary and urgently in response' (Flynn et al., 2021). Looking specifically at perceptions of risk, a YouGov survey in 28 countries found that the majority of respondents believed that climate change is likely to lead to serious economic damage, the loss of cities due to sea-level rise and mass migration (Smith, 2019). Respondents in Asian and Pacific countries believed that climate change would lead to the extinction of the human race. Highly visible extreme weather events (e.g. the wildfires in Australia and the USA across 2019 to 2021) and frequency of unusual weather events may contribute to increasing public support for policies to reduce climate change. Personal experience is an important factor in people's beliefs, attitudes and perceptions of risk, alongside ideology, knowledge, emotion and other factors (Capstick et al., 2015). The growing understanding of the health impacts of environmental damage, alongside increasing public support for action, demand a stronger reflection of environmental science and sustainability measures in built environment health guidance.

3.3.3 Under-represented Groups and Urban Development

There is growing evidence that urban development adversely affects the health of multiple population groups that are under-represented in decision-making. In recent years, scholars and practitioners have exposed the unfair health and wellbeing impacts caused by design that excludes some groups, including with regard to race, gender, age, disability and other characteristics. These are not new concerns, but they are receiving increasing attention in research and practice. For instance, Leslie Kern's (2020) book *Feminist City* describes how cities have not met the needs of women in terms of safety, social spaces, transport and other factors. There is growing awareness of the concept of gender mainstreaming in urban planning practice, evidenced by recent publications from local governments and professional institutions (City of Vienna, 2013; Bicquelet-Lock et al., 2020).

Low-income and minority ethnic groups have been under-represented in decision-making processes for urban change with negative health consequences. The health impact of urban regeneration (also called revitalisation and renewal) processes has received significant focus in recent years as it has become clear that the intention to increase residents' quality of life is not necessarily realised. One consequence of urban regeneration relates to the potential for low-income population groups to be permanently displaced from communities where they have lived and worked, potentially for generations. Changes in the urban environment that result in increasing residents of higher socio-economic or ethnic status are known as gentrification (Anguelovski et al., 2019). Gentrification is closely linked to housing affordability and tenure security, among wider neighbourhood factors that influence health.

Melody Tulier et al. (2019) examine studies of the health impacts of gentrification in the USA. They describe gentrification as

> an interactive process in formerly declining, under-resourced, predominantly minority neighborhoods involving economic investment and increasing sources of capital infusion and in-migration of new residents, generally with a higher socio-economic status. The process is dynamic, uneven, and occurs in stages. (p. 1)

They identify three potentially overlapping conceptualisations of the gentrification process: socio-economic upgrading, political conflict and urban restructuring and stages of gentrification. Within these overarching conceptual frameworks, multiple mechanisms for health impact are studied at different levels, including: (1) changes to neighbourhood attributes (e.g. infrastructure, economic opportunities/development or social cohesion), (2) changes to individual resources (e.g. financial status), (3) changes to both neighbourhood and individual mechanisms, and (4) the role of political and economic institutions. In unpicking and synthesising these diverse ways of conceptualising and measuring the health impact of gentrification, Tulier et al.'s review demonstrates the complexity of this topic and the associated challenges for policy-makers and researchers in identifying appropriate interventions. A key area for clearer conceptualisation of gentrification relates to 'the power dynamics inherent in gentrification, the differential valuing of individuals based on class and race, in addition to the upstream structural factors that drive gentrification and engender class and racial conflict' (p. 7). Drawing on Krieger (2001), the authors highlight that unclear investigation of these power dynamics risks failure to identify responsible actors and interventions. Given the potentially long timescales between exposures and outcomes in gentrification processes, Tulier et al. also advocate a life-course approach that considers levels and periods of exposure, timing and embodiment.

3.4 Systems Thinking for Urban Health

The theories and concepts that informed the THRIVES framework, such as ecosocial epidemiology and just sustainabilities (described in the next section), recognise the need for holistic perspectives to complex problems. Systems thinking provides another valuable set of tools for considering the impact of urban development on health. Systems theories, specifically systems thinking as described by Donella Meadows (2008), refocuses attention from detailed analysis of constituent parts to examination of the behaviour of the whole system. A system is an 'interconnected set of elements that is coherently organized in a way that achieves something' such as a forest, a city or a national economy (ibid., p. 11). Systems thinking is about examining problems from different viewpoints and 'expand[ing] the boundaries of our mental models' (Sterman, 2006, p. 511). A systems thinking approach to urban health provides understanding of the complex urban health system (i.e. how humans are affected by the urban environment) and the policy and decision-making processes to improve urban health.

Scholars have described the characteristics of complex systems generally and as they relate to urban health (Gatzweiler et al., 2018; Rydin et al., 2012; Glouberman et al., 2006). There are seven characteristics of complexity in urban health systems: dynamic, number of elements, interconnected, non-linear structure, feedback, counter-intuitive and emergent behaviour (Luke & Stamatakis, 2012; Glouberman et al., 2006; Sterman, 2000). Table 3.2 builds on previous research to describe these aspects of complex systems and how they relate to healthy urban policy and decision-making (Pineo, 2019; Pineo et al., 2018b). It is necessary to understand the general characteristics of complex systems in order to identify interventions to improve health.

There are several implications of systems thinking that inform the THRIVES framework. First, the dynamic and counter-intuitive aspects of urban health require the boundaries of health impact to become wider in terms of space and time. Health impacts of urban development should be considered for occupants, people living in neighbouring communities or even across the globe, across timespans of months, years and decades. Second, the interconnected characteristic means that design, planning and engineering solutions need to be integrated, not addressed in isolation. Third, the high number of elements and feedback behaviour in systems (among other factors) mean that no single disciplinary or development stakeholder

Table 3.2 Characteristics of complex urban health systems

Characteristic	Description in urban health terms	Example in urban health system
Dynamic	Health and wellbeing impacts or exposures change over time (possibly in unpredictable ways)	Air pollution has long-term trends (increasing over time), seasonal trends and extremes (spikes).
Number of elements	High number of variables within system	Transport system includes many elements which interact to create effects such as a walkable community.
Interconnected	Multiple interactions across and within systems	Transport emissions affect health through air pollution whilst contributing to climate change which has additional health impacts.
Non-linear structure	Non-linear relationship between exposure and health and wellbeing impact—effects are rarely proportional to causes	Impact of vehicle speed on pedestrian injury/death does not change proportionately as speed increases.
Feedback	System elements interact recursively (in feedback loops) to change the behaviour of the system	Increasing road capacity usually has the unintended effect of increasing traffic congestion by attracting more drivers.
Counter-intuitive	Health and wellbeing impacts are distant in space and time to exposures	Presence of many fast-food outlets in a community may result in increased obesity levels over time.
Emergent behaviour	Health and wellbeing effects are greater than the sum of individual effects within the system	A park or 20 mph limit is not sufficient on its own to support physical activity, but is effective if combined with other elements (e.g. pavements and mixed land uses).

Source: Adapted from Pineo (2019), Pineo et al. (2018b)

perspective will have a grasp on the challenges and solutions. Multiple disciplines and knowledge sources need to be involved in urban development processes. Systems thinking approaches for urban policy and design are further described in Chap. 8.

3.5 Interpreting the THRIVES Framework

Up to this point, this book has set out the need for a new understanding of built environment health impacts, arguing that despite its importance, professionals have received very little training about how to design and plan healthy places that are inclusive, equitable and sustainable. This section explains the key concepts and terms in the THRIVES framework which are underpinned by the theory and literature set out earlier. There are three key messages communicated through THRIVES, as shown in Fig. 3.4.

The first key message is that health impacts that are created by urban development often occur in a place or time that is distant to the development itself. For

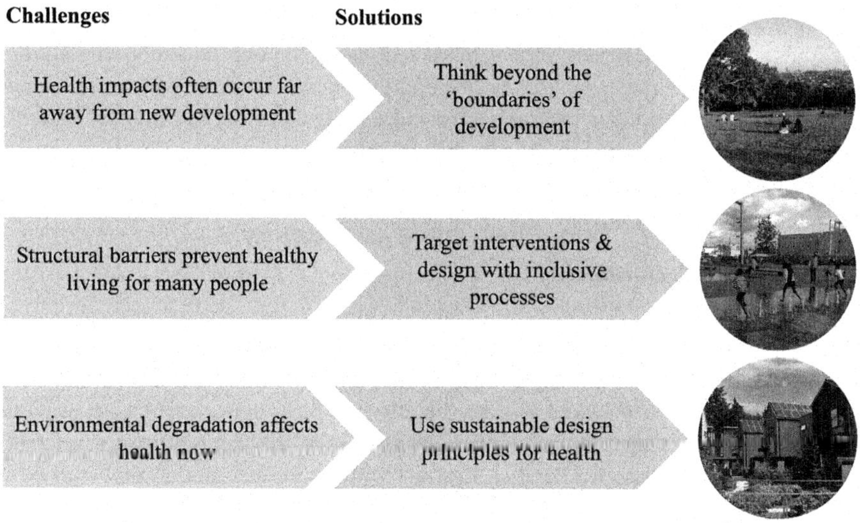

Challenges **Solutions**

Health impacts often occur far away from new development → Think beyond the 'boundaries' of development

Structural barriers prevent healthy living for many people → Target interventions & design with inclusive processes

Environmental degradation affects health now → Use sustainable design principles for health

Fig. 3.4 Key challenges and solutions for healthy urbanism highlighted by the THRIVES framework. (Source: Author)

example, the decision to use a building material that emits pollutants could harm the health of people in the supply chain, during installation or for building occupants and neighbours. There may be short-, medium- and long-term impacts of any of these effects. Architects and engineers therefore need more information about the health impacts of building materials to enable them to specify materials that are safe. In that sense, thinking beyond the boundary of the development involves consideration of the material supply chain and potential health impacts from the material over time.

Second, it was argued above that structural barriers prevent many people from living a healthy life. This could result from environmental discrimination and spatial inequalities. Policy and decision-makers can target improvements to the built environment where they are most needed; however, those processes of change should be done in cooperation with the affected communities. For example, local officials may identify residents who are experiencing overheating in specific property types (e.g. apartments or social housing), causing illness and excessive deaths during heatwaves. Officials may work with politicians, social housing providers and landlords to devise a programme to improve thermal comfort in properties. Funding to receive retrofit measures (e.g. external shades and shudders) may be targeted at those properties with prioritisation of vulnerable residents. Property managers would work with community representatives to design a plan for where, how and when building works would be completed.

Finally, it is clear that environmental degradation is affecting health now and this will increase over time. Built environment professionals can use sustainable design and planning principles to support health, which is effective for many other objectives beyond environmental sustainability, such as increasing physical activity. For example, a new large-scale mixed-use community (e.g. with homes, offices, retail and public space) can be designed using densities and layouts that support walking, cycling and public transport. This will support residents to use active travel for their daily commute, resulting in increased physical activity and reduced carbon emissions.

3.5.1 *Three Core Principles: Inclusion, Equity and Sustainability*

The centre of the THRIVES image (Fig. 1.2) is an illustration of landscapes that connect global to local environments with three core principles that define a healthy place: inclusive, equitable and sustainable. Moving these considerations to the centre of the diagram purposely reorients understandings of the determinants of health towards structural factors. This is in contrast with traditional models that start with individuals, their genes and 'lifestyle choices'. Choices about where and how to live are heavily constrained by societal and environmental factors, many of which are determined by the built environment.

The inclusion principle is defined holistically, considering the design and policy approaches that enable and support everybody to participate in society, regardless of age, disability, gender, ethnicity or other characteristics. Inclusive design has often been interpreted as creating greater accessibility for disabled people. A broader conceptualisation of inclusion is adopted here, in opposition to the historic tendency for places to be designed by and for only a portion of society. Building on previous definitions (EIDD, 2009; Heylighen et al., 2017), THRIVES defines an inclusive built environment as one that 'enables all members of society to conveniently participate in daily activities without feeling that they are disadvantaged by their personal characteristics or needs, and this is achieved through participatory processes' (Pineo, 2020, p. 10). This definition highlights residents' feelings of inclusion because these subjective emotions can be damaging or supportive of health and wellbeing (Marsh et al., 2020).

Health inequities were introduced in Chap. 1 as a key global health trend present in cities, countries and across nations. The THRIVES framework defines an equitable place as one that 'gives access to health-promoting environments to all residents and specifically considers and seeks to reduce barriers to access (be they physical, cultural, social or economic)' (Pineo, 2020, p. 11). Part of the solution to inequities created through urban environments is to adopt the public health principle of 'proportionate universalism' (Carey et al., 2015; Egan et al., 2016). Applying Gemma Carey and colleagues' approach to proportionate universalism for healthy urban environments would involve universal and targeted measures. A universal measure is applied for everybody in society, such as safe water. A targeted measure is applied for those with the most need, such as high-quality social housing. Carey et al. also promote the governance principle of 'subsidiarity', which means that residents are closely involved in determining solutions to the challenges they face. This aligns with the inclusion principle in THRIVES.

The final principle of sustainability could be seen as all encompassing, folding concepts of health, inclusion and equity within its broad meaning. Julian Agyeman's (2013) theory of 'just sustainabilities' views sustainability as being about the processes and outcomes involved in improving wellbeing, equity and justice for current and future generations, and doing so within ecosystem limits. For Agyeman, these principles are mutually supportive, and he argues that 'both social and environmental health are dependent, to a large extent, on greater justice and equality' (ibid., p. 18). Drawing upon previous definitions of sustainability and Agyeman's explicit linkage of social and environmental health, the THRIVES framework defines sustainable urban development as being 'supportive of the needs of the current (and immediately local) population without compromising the needs of future (or spatially distant) populations' (Pineo, 2020, p. 11). The three scales of health impact described by THRIVES offer a way to consider how environmental and social sustainability are linked at multiple scales through urban built environment decision-making.

To achieve healthy places, these three core principles (inclusion, equity and sustainability) must underpin policy and design. The concepts of inclusion and equity overlap and their interpretations vary. The THRIVES framework refers to equity in terms of overcoming unfair distribution of resources, whilst inclusion is about ensuring everybody can fully participate in society and daily activities, achieved through a process of participatory design and planning. All three core principles are weaved throughout the next four chapters by linking each principle to evidence-based design and planning strategies and practical case studies.

3.5.2 Three Scales of Health Impact: Planetary, Ecosystem and Human

The THRIVES framework moves out concentrically from the three core principles to three scales of health impact: planetary, ecosystem and local. These scales are interconnected, which is visually depicted in the framework by the arrows in the illustrated sections. The scales of health impact are aligned to the scales of decision-making (i.e. regional to building) and the example evidence-based urban design and planning goals. This alignment seeks to highlight that the planetary health and ecosystem health scales are best achieved through urban policy and large-scale infrastructure provision; however, there are also suitable building and neighbourhood scale initiatives. The local health scales may be supported by policy but are often implemented through design at the neighbourhood (inner local health circle) and building (outer circle) scales.

In THRIVES, planetary health focuses on climate change and other global environmental costs of urban development such as depletion of natural resources and biodiversity losses. Ecosystem health refers to the stability of the local ecosystems in which we live, made up of watersheds, forests, farmlands and other living and non-living environments that, in complex ways, impact human health and wellbeing (Dakubo, 2011). At the local health scale, neighbourhood and building design are associated with positive and negative health impacts. The next four chapters in this book will move through each of the scales of health impact (sub-dividing local health into chapters on neighbourhoods and buildings).

3.6 Understanding the Epidemiological Evidence Base

The knowledge base about how design and planning activities influence health is informed by epidemiology and medical statistics, two fields that are foreign to built environment curriculum. Epidemiology is the study of the determinants and distribution of diseases at the population level. It gained recognition among the general public during the COVID-19 pandemic; however, epidemiology is not a field that most people understand. For built environment professionals, epidemiology can help answer whether a particular policy or design measure 'works' to improve health and wellbeing. Unfortunately interpreting epidemiological study results is challenging, even for medical practitioners, and 'cherry-picking' of evidence frequently occurs (Greenhalgh, 2019). This section briefly introduces key concepts in epidemiology to enable critical thinking and evidence-based practice in healthy urbanism.

Information appraisal and synthesis are key skills for urban planners and others who pull together wide sources of evidence about a place and its inhabitants. Consider two examples that the author has encountered in professional practice. Design teams on separate projects wrote that their proposals would improve health through provision of on-site beehives and indoor potted plants. The beehives were intended to reduce allergies among people who ate the honey produced, while the plants were intended to reduce indoor air pollutants. On reading the scholarly literature, there is very weak evidence for both of these anticipated outcomes, although both measures could benefit health for other reasons. It is important to consider the strength of evidence about the built environment and health when deciding how to act upon research results. This means applying academic critical thinking and study appraisal skills which are transferable to wider literature, including newspaper articles and industry reports. Not all summaries of health evidence are accurate, and readers should use their knowledge to draw their own conclusions about the implications of research findings.

3.6.1 Components of Epidemiological Studies

As epidemiologists try to understand what makes people healthy or ill at the population level (so-called determinants of health), they focus on investigating associations between exposures and outcomes. Drawing upon the aforementioned example, eating honey is an exposure and allergy symptoms are an outcome. Exposures can be environmental (e.g. air pollution), behavioural (e.g. smoking or physical activity), or a personal characteristic (e.g. blood group), and they may also be referred to as a 'risk factor'. The amount of exposure is called the dose, while the length of exposure can be measured in short-term (acute) and long-term (chronic) timescales. Outcomes include any disease (referred to as morbidity), health or wellbeing state/event or death (mortality). Not all individuals are equally affected by the same dose of a particular exposure, referred to as sensitivity, which may be caused by differences in age, genes, illness, diet and many other factors. Dose–response relationship refers to the quantified measure of the body's reaction to a given dose.

The results of an epidemiological study include measures of effect such as risk, rates, odds ratios and life expectancy. Studies may also produce estimates about the impacts of treatments/diseases through comparable health metrics to assist policy and decision-making. Two widely used metrics are disability adjusted life years (DALYs) and quality adjusted life years (QALYs). DALYs allow public health experts to describe the overall burden of disease, rank diseases and compare the burden over time and across places. One DALY is described by the WHO to mean one lost year of 'healthy' life. A QALY is used for health economic evaluations and can help understand the impact of treatments. A QALY should encapsulate the impact of a treatment on a patient's length and health-related quality of life (Whitehead & Ali, 2010).

To see these terms in context, consider a study about the health benefits of a green infrastructure project in Northern Ireland, the Connswater Community Greenway. The study quantified the health benefit in terms of life expectancy for several common chronic diseases because the project aimed to increase physical activity (a key risk factor for chronic diseases) among the local population. The researchers also estimated the DALYs saved as a result of this project and the cost-effectiveness of the investment for health (Dallat et al., 2014). The study found that if 10% of physically inactive local residents became active, this could prevent 886 incident cases and 75 deaths from several common chronic diseases by 2050. The cost would be £4469.45 per DALY, which the authors reported to be cost-effective as a healthcare treatment.

3.6.2 Advantages and Disadvantages of Epidemiological Study Designs

Not all study findings can inform built environment practice or establish that a particular exposure causes a particular outcome. For example, demonstrating causality requires a number of criteria to be satisfied, including the validity of the study, whether the exposure occurred before the outcome and that the results were not caused by some other factor (Elwood, 2017). Traditional epidemiological studies can be divided into observational (non-experimental) and interventional (experimental), of which the latter category contains the randomised controlled trial (RCT) (Beaglehole, 1993). Within these two categories, studies may gather data from individuals or groups in different ways, further distinguishing types of studies such as ecological, cross-sectional, cohort and case control studies (all observational). Each of these study designs has advantages and disadvantages that affect how easy they are to conduct and how they can be interpreted.

It is difficult to find randomised interventional studies with built environment exposures. This is understandable because there are many challenges with randomly allocating some people to receive a built environment intervention whilst others do not. One such study was Philippa Howden-Chapman et al.'s (2008) study of the effectiveness of improved home heating (heat pump, wood pellet burner and flued gas) on the health of children with asthma in New Zealand. By ensuring that an RCT study contains a large enough random sample of people in the relevant population (e.g. households containing children with asthma), differences in the various outcomes can be attributed to the intervention rather than other factors that might affect health. The home heating RCT found a number of benefits from the heating systems, including 1.80 fewer days off school and fewer reports of poor health, leading the authors to conclude that this was an effective adjunct to pharmaceutical treatments for asthma in children. Notably, households in the control group received a new heater after the data collection period ended, which is ethically appropriate when an intervention is shown to work.

Observational studies are far more common in the built environment and health literature and they can be used for descriptive and analytical purposes. Ecological studies are a common example because they can use existing (sometimes 'open') datasets. The studies by Marmot et al. (2010, 2020) described in Chap. 1 were ecological analyses of life expectancy (outcome) and neighbourhood deprivation (exposure) in which data were analysed at the neighbourhood level. Another example relates to air pollution exposure and COVID-19 deaths. As described in Box 3.1, multiple ecological analyses of these factors were performed in Europe, China and the USA using data at the city or county scale. Ecological studies are useful for exploring understudied relationships between exposures and outcomes (as would be the case with a novel disease such as COVID-19), yet they have a significant disadvantage known as the ecological fallacy. This means that any association (or causal relationship) that can be observed between variables at an aggregate level does not necessarily reflect the relationship that exists at an individual level (Porta, 2014). In

Table 3.3 Strengths and weaknesses of traditional epidemiological study designs (Bonita et al. 2006) and examples from urban health

Study type	Strengths	Weaknesses	Examples in urban health
Ecological: Data collected at group level, typically at single point in time	Testing hypotheses, exploring rare diseases, can be low cost and fast	Ecological fallacy, potential differences in data across areas, time relationship usually unknown, high risk of confounding	Income, race, household overcrowding and COVID-19 at county level (Chen & Krieger, 2020)
Cross-sectional: Data collected at individual level at single point in time	Testing multiple exposures and effects, can be low cost and fast	Time relationship unknown, not suited to rare causes/diseases, risk of recall and selection bias, confounding	Neighbourhood environment characteristics (e.g. access to parks, residential density, nearby supermarket and fast-food outlets) and obesity (Saelens et al., 2012)
Case-control: Participants selected from same underlying population then asked about their exposures.	Can use existing datasets, suitable for rare diseases and multiple exposures	High risk of recall and selection bias	Built environment factors associated with child pedestrian motor-vehicle collision (PMVC) rates (Rothman et al., 2017)
Cohort (or longitudinal): Participants selected based on their exposure status and followed up over time.	Low risk of selection bias, measurement of time relationship, suitable for rare causes	High risk of statistical chance and loss to follow-up, takes a long time to gain results and is costly	Neighbourhood design, perceptions of crime and safety and physical activity, comparing people before and after moving home (Nightingale et al., 2019)
Randomised controlled trials: Participants are selected from a population sample and randomised to receive the intervention or be in the control group and they are followed up over time.	Clear temporal sequence, high-quality data on multiple outcomes, suitable for known and unknown confounders	Expensive, time-consuming, regulatory impediments, representativeness and generalisability, multiple biases	Effective home heating (heat pump, wood pellet burner, flued gas) and the health of children with asthma (Howden-Chapman et al., 2008)

other words, a doctor could not understand an individual's risk of dying from COVID-19 on the basis of their air pollution exposure using the studies reporting county-scale associations.

There are strengths and weaknesses across study types and it is often best to have evidence from many studies of different designs to gain understanding of built environment health impacts. Table 3.3 is informed by Bonita et al. (2006) and describes traditional epidemiological study designs, their strengths and weaknesses and example applications in urban health. Studies that analyse the results of multiple RCTs (a meta-analysis) are considered to be a strong form of evidence for determining whether an intervention works, yet systematic reviews that synthesise evidence from multiple studies (often observational) and umbrella reviews that compare evidence from multiple systematic reviews are more common in the built environment.

The complexity of the built environment and health reduces the applicability of traditional epidemiological methods. Obesity provides a useful example because it can be influenced by many environmental factors that affect physical activity and diet, and potentially also through stress, sleep and air pollution (Lam et al., 2021). In obesity research, the diversity of measures used for built environment factors (e.g. residential density and walkability) hamper meta-analysis because the results cannot be aggregated, but they are important to recognise context. The cumulative impact of multiple environmental factors is understudied, with many analyses looking at a single-exposure and a single-outcome, when in the built environment, people are exposed to multiple factors simultaneously (Mackenbach et al., 2014). Cross-sectional studies are common, limiting potential for causal inference because these studies only have data from a single point in time, meaning that it is not possible to determine the sequence of events between exposure and outcome. In relation to obesity, Rutter (2018) argues that there is a need to move beyond studies of individual interventions and biomedical approaches (e.g. the randomised controlled trial) towards systems approaches that recognise the complexity of obesity interventions, which are subject to adaptations at the individual and system level. This holds true for other built environment and health topics and new approaches are being developed, such as those using systems science.

3.6.3 Study Appraisal Techniques

It is unlikely that built environment readers of this book will expect to perform quality appraisals of epidemiological studies. However, understanding the limits of the existing urban health evidence base and being able to question the findings of a particular study can empower readers to form their own judgements about new studies. The evidence base is constantly changing and growing, but not all studies are considered equal. Box 3.2 contains a set of criteria for study appraisal that should be suitable to students and practitioners of built environment disciplines.

When interpreting epidemiological studies, it is important to consider chance, bias, confounding and reverse causality before accepting a risk factor as a possible cause of disease (Bonita et al., 2006). Determining whether any observed differences between groups within a study could be due to chance requires the use of statistics. Studies will report confidence intervals and p-values to help readers interpret the strength of the associations observed. For example, the RCT about home heaters and children with asthma reported 1.80 fewer days off school for the intervention group (Howden-Chapman et al., 2008). Alongside that figure, they included the 95% confidence interval which ranged between 0.11 and 3.13. This means that given the population size, the authors are 95% certain that the true number of days off school is between 0.11 and 3.13. If this 95% confidence interval had crossed zero, the direction of effect would be unknown, meaning that the true number of days off school may not have been different between the intervention and control group. Different forms of bias are introduced as a result of the study design. Bias refers to any systematic problem that leads to inaccurate measurement of an exposure, confounder or outcome. There are many types of bias. For example, selection bias is about the study population being systematically different to the population that the research is about. Confounding refers to a factor that is associated with a disease and a possible risk factor and which can lead to a misleading association between that possible risk factor and the disease. A confounder can be accounted for in the analysis, but it has to be measured. Common confounding factors include socio-economic status and smoking. Finally, reverse causality refers to situations where the studied outcome actually caused the risk factor, rather than the other way around. To ensure that reverse causality has not occurred, the exposure or risk factor should be measured before the outcome. The overall impact of these factors (chance, bias, confounding and reverse causality) is that epidemiological studies need to be interpreted critically and with caution. Unfortunately, journalists and even study authors may mispresent study results. Most built environment professionals will find evidence reviews (e.g. in systematic reviews or professional reports) more accessible, but it is still useful to understand the basic principles and apply study appraisal skills (see Box 3.2).

Box 3.2 Simple study appraisal techniques

Academic and practising physician Trisha Greenhalgh's (2019) authoritative book *How to Read a Paper: The Basics of Evidence-Based Medicine and Healthcare*, now in its sixth edition, explains how to use epidemiological and other research evidence in medical practice. Greenhalgh catalogues the typical decision-making practices that continue to be used in healthcare, such as anecdote and cost-minimisation, rather than rational analyses of evidence. She points out that there is push-back against evidence-based medicine partly because practitioners feel that their experiences and patients' circumstances are not necessarily covered by research evidence. In the built environment industry, an Australian study found that practitioners relied primarily on feedback from past projects and colleagues' insights or advice, over knowledge from education or research (Criado-Perez et al., 2020). The problem with these modes of practice is that lessons learned about bad outcomes and risks are not shared widely, leaving people to make the same mistakes over time.

Evidence-based practice in any field is not about uncritically accepting the findings of all research studies. Greenhalgh (2019) claims that 99% of studies 'belong in the bin' because of methodological flaws (p. 29). Practitioners

(continued)

Box 3.2 (continued)

need a set of study appraisal skills to understand how to interpret and make sense of research findings. The following are a set of questions to evaluate studies about the built environment and health, informed by Greenhalgh (2019) and Elwood (2017).

To begin with, it is useful to ask general points about the source of information: Who produced the evidence and are there any relevant conflicts of interest? Has the evidence been evaluated, such as through academic peer-review? How recent is the evidence?

The following questions may then be appropriate, depending on the readers' level of understanding: What was the study design and was it appropriate for the research questions? What are the limitations? What was the study population (e.g. age, illness and location) and sample size? Can the results be applied to other populations? Are there potential problems with chance, bias, confounding or reverse causality? What does this study add to the existing evidence base?

3.6.4 *Measuring the Impact of Healthy Place-Making*

Indicators provide a valuable tool to inform healthy urban development and policy and they are typically informed by epidemiological evidence about the built environment impact on health. Indicators are summary measures of complex phenomena. There are many indicators covering topics related to quality of life, wellbeing and health. These measures can be used for the following purposes (Pineo et al., 2018b):

1. to understand a place and its impact on health to inform the development, monitoring and adjustment of urban policies;
2. to compare places or the success of policy programmes (e.g. within a city, across cities or internationally);
3. to support funding bids and/or decisions and
4. to involve the community in agreeing priorities.

Chapters 4, 5, 6 and 7 provide a selection of indicators to measure each goal in the THRIVES framework. Chapter 8 outlines key uses of indicators in healthy urbanism including health impact assessment, monitoring and evaluation.

3.7 Conclusion

This chapter described several gaps in existing conceptualisations of health and place that led to the development of THRIVES. Several field advancements and societal shifts were described, including new knowledge about the structural barriers to health associated with the built environment, the health disadvantages experienced by groups that are under-represented in urban design and planning process, and the urgency of environmental breakdown caused by climate change and other degradation. Existing theories and approaches were used to re-interpret healthy urban design and planning priorities through the THRIVES framework. The scales of health impact form the structure for the next four chapters. Each chapter sets out evidence for the health impacts of the respective design and planning goals. Readers will find ample epidemiological evidence to support action alongside international case studies of healthy urbanism.

There are many synergies between healthy design and planning measures and other environmental, social or economic objectives within a city. These synergies may be obvious and intentional, but often they are part of a wide range of potential effects of policy or design decisions for which there are varying degrees of understanding. Systems thinking methods can help identify some of the potential unintended effects of design and policy measures, which may be synergistic with the original goal or other goals (in which case they are called 'co-benefits') or they may produce unwanted effects (and may be referred to as tensions or disbenefits). The following four chapters identify synergies and tensions for design and planning goals in each of the scales of health impact in THRIVES.

References

Agyeman, J. (2013). *Introducing just sustainabilities: Policy, planning and practice, Just sustainabilities*. Zed Books.

Agyeman, J., Bullard, R. D., & Evans, B. (Eds.). (2003). *Just sustainabilities: Development in an unequal world*. Earthscan.

Aldridge, R. W., Lewer, D., Katikireddi, S. V., Mathur, R., Pathak, N., Burns, R., Fragaszy, E. B., Johnson, A. M., Devakumar, D., Abubakar, I., & Hayward, A. (2020). Black, Asian and Minority Ethnic groups in England are at increased risk of death from COVID-19: Indirect standardisation of NHS mortality data. *Wellcome Open Res., 5*, 88.

Anguelovski, I., Triguero-Mas, M., Connolly, J. J., Kotsila, P., Shokry, G., Pérez Del Pulgar, C., Garcia-Lamarca, M., Argüelles, L., Mangione, J., Dietz, K., & Cole, H. (2019). Gentrification and health in two global cities: A call to identify impacts for socially-vulnerable residents. *Cities Health, 4*, 1–10.

Bailey, Z. D., Krieger, N., Agénor, M., Graves, J., Linos, N., & Bassett, M. T. (2017). Structural racism and health inequities in the USA: Evidence and interventions. *The Lancet, 389*, 1453–1463.

Barton, H. (2005). A health map for urban planners: Towards a conceptual model for healthy, sustainable settlements. *Built Environment, 31*, 339–355.

Barton, H. (2017). *City of well-being: A radical guide to planning*. Routledge; Taylor & Francis Group.

Barton, H., & Grant, M. (2006). A health map for the local human habitat. *The Journal of the Royal Society for the Promotion of Health, 126*, 252–253.

Barton, H., Grant, M., & Guise, R. (2021). *Shaping neighbourhoods: For local health and global sustainability* (3rd ed.). Routledge, Taylor & Francis Group.

Beaglehole, R. (1993). *Basic epidemiology*. World Health Organization.

Berke, E. M., & Vernez-Moudon, A. (2014). Built environment change: A framework to support health-enhancing behaviour through environmental policy and health research. *Journal of Epidemiology and Community Health, 68*, 586–590.

Berkowitz, R. L., Gao, X., Michaels, E. K., & Mujahid, M. S. (2020). Structurally vulnerable neighbourhood environments and racial/ethnic COVID-19 inequities. *Cities Health, 0*, 1–4. https://doi.org/10.1080/23748834.2020.1792069

Bicquelet-Lock, A., Divine, J., & Crabb, B. (2020). *Women and planning: An analysis of gender related barriers to professional advancement*. Royal Town Planning Institute.

Bonita, R., Beaglehole, R., & Kjellström, T. (2006). *Basic epidemiology* (2nd ed.). World Health Organization.

Bullard, R. D. (2005). *The quest for environmental justice: Human rights and the politics of pollution*. Sierra Club Books.

Bullard, R. D. (2007). Equity, unnatural man-made disasters, and race: Why environmental justice matters. In C. Wilkinson & R., R. Freudenburg, W. (Eds.), *Equity and the environment, research in social problems and public policy* (pp. 51–85). Emerald Group Publishing Limited.

Capstick, S., Whitmarsh, L., Poortinga, W., Pidgeon, N., & Upham, P. (2015). International trends in public perceptions of climate change over the past quarter century. *WIREs Climate Change, 6*, 35–61.

Carey, G., Crammond, B., & De Leeuw, E. (2015). Towards health equity: A framework for the application of proportionate universalism. *International Journal for Equity in Health, 14*, 81.

Chen, J., & Krieger, N. (2020). *Revealing the unequal burden of COVID-19 by income, race/ethnicity, and household crowding: US county vs. ZIP code analyses* (HCPDS Working Paper, Volume 19, Number 1). Harvard T.H. Chan School of Public Health, Boston, MA.

City of Vienna. (2013). *Gender mainstreaming in urban planning and urban development*. MA 18—Urban Development and Planning, Vienna.

Cole, M. A., Ozgen, C., & Strobl, E. (2020). Air pollution exposure and Covid-19 in Dutch municipalities. *Environmental and Resource Economics, 76*, 581–610.

Conticini, E., Frediani, B., & Caro, D. (2020). Can atmospheric pollution be considered a co-factor in extremely high level of SARS-CoV-2 lethality in Northern Italy? *Environmental Pollution, 261*, 114465.

Corburn, J. (2013). *Healthy city planning: From neighbourhood to national health equity* (Planning, History and Environment Series). Routledge.

Criado-Perez, C., Collins, C. G., Jackson, C. J., Oldfield, P., Pollard, B., & Sanders, K. (2020). Beyond an 'informed opinion': Evidence-based practice in the built environment. *Architectural Engineering and Design Management, 16*, 23–40.

Dahlgren, G., & Whitehead, M. (1991). *Policies and strategies to promote social equity in health*. Institute for Futures Studies.

Dakubo, C. Y. (2011). *Ecosystems and human health: A critical approach to ecohealth research and practice*. Springer.

Dallat, M. A. T., Soerjomataram, I., Hunter, R. F., Tully, M. A., Cairns, K. J., & Kee, F. (2014). Urban greenways have the potential to increase physical activity levels cost-effectively. *European Journal of Public Health, 24*, 190–195.

Egan, M., Kearns, A., Katikireddi, S. V., Curl, A., Lawson, K., & Tannahill, C. (2016). Proportionate universalism in practice? A quasi-experimental study (GoWell) of a UK neighbourhood renewal programme's impact on health inequalities. *Social Science & Medicine, 152*, 41–49.

EIDD. (2009). EIDD Stockholm declaration 2004.

Elwood, M. (2017). Critical appraisal of epidemiological studies and clinical trials. In *The diagnosis of causation* (4th ed.). Oxford University Press.

Flynn, C., Yamasumi, E., Fisher, S., Snow, D., Grant, Z., Kirby, M., Browning, P., Rommerskirchen, M., & Russell, I. (2021). *People's climate vote results*. United Nations Development Programme and University of Oxford.

Gatzweiler, F. W., Reis, S., Zhang, Y., & Jayasinghe, S. (2018). Lessons from complexity science for urban health and well-being. *Cities Health, 1*, 210–223.

Giles-Corti, B., Vernez-Moudon, A., Reis, R., Turrell, G., Dannenberg, A. L., Badland, H., Foster, S., Lowe, M., Sallis, J. F., Stevenson, M., & Owen, N. (2016). City planning and population health: A global challenge. *The Lancet, 388*, 2912–2924.

Glouberman, S., Gemar, M., Campsie, P., Miller, G., Armstrong, J., Newman, C., Siotis, A., & Groff, P. (2006). A framework for improving health in cities: A discussion paper. *Journal of Urban Health, 83*, 325–338.

Greenhalgh, T. (2019). *How to read a paper: The basics of evidence-based medicine and healthcare* (6th ed.). John Wiley & Sons Ltd.

Heylighen, A., Van der Linden, V., & Van Steenwinkel, I. (2017). Ten questions concerning inclusive design of the built environment. *Building and Environment, 114*, 507–517.

Howden-Chapman, P., Pierse, N., Nicholls, S., Gillespie-Bennett, J., Viggers, H., Cunningham, M., Phipps, R., Boulic, M., Fjällström, P., Free, S., Chapman, R., Lloyd, B., Wickens, K., Shields, D., Baker, M., Cunningham, C., Woodward, A., Bullen, C., & Crane, J. (2008). Effects of improved home heating on asthma in community dwelling children: Randomised controlled trial. *BMJ, 337*, a1411–a1411.

Kern, L. (2020). *Feminist city: Claiming space in a man-made world.* Verso.

Krieger, N. (1994). Epidemiology and the web of causation: Has anyone seen the spider? *Social Science & Medicine, 39*, 887–903.

Krieger, N. (2001). Theories for social epidemiology in the 21st century: An ecosocial perspective. *International Journal of Epidemiology, 30*, 668–677.

Lam, T. M., Vaartjes, I., Grobbee, D. E., Karssenberg, D., & Lakerveld, J. (2021). Associations between the built environment and obesity: An umbrella review. *International Journal of Health Geographics, 20*, 7.

Luke, D. A., & Stamatakis, K. A. (2012). Systems science methods in public health: Dynamics, networks, and agents. *Annual Review of Public Health, 33*, 357–376.

Lundberg, O. (2020). Next steps in the development of the social determinants of health approach: The need for a new narrative. *Scandinavian Journal of Public Health.* https://doi.org/10.1177/1403494819894789.

Lynch, K. (1981). *A theory of good city form.* MIT Press.

Mackenbach, J. D., Rutter, H., Compernolle, S., Glonti, K., Oppert, J.-M., Charreire, H., De Bourdeaudhuij, I., Brug, J., Nijpels, G., & Lakerveld, J. (2014). Obesogenic environments: A systematic review of the association between the physical environment and adult weight status, the SPOTLIGHT project. *BMC Public Health, 14*, 1.

Marmot, M., Allen, J., Boyce, T., Goldblatt, P., & Morrison, J. (2020). *Health equity in England: The Marmot Review 10 years on.* Institute of Health Equity, London.

Marmot, M., Allen, J., Goldblatt, P., Boyce, T., McNeish, D., Grady, M., & Geddes, I. (2010). Fair society, healthy lives: The Marmot review; strategic review of health inequalities in England post-2010. *Marmot Review*, London.

Marsh, H. W., Huppert, F. A., Donald, J. N., Horwood, M. S., & Sahdra, B. K. (2020). The well-being profile (WB-Pro): Creating a theoretically based multidimensional measure of well-being to advance theory, research, policy, and practice. *Psychological Assessment, 32*, 294–313.

Meadows, D. H. (2008). *Thinking in systems: A primer.* Chelsea Green Pub.

Nieuwenhuijsen, M. J. (2016). Urban and transport planning, environmental exposures and health-new concepts, methods and tools to improve health in cities. *Environmental Health, 15*, S38.

Nightingale, C. M., Limb, E. S., Ram, B., Shankar, A., Clary, C., Lewis, D., Cummins, S., Ellaway, A., Giles-Corti, B., Whincup, P. H., Rudnicka, A. R., Cook, D. G., & Owen, C. G. (2019). The effect of moving to East Village, the former London 2012 Olympic and Paralympic Games Athletes' Village, on physical activity and adiposity (ENABLE London): A cohort study. *The Lancet Public Health, 4*, e421–e430.

Northridge, D. M. E., Sclar, D. E. D., & Biswas, M. P. (2003). Sorting out the connections between the built environment and health: A conceptual framework for navigating pathways and planning healthy cities. *Journal of Urban Health, 80*, 556–568.

Pineo, H. (2019). *The value and use of urban health indicator tools in the complex urban planning policy and decision-making context* (Doctoral Thesis). University College London, London.
Pineo, H. (2020). Towards healthy urbanism: Inclusive, equitable and sustainable (THRIVES)—an urban design and planning framework from theory to praxis. *Cities Health, 0*, 1–19. https://doi.org/10.1080/23748834.2020.1769527
Pineo, H., Glonti, K., Rutter, H., Zimmermann, N., Wilkinson, P., & Davies, M. (2018a). Urban health indicator tools of the physical environment: A systematic review. *Journal of Urban Health, 95*, 613–646.
Pineo, H., Moore, G., & Braithwaite, I. (2020b). Incorporating practitioner knowledge to test and improve a new conceptual framework for healthy urban design and planning. *Cities Health, 0*, 1–16. https://doi.org/10.1080/23748834.2020.1773035
Pineo, H., Zimmermann, N., Cosgrave, E., Aldridge, R. W., Acuto, M., & Rutter, H. (2018b). Promoting a healthy cities agenda through indicators: Development of a global urban environment and health index. *Cities Health, 2*, 27–45.
Pineo, H., Zimmermann, N., & Davies, M. (2019). Urban planning: Leveraging the urban planning system to shape healthy cities. In S. Galea, C. K. Ettman, & D. Vlahov (Eds.), *Urban health* (pp. 198–206). Oxford University Press.
Porta, M. (2014). *A dictionary of epidemiology*. Oxford University Press.
Raworth, K. (2017). *Doughnut economics: Seven ways to think like a 21st-century economist*. Random House Business Books.
Rayner, G., & Lang, T. (2012). *Ecological public health: Reshaping the conditions for good health*. Routledge.
Rothman, L., Howard, A., Buliung, R., Macarthur, C., Richmond, S. A., & Macpherson, A. (2017). School environments and social risk factors for child pedestrian-motor vehicle collisions: A case-control study. *Accident; Analysis and Prevention, 98*, 252–258.
Rutter, H. (2018). The complex systems challenge of obesity. *Clinical Chemistry, 64*, 44–46.
Rydin, Y., Bleahu, A., Davies, M., Dávila, J. D., Friel, S., De Grandis, G., Groce, N., Hallal, P. C., Hamilton, I., Howden-Chapman, P., Ka-Man Lai, C. J., Lim, J. M., Osrin, D., Ridley, I., Scott, I., Taylor, M., Wilkinson, P., & Wilson, J. (2012). Shaping cities for health: Complexity and the planning of urban environments in the 21st century. *The Lancet, 379*, 2079–2108.
Saelens, B. E., Sallis, J. F., Frank, L. D., Couch, S. C., Zhou, C., Colburn, T., Cain, K. L., Chapman, J., & Glanz, K. (2012). Obesogenic neighborhood environments, child and parent obesity. *American Journal of Preventive Medicine, 42*, e57–e64.
Schulz, A., & Northridge, M. E. (2004). Social determinants of health: Implications for environmental health promotion. *Health Education & Behavior, 31*, 455–471.
Smith, M. (2019). International poll: Most expect to feel impact of climate change, many think it will make us extinct. *YouGov.co.uk*.
Sterman, J. D. (2000). *Business dynamics: Systems thinking and modeling for a complex world*. Irwin/McGraw-Hill.
Sterman, J. D. (2006). Learning from evidence in a complex world. *American Journal of Public Health, 96*, 505–514.
Tulier, M. E., Reid, C., Mujahid, M. S., & Allen, A. M. (2019). "Clear action requires clear thinking": A systematic review of gentrification and health research in the United States. *Health & Place, 59*, 102173.
Watts, N., Amann, M., Arnell, N., Ayeb-Karlsson, S., Beagley, J., Belesova, K., Boykoff, M., Byass, P., Cai, W., Campbell-Lendrum, D., Capstick, S., Chambers, J., Coleman, S., Dalin, C., Daly, M., Dasandi, N., Dasgupta, S., Davies, M., Di Napoli, C., … Costello, A. (2021). The 2020 report of The Lancet Countdown on health and climate change: Responding to converging crises. *The Lancet, 397*, 129–170.
Whitehead, S. J., & Ali, S. (2010). Health outcomes in economic evaluation: The QALY and utilities. *British Medical Bulletin, 96*, 5–21.
WHO. (1986). *Ottawa charter for health promotion*. World Health Organization, Geneva, Switzerland.

WHO. (2020). *Personal interventions and risk communication on air pollution*. WHO, Geneva, Switzerland.

Wu, X., Nethery, R. C., Sabath, M. B., Braun, D., & Dominici, F. (2020). Air pollution and COVID-19 mortality in the United States: Strengths and limitations of an ecological regression analysis. *Science Advances, 6*, eabd4049.

Zhu, Y., Xie, J., Huang, F., & Cao, L. (2020). Association between short-term exposure to air pollution and COVID-19 infection: Evidence from China. *Science of the Total Environment, 727*, 138704.

Chapter 4
Planetary Health

4.1 Introduction

The term 'planetary health' may mislead some readers to think that the topic is more concerned with the health of the planet than people; however, that is not the case. The concept is rooted in the related fields of ecological public health, One health and ecohealth (among others) which have slightly different emphases but all focus on the interconnections between ecological systems and human health (Buse et al., 2018). The planetary health agenda is focused on the health impacts of anthropogenic environmental change and through research and policy action it aims to: understand the interlinkages between health, civilisation and environmental degradation; adopt transformative actions; change governance systems through integrated policy-making; and reduce inequities (Whitmee et al., 2015). In the THRIVES framework, the term 'planetary health' references a scale of health impact that is global in nature but can be strongly influenced by decisions taken through the design of cities and urban infrastructure.

Better management of the built environment is an important action to reduce the health harms that result from environmental degradation. The concept of 'planetary boundaries' refers to thresholds within which earth's systems need to operate to avoid catastrophic changes that would not be safe for human survival (Rockström et al., 2009). Table 4.1 summarises some of the health effects (both positive and negative) related to a selected set of planetary boundaries from Butler (2016) with an explanation of how these relate to built environment decisions. The health benefits describe how humans currently gain from activities that threaten transgression of a planetary boundary, such as burning fuel for heating. The health disbenefits are the impact of approaching or exceeding the boundary. Although Table 4.1 is not comprehensive, it showcases the wide range of environmental decisions that relate to planetary boundaries from building design to urban form and energy systems. Researchers continue to investigate the poorly understood interconnections across

H. Pineo, *Healthy Urbanism*, Planning, Environment, Cities,
https://doi.org/10.1007/978-981-16-9647-3_4

Table 4.1 Example positive and negative health effects of selected planetary boundaries with examples of built environment contributions

Planetary Boundary	Health benefits	Health disbenefits	Relationship to built environment decisions
Atmospheric aerosol loading	Thermal comfort and nutrition through heating and cooking systems, use of electrical lighting and appliances	Respiratory and cardiac disease, mercury accumulation in marine animals	Energy production systems, building design and orientation
Loss of biological diversity	Reduced predation and increased food production	Loss of potential pharmaceuticals and products, more human and animal disease (e.g. transmitted by insect and mammalian vectors)	Land use patterns, mining for construction materials
Chemical pollution	Use of many technologies and materials	Endocrine disruption, cancer, birth defects, neurological conditions, mining disasters	Technologies and materials used in construction
Climate change	Industrialisation (i.e. transition from manual labour for manufacturing and farming)	Heat stress, extreme weather, infectious diseases, food scarcity, population dislocation, conflict, mental health effects	Land use (promote mixed-use and connected development), energy systems, building design

Source: Adapted from Butler (2016)

these boundaries and the associated health effects. Although the evidence-base is still developing, built environment professionals should take action now. Sustainable planning and design practices that have been developed since the 1990s can be used to reduce the negative health impacts that arise from transgressing planetary boundaries.

As introduced in Chap. 1, the health threat of climate change has been described as the biggest of the twenty-first century and one that will have the greatest effect on people who have contributed the least emissions, such as low-income populations (Costello et al., 2009). Just as was observed with COVID-19, the negative consequences of climate change will exacerbate existing inequalities, cutting away at gains towards international goals of poverty reduction and increasing the quality of life for people around the world. The *Lancet* Countdown project is committed to monitoring and reporting the health effects of climate change. Their 2021 report highlights that despite international agreements to limit emissions, the period 2015–2020 saw continued growth in carbon dioxide emissions and the five hottest years on record (Watts et al., 2021). Equally concerning is that the conditions monitored for climate change impacts, exposures, and vulnerabilities have all worsened since the project began in 2016, with the effects being disproportionately worse for disadvantaged populations. The figures are shocking, and they provide valuable evidence about the health and economic impacts of environmental degradation. For example, in 2018, there were 296,000 heat-related deaths (primarily occurring in Japan, eastern China, northern India, and central Europe) with the cost of these deaths in Europe alone being equivalent to 1.2% of regional gross national income (ibid.). These findings represent 2 of 43 indicators setting out the scale and urgency of climate change. There is also growing research on the health effects of climate change that are considered to be further upstream, such as those related to social instability. Impacts that are both short-term and long-term, such as disasters and resource scarcity respectively, can lead to social instability that threatens health (Sellers et al., 2019). In addition, the mental health effects of the climate crisis are increasingly understood. A recent report explains that climate change exacerbates mental distress, that severe distress is observed after extreme weather events and that warmer temperatures are linked to increased suicides, among other effects (Lawrance et al., 2021).

Reducing emissions and adapting to the impacts of climate change will require action from many sectors, tiers of government and community groups. As authors for the Intergovernmental Panel on Climate Change have pointed out, transformative systemic change in society is required, yet very few countries, cities or communities can claim to be implementing pathways that will achieve the required carbon reductions (de Coninck et al., 2018). The Planetary Health Alliance has created a set of 12 cross-cutting principles (Box 4.1) that should inform all planetary health education (Stone et al., 2018). These principles emphasise the urgency and multi-level health impacts of environmental change, alongside knowledge needed to enact change. The issues covered in Box 4.1 are well integrated with those set out in the first three chapters of this book, including introductions to systems thinking, just sustainabilities, environmental justice, and governance principles for health, among others. The case studies in this chapter and the following will help bring to life the importance and meaning of these cross-cutting principles as they apply to practical urbanism examples.

Box 4.1 Twelve cross-cutting principles for planetary health

The following twelve principles were developed by the Planetary Health Alliance's Planetary Health Education Brainstorm Group with the intention of informing all planetary health education (Stone et al., 2018).

1. *A planetary health lens*: Viewing the world with an understanding of the important interconnections and cause–effect relationships between environmental change and human health.
2. *Urgency and scale*: Examining the complexity of environmental change and human health impacts with reference to multiple scales (geographical and temporal) and other contextual socio-economic, political and cultural factors that shape potential challenges and solutions.

(continued)

Box 4.1 (continued)

3. *Policy*: Being familiar with policy actors at global and local levels and providing them with quantified health effects of anthropogenic environmental change to inform cross-sector policy action.
4. *Organising and movement building*: Recognising that community organisations can have significant influence in local and global political processes, and that this is important to identify workable solutions for planetary health challenges.
5. *Communication*: Learning effective communication methods (including the importance of listening) to translate complex research evidence to inform action across sectors, scales and geographies.
6. *Systems thinking and transdisciplinary collaborations*: Gaining skills to understand complex challenges and integrate knowledge across disciplines to develop new solutions that overcome existing gaps in research methods and policy.
7. *Inequality and inequity*: Appreciating the theory and practice of equality and equity (including marginalisation, vulnerability and resilience) to inform research and decision-making.
8. *Bias*: Identifying potential biases and vested interests (political, social or economic) that determine research and perceptions of the health effects of environmental change.
9. *Governance*: Understanding processes of decision-making and policy implementation, including challenges related to capacity, politics, institutional issues and governance across scales/regions.
10. *Unintended consequences*: Recognising that the consequences of environmental change and the impacts on health will sometimes be unexpected, requiring adoption of adaptive management and resilience building.
11. *Global citizenship and cultural identity*: Seeing one's own place in global and local communities and using this to define the values and practices of future generations in positive ways.
12. *Historical and current global values*: Gaining knowledge of the past in terms of milestones in the field, ignored or marginalised perspectives and patterns that will influence future action.

4.2 Achieving Planetary Health Framework Goals

It is clear from Table 4.1 that many built environment policy and design decisions will affect planetary health, not only those that are covered in this chapter. THRIVES explicitly recognises the interconnections across the three scales of health impact (planetary, ecosystem and local) that are aligned to the design and planning goals and scales of decision-making via the concentric circles in Fig. 1.2. Actions taken across these scales and goals can be supportive of health (e.g. building design can be energy efficient to reduce carbon emissions); however, the framework attempts to show that there is potential for greater impact when issues of energy, resource efficiency and biodiversity are considered at regional and city scales.

4.2.1 *Zero Carbon*

Carbon dioxide (CO_2) emissions are the primary cause of climate change and, among other greenhouse gases (GHGs), they can be reduced through urban built environments, including energy systems, land use, transport and buildings. The global breakdown of GHGs across sectors, including energy use in buildings (17.5%) and transport (16.2%), demonstrates the significant potential for climate change mitigation by reducing built environment emissions (Fig. 4.1). Analysis

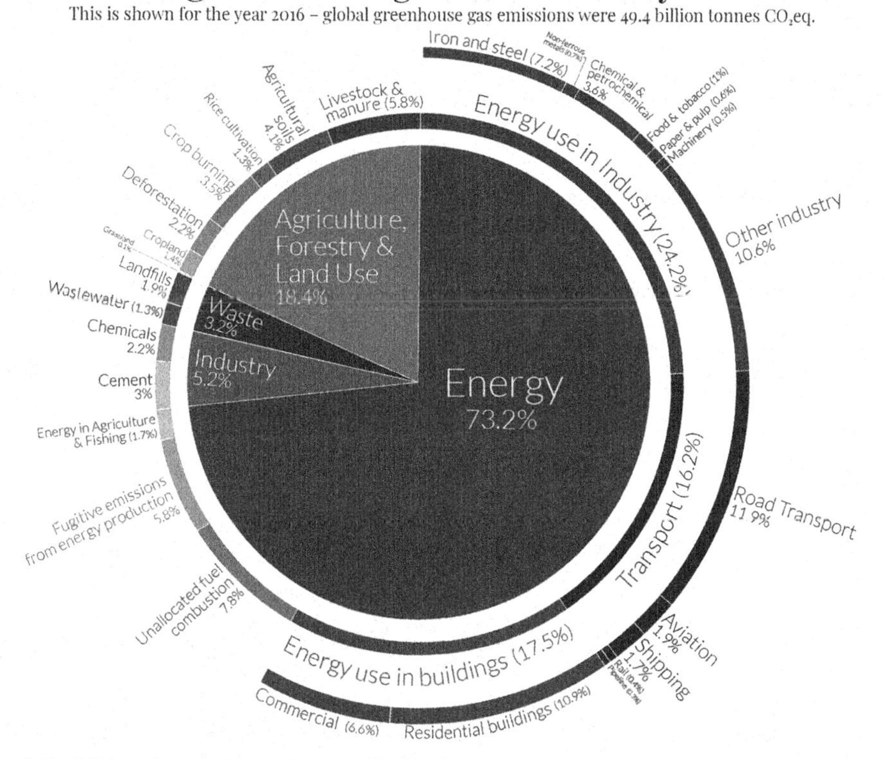

Fig. 4.1 Greenhouse gas emissions by sector. (Source: Hannah Ritchie (2020), available from Ritchie and Roser (n.d.) using data from Climate Watch, the World Resources Institute (2020))

from the *Lancet* Countdown shows that the carbon intensity of the global energy system has been roughly unchanged over the past 30 years, and in 2017, it was 0.4% higher than 1990 levels (Watts et al., 2021). There are regional differences whereby in 2018 the carbon intensity levels in the USA and northern and western Europe were 12% and 20%, respectively, lower than 1990 levels. China's carbon intensity reduced by 4% from 2013 to 2017 but remains relatively high. The report also highlights more positive findings, such as increases in renewable and low-carbon energy sources, accounting for 33% of total energy generation in 2017. Unfortunately, the pace and scale of decarbonisation has not been sufficient and leading climate scholars are concerned that achieving the goal of net zero CO_2 emissions by 2050 may not be possible, and would not necessarily keep warming below the 2 degrees target (Deutch, 2020; de Coninck et al., 2018). Achieving 'net zero' means 'eliminating all CO_2 and eventually all other GHG emissions…from every end-use sector' including industry, transport, buildings and land use (Deutch, 2020, p. 2238).

The general principles for reducing carbon emissions through the built environment are clearly described by the energy hierarchy, often depicted as an inverted pyramid showing the prioritisation of energy reduction goals. The energy hierarchy in Fig. 4.2 is adapted from the Institution of Mechanical Engineers (2020) and the version used in London's Energy Assessment Guidance (Mayor of London, 2020a). Reducing energy demand can be achieved through urban form and building design (see Box 4.2). Energy supply and building management systems and technologies can be designed to use energy efficiently, reducing overall consumption. Energy sources should be renewable or low-carbon and these would preferably be on-site (e.g. solar panels) or near end-users (e.g. a decentralised heating system). Finally, London's Energy Assessment Guidance allows for offsetting, which may be appropriate for specific developments and usually operates through payment into a fund that is allocated according to city priorities.

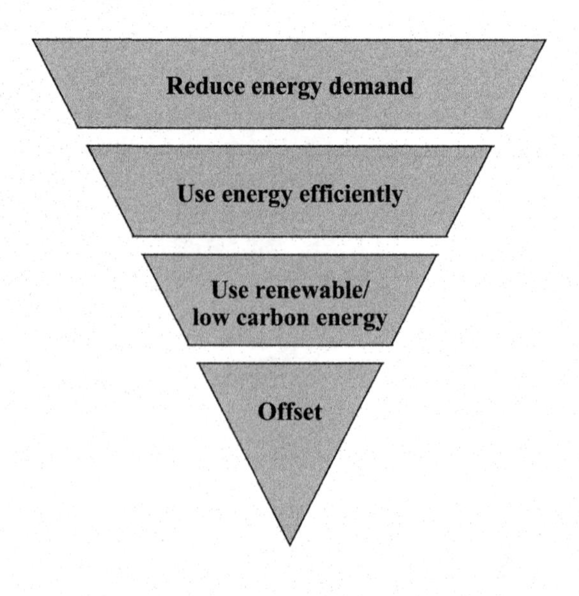

Fig. 4.2 An example of the energy hierarchy. (Source: Adapted from Institution of Mechanical Engineers, 2020 and Mayor of London, 2020a)

Box 4.2 Urban design and policy measures to reduce carbon emissions
Large-scale infrastructure projects, city-wide policies and buildings can significantly reduce carbon emissions. The general principles for carbon reduction in the built environment are to reduce the use of energy (e.g. for moving people and goods or to heat buildings), to use energy efficiently and to switch to low and zero carbon energy systems for electricity, heating and cooling.

Specific evidence-based design and planning policy measures that will support carbon reductions and improve health include the following:

- Reduce the need to travel via motor vehicles via provision of public transport, mixed-use development and active mobility infrastructure.
- Plan for compact urban form to support decentralised energy systems, low carbon transport and infrastructure, and leaving undeveloped land for carbon sinks.
- Install decentralised low or zero carbon energy systems.
- Use low carbon or recycled construction materials.
- Encourage food vendors to use local suppliers.
- Design buildings and masterplans to reduce reliance on mechanical heating and cooling, such as orientation and layout, natural ventilation, daylighting and solar shading.
- Reduce energy use within buildings by situating stairs in prominent locations, using low energy or passive lighting and sensors.
- Use a sustainable building/neighbourhood standard or principles to reduce embodied carbon and energy demand across the life cycle.

For further information, see analyses by Cedeño-Laurent et al. (2018b), Pizarro (2009) and Younger et al. (2008).

The sustainability agenda was established in the early 1990s and over the decades much expertise has accrued in industry and government about low carbon design and planning. Steffan Lehmann's (2014) book *Low Carbon Cities* argues that decarbonisation will result in cities that are more 'efficient, competitive, liveable and sustainable', providing global examples of progress and success aligned with his green urbanism framework from Korea, India, Australia and more (p. 9). Environmental or 'green' design and planning approaches are usually supportive of human health, yet the opportunities are often missed to explicitly link these agendas in urban policy. Box 4.2 sets out evidence-based design and planning measures that support carbon reductions and health. Many of these measures are universally applicable, but they should be applied with respect to local contexts. For example, compact urban form will support climate change mitigation and adaptation in hot-dry climates because shadows that cast across buildings which are close together can reduce the need for mechanical cooling systems (while compact form reduces transport emissions); however, in hot-humid climates, this form could reduce air circulation around dwellings and exacerbate indoor overheating (Pizarro, 2009). It is therefore important to use local experience and research to inform design and policy standards for carbon reduction.

4.2.2 Biodiversity

Biodiversity is declining rapidly with significant impacts for human health and wellbeing. The Convention on Biological Diversity (2006) defines biological diversity as 'the variability among living organisms from all sources including, inter alia, terrestrial, marine and other aquatic ecosystems and the ecological complexes of which they are part; this includes diversity within species, between species and of ecosystems'. The component parts of biodiversity can be grouped at three levels: genetic, species and ecosystem diversity. The Intergovernmental Science Policy Platform on Biodiversity and Ecosystem Services reported that biodiversity losses are declining at an unprecedented pace with multiple health effects, including threats related to: natural products for medicines, infectious diseases, food production, clean water and other factors (IPBES, 2019). Loss of nature also threatens the livelihoods of many indigenous and urban/rural poor populations. The IPBES analysis quantifies the risks to humans through multiple pathways. For example, 75% of global food crop types require animal pollination (including coffee, cocoa and almonds), yet humans are responsible for a quarter of assessed animal and plant populations being threatened and the extinction of over 680 vertebrate species since 1500 (ibid.). Biological diversity underpins ecosystem functions that humans rely upon for health, including through important spiritual and cultural values, climate regulation and disaster risk reduction (Romanelli et al., 2015).

Although the distributed health effects of biodiversity loss have been well-documented, the shorter-term public health benefits of designing for biodiversity and ecological restoration have not been adequately quantified, although the evidence base is growing (Breed et al., 2020). There are direct and indirect health benefits of urban nature (including in residential gardens), such as those associated with reducing the urban heat island effect and the restorative effects of being in greenspace that may reduce stress (Shanahan et al., 2015). Studies have demonstrated that exposure to areas with increased biodiversity can support human wellbeing. An analysis of different urban and peri-urban greenspaces in Italy found a positive correlation between biodiversity levels and visitors' wellbeing (Carrus et al., 2015). Another study in Sheffield, England, found associations between biodiversity measures in urban greenspaces, including area, plant richness and number of habitats, with psychological measures for recovery from mental fatigue, the opportunity for reflection, emotional attachment and personal identity (Fuller et al., 2007). The benefits of designing for biodiversity may not always have quantifiable short-term local health benefits, but this does not reduce their importance.

As with energy, there is a 'mitigation hierarchy' for biodiversity that can help guide decision-making on specific urban policies and projects. The hierarchy is to avoid, minimise, restore and offset the predicted biodiversity impacts on projects with the aim of achieving 'no net loss' (Gardner et al., 2013). A key challenge is that biodiversity accounting does not accurately measure cumulative impacts that consider appropriate baselines, trends and uncertainty, meaning that 'no net loss' is not assured (Bull et al., 2016). Offsetting ecological losses created through new development is unlikely to fully compensate for damage; however, it may be seen as the only option in certain circumstances. Birkeland and Knight-Lenihan (2016) describe international approaches to offsetting in relation to urban development. A practical approach uses the concept of 'habitat/condition-area' to inform a 'bio-banking' system, whereby the biodiversity condition of the development site is assessed, and the developer pays a proportionate offset fee into a fund managed by the local authority (or another body) for ecological restoration on another site. This approach is criticised due to the challenges with accurately measuring the site's condition. Another approach is called 'conservation benefit matching' which involves adopting the mitigation hierarchy at multiple tiers of government and the use of high-quality data about local conditions to determine whether offsets will meet national goals. An example of how this could work in practice is the Manawaüt-Wanganui Regional Council's One Plan in New Zealand which 'proposes protecting and maintaining indigenous biodiversity and enhancing it where appropriate' through set outcomes and timelines associated with the mitigation hierarchy (ibid., p. 55).

Urban development can support biodiversity (see example in Box 4.3), but significant improvements to current practice are needed to avoid net biodiversity

Box 4.3 Biodiversity in London's Queen Elizabeth Olympic Park
The Queen Elizabeth Olympic Park (QEOP) was planned as London's newest park situated on the site of the 2012 London Olympic and Paralympic Games. The London Legacy Development Corporation (LLDC) was established in 2012 as a mayoral development corporation under the power of the Localism Act 2011 to lead redevelopment of the site. Totalling 560 acres, the QEOP is a large-scale project comprising '35 km of pathways and cycleways, 6.5 km of waterways, over 100 hectares (ha) of land capable of designation as Metropolitan Open Land, 45 ha of Biodiversity Action Plan Habitat, 4000 trees, playgrounds and a Park suitable for year-round events and sporting activities' (LLDC, 2013). There were also plans for up to 6800 new homes and 91,000 square metres of new commercial space.

Biodiversity actions were part of the wider principles to improve health and wellbeing for residents, which included active design, walkable communities, access to nature and biodiversity, public access to the sports venues, integration of green paths within the residential developments and provision of affordable and social housing. There was a 2008 Biodiversity Action Plan (BAP) for the Olympic Park during the games, and an updated 2013 plan for the legacy development. The 2008 BAP was viewed as a great success, having provided more than 25 ha of BAP habitat and a precedent for the BAP process, involving stakeholder and community engagement. The BAP's vision was to extend existing networks of greenspaces (linking the Lee Valley Regional Park in the north with the Thames to the south), to connect new and existing parks with residential communities and to create new wildlife-rich parks that would help London adapt to climate change. The 2013 BAP sets out the Habitat Action Plan with targets and associated monitoring indicators. The Plan contains detailed requirements for different habitats on the park and associated management requirements. New residential developments were required to meet specific BAP targets for the QEOP. Figure 4.3 shows different habitats on the park and adjacent developments over time.

Fig. 4.3 Biodiversity elements at Queen Elizabeth Olympic Park. Top left: Wildflower meadow at 2012 London Olympic and Paralympic Games. Bottom left: Bug hotel in Chobham Manor in 2021. Top right: Grasslands and wet woodlands at centre of QEOP in 2021. Bottom right: Wetland and trees at East Village in 2019. (Source: Author)

decline. Janis Birkeland's (2008) positive development concept sets out that a net positive development would return to ecosystems and the local community more than it takes, considering impacts across space and time. Benefits for the 'public estate' mean equitable and universal access to the 'means of survival, health and wellbeing', such as shelter, food, clean water and air (Birkeland, 2017, p. 85). A key point of her argument is that existing standards and guidelines seek to incrementally boost performance against business-as-usual practice rather than requiring the radical and transformative change that is needed to avoid catastrophic biodiversity loss. She has advocated a biodiversity credit scheme (BCS) used by the Australian Green Building Council that includes strategies such as: protecting ecological value (for example, by reusing previously developed land); minimising ecological impact (for example, by retaining on-site habitats); enhancing ecological value and biodiversity (for example, by prioritising on-site biodiversity actions and allowing off-site actions that create additional benefits); connecting ecological networks (for example, through habitat corridors); and finally, managing on-site and restoring off-site enhancements through adaptive management approaches using monitoring indicators (Birkeland, 2018).

Measures to enhance biodiversity through urban policy include growth strategies that avoid low-density sprawling development and support habitat corridors or 'stepping stones' that link ecosystems across and beyond the boundaries of a city. At the scale of individual developments, Georgia Garrard et al.'s (2018) Biodiversity Sensitive Urban Design (BSUD) framework can support design teams to achieve on-site biodiversity benefits for human and ecosystem health. The framework includes five principles (Box 4.4) and a sequential process for adopting these principles is urban development. The BSUD process involves the following stages: documenting biodiversity values on the site, identifying biodiversity objectives, identifying BSUD actions using the five principles (informed by the wider development objectives), assessing the contribution of the BSUD actions using metrics and finally, deciding the appropriate BSUD to meet biodiversity and development objectives. There are many potential BSUD actions that can be adopted on a specific site across design, construction and occupation and these should be informed by local context, including the need to resolve trade-offs with other development objectives.

Box 4.4 A framework for Biodiversity Sensitive Urban Design

There are five principles in Garrard et al.'s (2018) Biodiversity Sensitive Urban Design (BSUD) framework that are based on existing evidence of biodiversity and urbanisation. These principles should inform all design projects and be accompanied with community stewardship initiatives. The BSUD principles are summarised as follows:

1. *Maintain and introduce habitat*: Prioritising new development in areas of low ecological value, retaining and protecting vegetation during construction, using native species and increasing vegetation complexity in landscape strategies, and adding green infrastructure or habitat walls.
2. *Facilitate dispersal*: Creating infrastructure for animal movement and habitat connectivity (avoiding the spread of invasive species and pests).
3. *Minimise threats and anthropogenic disturbances*: Using landscaping to contain pets and minimise weeds and exotic predators, adopting sustainable drainage techniques (e.g. swales and rain gardens) to reduce runoff, reducing noise and light pollution through measures such as sound barriers, and dimming lights and temporary road closures.
4. *Facilitate natural ecological processes*: Aiming mitigation strategies and resources at target species, enhancing pollinator habitat and planning for events such as fire or flooding.
5. *Improve potential for positive human-nature interactions*: Including 'cues to care' (Nassauer, 1995) that help facilitate local stewardship of biodiversity, designing for opportunities to interact with nature and considering potential conflicts or disservices between biodiversity, safety and other local requirements.

4.2.3 Resource Efficiency

The pace and scale of environmental degradation is risking the viability of human life on earth and increasing inequalities, insecurity and poverty. The UN Environment Programme (UNEP, 2021) report on *Making Peace with Nature* sets out the impact of consumption practices from 1970 to 2020 noting that while the economy grew five times larger, resource use tripled, carbon emissions doubled and 1.3 billion people are living in poverty. The pollution created through current industrial practices and urban infrastructure is killing millions of people annually. For example, the UNEP report notes that approximately 400 million tons of heavy metals, solvents, toxic sludge and other industrial wastes pollute the world's waters every year. Plastic waste has increased tenfold in marine ecosystems since 1980 and it pollutes other ecosystems, creating human health problems. Endocrine-disrupting chemicals (EDCs) are found in many products, including plastics, and they can be ingested through contaminated water and food causing numerous disabilities and diseases. An analysis of the health and economic impact of EDCs in Europe estimated that the total cost is 157 billion euros annually, or 1.23% of GDP (Trasande et al., 2015). The way that resources are used for energy, industrial processes and infrastructure is an important consideration in the built environment impact on health.

Resource efficiency refers to minimising the use and waste of natural resources and raw materials, including fuels, minerals, metals, food, soil, water, air and others. The concept is promoted in SDG 12 which aims to 'ensure sustainable consumption and production patterns' through efficient use of resources, waste management and avoiding chemical pollutants (UN General Assembly, 2015, p. 22). The model that is driving current thinking in resource efficiency is the circular economy, which applies systems thinking principles to move away from a linear economy, described as 'take-make-waste' (Van Buren et al., 2016) to one that eliminates waste altogether, thus closing resource loops. The experience that inspired Dame Ellen MacArthur to drive the circular economy agenda through her eponymous foundation puts the existential threat of linear resource use in stark light. MacArthur (2015) tells the story of how, in 2005, she succeeded in being the fastest solo sailor to circumnavigate the globe. The supplies that she had on her roughly ten-week journey were finite and she depended on them for her survival. The sense of risk was particularly acute when she was in parts of the Southern Ocean where it would have taken days for another person to reach her. After the race, she began looking into the stocks of finite materials left on earth, such as coal, which at that time was predicted to last for a further 118 years—a tiny fraction of earth's history. This realisation led her to leave her career in solo sailing to work on the circular economy. This personal story about humanity's future told through an individual's experience brings the vast challenge of resource consumption down to a clear imperative. Circular economy frameworks emphasise a full closure of resource loops from avoiding the use of raw materials through to recovering energy from materials (Van Buren et al., 2016).

Circular economy frameworks can be applied at micro, meso and macro scales, for example, to building materials, construction practices and land use management.

Fusco Girard and Nocca (2019) reviewed circular economy strategies in 14 cities, primarily in Europe but also in Kawasaki, Japan. They identified a wide range of actions that can be taken from urban planning and governance through to building design. Urban planning can be used to enable efficient resource flows within a city through waste planning, mixed land uses that facilitate low-energy transport of goods and people, provision of infrastructure for sharing of common assets (e.g. libraries and public space), and regulations that stimulate material reuse and recovery. Multiple governance strategies are proposed for circular cities, including collaborative, adaptive, experimental and reflexive governance models. These approaches emphasise the importance of ongoing monitoring and partnership across sectors and the public, which would begin to address some of the key implementation challenges identified by Williams (2019) such as lack of political support and siloed-thinking. Adaptable and modular buildings can become self-sufficient in terms of resources (see Bullitt Center case study later in this chapter) and built of dismountable materials that can be repurposed or recycled. The Park 20|20 project in the Netherlands shows how an urban development can integrate circular economy principles across its life cycle stages, making adaptation and recovery of building components more feasible. The development used 'resources passports' which allow for 'tracking of materials and their corresponding residual value along the lifecycle of a building' (Leising et al., 2018).

Goals for the circular economy go beyond resource management. The Ellen MacArthur Foundation (2015) focuses on creating value through optimised resource use models. 'Circular cities' approaches consider urban planning and governance, social systems and the economy (Fusco Girard & Nocca, 2019). This holistic view is encapsulated in Van Buren et al.'s (2016) definition which states:

> A circular economy aims for the creation of economic value (the economic value of materials or products increases), the creation of social value (minimization of social value destruction throughout the entire system, such as the prevention of unhealthy working conditions in the extraction of raw materials and reuse) as well as value creation in terms of the environment (resilience of natural resources). (p. 3)

Including social value as a key component of circular economy models makes them particularly suited to supporting health and wellbeing at multiple spatial and temporal scales. Reducing waste and pollution will promote health through the planetary health mechanisms outlined earlier, such as avoiding catastrophic climate change and ensuring ecosystems can provide people with water, food and clean air. In the shorter term, there are cultural and economic benefits that will directly support health and wellbeing. The adaptive reuse of buildings with cultural heritage value is one example (Fusco Girard & Nocca, 2019). Through the restoration of such buildings, local communities will benefit from jobs and strengthened place attachment, social cohesion and local identity—factors which can then positively reinforce productivity and the economy.

Adapting rather than demolishing buildings that reach obsolescence is a part of circular economy strategies to reduce the use of materials, energy and construction-related pollution (Wilkinson et al., 2014; Williams, 2016). Adaptable building

Box 4.5 Unintended health effects of building reuse
Conserving materials by reusing buildings can create carbon savings and the avoidance of using new materials. However, it is important that the refurbishment is completed to a high standard to avoid potential negative health impacts. One example of this challenge can be found in the UK where the conversion of non-residential buildings (e.g. offices) to residential properties has increased following changes to the Permitted Development Rights since 2013. Ben Clifford et al. (2018, 2020) have documented some of the detrimental effects of these conversations, including potential health impacts. In their analysis of 639 converted buildings, they found that these properties had the following characteristics: primarily located in commercial or industrial areas with few residential amenities; only 22% of dwelling units would meet the Nationally Described Space Standards; single aspect windows in 72% of dwelling units; and only 3.5% of units had access to private amenity space, such as a balcony or garden (Clifford et al., 2020). The primary policy goal of extending PDR in the UK is to increase the housing supply, particularly for affordable housing. However, the characteristics of this housing may create numerous health and wellbeing impacts for occupants, resulting in increased costs to the public sector. To avoid these unintended consequences and gain the co-benefits of building reuse, the location and design of properties must be adequately considered alongside goals to increase housing numbers.

design reduces the need for costly and environmentally damaging adaptation (or demolition) processes during a building's lifespan. However, many adaptations will continue to occur with buildings and infrastructure that were not designed for that purpose. To facilitate building adaptation planners and design teams will need to consider how a building will function in its location over time. Architects will need to create flexible structures within which internal layouts can shift depending on occupants' needs for space and other functional requirements. Adaptations that involve a change of use can produce quirky and exciting spaces, such as the sought-after conversions of nineteenth-century industrial buildings into flats; however, caution is needed to avoid unintended consequences (see Box 4.5).

4.3 Monitoring Indicators for Planetary Health

The indicators in Table 4.2 were gathered from a systematic review of urban health indicators, meaning that the included studies' authors believed there was a relationship between the physical environment factor being measured and health outcomes (Pineo et al., 2018a). Readers can access the indicator sources in the review to find reported data in their region or city. There were 145 indicator tools from 28 countries, including 8006 indicators. The selection of indicators in Table 4.2 should be seen as a starting point that can be used in combination with others provided in this book and elsewhere to inform different stages of policy and design processes.

Table 4.2 Examples of monitoring indicators for built environment impact on planetary and human health

Goal	Monitoring indicators
Zero carbon	Consumption of electricity per household in megawatts/hour
	% of electricity generated from renewable sources
	Number (or %) of new (or refurbished) building stock certified with a sustainable building standard (e.g. BREEAM and LEED)
Biodiverse	Total hectares of greenspace per 1000 residents
	Native vegetation cover
	% of area covered by trees
Resource efficient	Annual average utility bills
	Volume of water consumption per capita

There are many other sources of relevant indicators and data for planetary health. Several examples include the following:

- The *Lancet* Countdown produces valuable data about global and regional trends in planetary health topics. For example, their indicator on clean household energy measures the proportion of the global population that uses clean fuel and technology for cooking (63% in 2018) and the use of zero-emissions energy in homes (26% in 2017) using data from WHO and the International Energy Agency (Watts et al., 2021).
- Indicators produced in life-cycle assessments, social return on investment tools, sustainable construction standards and occupant surveys (see Coleman et al., 2018; RIBA et al., 2017).
- In relation to biodiversity, Garrard et al. (2018) describe proxy measures of cover and proportions of native and non-native species alongside a more appropriate measure of 'viability' which can be assessed with various methods including population viability analysis (PVA).
- For the circular economy, Fusco Girard and Nocca (2019) compiled relevant indicators across resource use and wider topics related to health, jobs and other impacts.

These selected indicators represent a wider range of potential metrics for planetary health. See Chap. 8 for further information on applying indicators to urban policy and development.

4.4 Bullitt Center: A Deep Green Building for Planetary Health

The city of Seattle, Washington, USA, has long been recognised as a green city, not just from the evergreen trees that permeate the urban fabric leading to its nickname the Emerald City, but also from its position as a pioneer of sustainable development policies. It was an early adopter of climate change action both through city policies,

Fig. 4.4 Bullitt Center, note prominent roof-top solar array. (Source: Author, 2019)

such as a resolution officially recognising the impact of global warming in 1992 (Rice, 2010) and through the work of community movements for sustainability and health, focused on indicators (Holden, 2007, 2008). Like other North American cities, historically Seattle adopted engineering and planning approaches that also caused harm to the local landscape (see Karvonen, 2010), yet the city remains an example of leadership in sustainable development in many respects. A success story of the city's collaborative and ambitious approach to sustainability is the Bullitt Center (Fig. 4.4), an award-winning six-storey ($51,000ft^2$) office building located one mile East of downtown and the Puget Sound that opened on Earth Day in 2013. This case study describes the project's successes, including the design and permitting process through to post-occupancy monitoring.

Dorothy Bullitt, a prominent Seattle businessperson, created the Bullitt Foundation in 1952 as a charitable organisation that invested in civic and cultural services (Bullitt Foundation, n.d.-a). From 1992, the organisation focused on environmental conservation and restoration in the Pacific Northwest region, led by its president Denis Hayes. The Bullitt Center project was driven by Hayes as more than just the Foundation's headquarters, but as a 'touchstone' and proven example for the organisation's Deep Green Buildings programme, which 'promotes huge leaps—as opposed to incremental shifts—in the built environment' towards buildings that mitigate against climate change and adapt to its impacts (Bullitt Foundation, n.d.-b). The Deep Green concept focuses on operational performance beyond design measures or commitments, taking a holistic view of sustainability; encompassing productivity, health and even beauty; and viewing the building as a living system.

The Bullitt Center's achievements span the breadth of planetary, ecosystem and local health impacts set out in the THRIVES framework. The building produces more electricity than it uses through the roof-top solar PV panels and it collects all of its on-site water demand through rainwater. Sustainable design measures include: a well-insulated envelope and windows, exterior shading, daylighting, radiant heating served by a ground source heat pump, natural ventilation, heat recovery ventilation, a regenerative elevator, smart building management systems, solar panels, composting toilets (later replaced with a vacuum flush system), greywater cleaning system, rainwater harvesting, promoting stair use and sustainable materials. According to one evaluation, the building 'has about the same impact on the greater environment as the Douglas fir forest that occupied this site 150 years ago' (Peña et al., 2017, p. 164). Each of the building's sustainability measures can be analysed and mapped to THRIVES to consider its health impacts at different scales. For example, the composting toilets used 96% less water than standard flush toilets and produced zero waste. The composting system turned human excreta and toilet paper deposits into two bi-products: (1) a liquid called leachate which was taken off site and treated aerobically before being released into regional constructed wetlands; and (2) human biosolids that were removed and made into GroCo Compost sold commercially as a soil amendment (H Burpee 2021, personal communication, 8 November). This is a practical example of a circular economy system that meets the resource efficiency, biodiversity and water goals of THRIVES.

The ambitious goals at Bullitt Center were aided by highly collaborative processes in design, permitting and post-occupancy evaluation. Interviews conducted with city planners and architects demonstrated that collaborative approaches helped this building push the limits of what was possible, creating new pathways for sustainable development in Seattle and elsewhere (Pineo & Moore, 2021). Project team members were carefully screened with regard to their ability to work with other organisations and professions in what the Urban Land Institute (2015) described as a 'deeply integrated design process' that proactively identified and solved problems before construction commenced (p. 2). Academics at the University of Washington (UW) informed the Request for Proposals, technical studies and evaluation, later becoming tenants. Close collaboration with city officials resulted in a key solution for overcoming legal challenges to the innovative technologies and design measures. The Living Building Pilot Programme was created by the City of Seattle to allow the Bullitt Center and similar projects to adopt radical solutions that depart from building codes and other regulations. Examples of how the project did not meet city regulations included the solar panels overhanging the sidewalk (Fig. 4.4), rainwater consumption, the greywater infiltration system (Fig. 4.5, described in Chap. 5 in detail) and the provision of composting toilets in a commercial building (ibid.). Adopting these new systems had long-term effects on the city's permitting process to improve sustainability and health.

A legacy of the Bullitt Center project was that Seattle's Department of Construction & Inspections implemented the Living Building Pilot Programme.

Fig. 4.5 Greywater infiltration system at Bullitt Center, Seattle. Left: Street-level bioswales with 20 feet of gravel below ground. Right: Constructed wetland on the third floor of the building filters greywater from sinks and showers before discharging to the bioswales. (Source: Author, 2019)

This programme and the 2030 Challenge Pilots are part of the city's Climate Strategies to become carbon neutral by 2050. Roughly one-third of Seattle's carbon emissions are produced from buildings and city leaders recognised a need to incentivise transformative changes. Developers can request departures from the Seattle Land Use Code through Design Review if they adopt higher environmental standards through the Living Building Challenge standard or the 2030 Pilot, aimed at new build and retrofit, respectively (Seattle Department of Construction & Inspections, n.d.). The city provides incentives, including allowances, for greater floor area and building heights. Pilot programmes provide an important mechanism to de-risk innovative design measures and technologies, allowing developers and city officials to trial and monitor approaches before scaling them up to wider projects (Pineo & Moore, 2021).

A key accolade for the Bullitt Center was its certification to the Living Building Challenge (version 2.0), created by the International Living Future Institute, which now has office space in the building. Certification with this sustainable building standard requires on-site energy production that matches energy use, rainwater harvesting and treatment to meet the building's water needs year-round, and the exclusion of any toxic materials as set out in the 'Red List', among other ambitious targets. The LBC standard (version 4.0) has seven 'petals' or categories: place, water, energy, health and happiness, materials, equity and beauty. Beauty and health are supported by biophilic design principles that are evident in the ample use of natural materials inspired by the Douglas fir forest that was present on the site

Fig. 4.6 Sustainable design elements at Bullitt Center, Seattle. Left: Modular carpet sourced from Interface's 'Urban Retreat' collection. Right: Educational plaques explaining sustainable design features and technologies. (Source: Author, 2019)

historically, such as the use of timber in the staircase and building structure. Biophilic design and acoustic comfort within the building were supported by soft furnishings, including a modular carpet in earth tones (Fig. 4.6) of which individual tiles can be replaced and recycled (Nayar, 2015), a prime example of the circular economy.

Monitoring and evaluation are important aspects of a pilot project. New systems and design measures may not work as expected, but with monitoring data, adaptations can be made to correct problems. The University of Washington's Integrated Design Lab has office space at the Bullitt Center and uses data from the building management system to study the effectiveness of the adopted technologies, gaining lessons for future building design. UW researchers, Robert Peña et al. (2017), describe the extent of data availability at the Bullitt Center:

> The building was designed and constructed for an exceptionally granular level of performance monitoring. Every electric circuit in the building is metered and recorded. The building management system collects data on environmental conditions and the operation of all its systems including window and exterior blind operation, floor slab, and indoor air temperatures, CO2 levels, and flows of heated and cooled water generated by the ground-source heat pump system. (ibid., p. 164)

The UW team analysed energy data to compare the design simulation and actual performance, which was challenging in this multi-tenant building. The predicted energy use intensity (EUI) was 16 kBTU/ft^2/yr and the actual EUI over the first three years of use exceeded that goal at roughly 11 kBTU/ft^2/yr, surpassing the goal of net zero and operating as net-positive. They used a sub-set of the data to perform more detailed analyses of operational performance to understand how and where energy was being used. Overall, they have observed a close alignment between design goals and operational performance. Using the monitoring data, they found that the automated night-flush cooling system did cool indoor air and floor slabs, maintaining thermal comfort during warm afternoons with little or no mechanical cooling. There were hiccups with the building management system that were initially identified with this data and later corrected. Peña et al. call for greater 'stewardship systems' in buildings that provide continuous feedback information to inform operational improvements.

Alongside the monitoring of technical systems, there have also been evaluations of occupants' experiences and perceptions of the building demonstrating its impacts on health, comfort and wellbeing. Heather Burpee and a team of UW researchers gathered data through surveys, activity trackers, indoor environment monitors and samples of the building's airborne and surface microbiome (Burpee et al., 2014, 2015; Gilbride et al., 2016). The first round of surveys focused on employees prior to their move into the Bullitt Center to gather baseline data about quality of life, physical activity and other factors. Later rounds of surveys included the original participants and new building occupants. It should be noted that the sample sizes for survey data were small (between 16 and 25 people) and data were gathered at the early stages of occupation. Environmental conditions in the building may have changed since their study, especially given the project's commitment to ongoing monitoring and adaptation.

There were many positive findings about the impacts of the Bullitt Center on health and other measures. The building's location and active-commuting amenities (e.g. storage and showers) may have supported an observed shift in transport-related physical activity, with time spent travelling by car decreasing by 12%, with increased time on public transport (65%) and cycling (58%) (Burpee et al., 2014). The 'irresistible staircase' (Fig. 4.7) was designed to support physical activity and reduce energy use through operation of the lift. The evaluation found that the stairs were

Fig. 4.7 Healthy design elements at Bullitt Center, Seattle. Left: The 'irresistible staircase' with views of the city. Right: Workspaces with natural light and timber. (Source: Author, 2019)

used for 68% of trips from the second floor (location of main building entrance) to the sixth floor, compared to 22–27% in typical office buildings (Burpee et al., 2015). If staff took two of these ascents per day, the monthly benefit would be 20% of recommended monthly aerobic activity (ibid.). The economic impact has been evaluated across many measures to provide evidence of the success of this ground-breaking project. The local supply chain was improved for sustainable building products, including for a window manufacturer and a vapour barrier producer (Cowan et al., 2014). Furthermore, an ecosystems services valuation estimated a financial value over a 250-year period (with a 4% discount rate) of US$18.45 million from six building features: site transportation measures, rainwater capture and reuse, the composting toilets, energy efficiency measures, the solar array and the use of Forest Stewardship Council (FSC) certified wood (ibid.).

The reflective learning process embedded into Bullitt Center helped to surface some problems that could be remedied as part of building management practices. As previously mentioned, the UW Integrated Design Lab was able to identify issues with the building management system's sequence of operations that were corrected (Peña et al., 2017). Early evaluations found that indoor temperatures were periodically outside of the thermal comfort range for 20% of occupants and this matched survey responses, with 20–29% expressing dissatisfaction (Burpee et al., 2015). Acoustic discomfort was another area that did not perform well in early evaluations and indeed this was not a key design goal (ibid.). Overall, the vast majority of occupants were satisfied or highly satisfied with the Bullitt Center as a workplace (ibid.). The value of these various sources of data about building performance and occupants' experiences is that problem areas can be fixed. Most recently, the composting toilets were replaced for a vacuum flush system after seven years of operation for a number of reasons (B Kahn 2021, personal communication, 11 November). With the aim of aiding other projects, a lessons learned paper was produced by the Bullitt Center (2021) outlining the ongoing value of composting toilets if they are designed and sized appropriately.

Educating building occupants, professionals and the wider public is a key part of the Bullitt Center's mission to drive transformative change in sustainable and healthy buildings. In partnership with UW's Center for Integrated Design and the Integrated Design Lab, there are many resources for gathering, interpreting and sharing the success stories and lessons learned on this project. The building is open to public tours to complement the wealth of online information. The design features and technologies are explained via on-site plaques that relate to more detailed information on the website (www.bullittcenter.org) (Fig. 4.6). This information may be part of environmental cues that affect occupant behaviour. Gilbride et al. (2016) examined concepts from environmental psychology relating to values (see Steg et al., 2014) to interpret cues in the Bullitt Center that may influence pro-environmental behaviour. In essence, the environmental psychology literature suggests that design measures that remind people of environmental goals may reinforce pro-environment values and behaviour. At the Bullitt Center, Gilbride et al. argue that the highly visible sustainable building systems (e.g. the roof-top solar array and the greywater infiltration system) and the availability of monitoring data may form 'situational cues framing goals in daily decision making thereby stimulating a person's environmental values and motivating environmentally friendly actions' (p. 8–2).

The Bullitt Center provides 'living proof' (Peña, 2014) of a high-performing sustainable building that also supports occupant health and wellbeing. While some might see its use of unconventional technologies as risky, the Bullitt Foundation saw the building as an opportunity to showcase the processes and outcomes of deep green buildings. According to their website, the building has 'operated in the black for six years' which is impressive evidence to help build the business case for sustainable and healthy workplaces (Bullitt Foundation, n.d.-a). This case study showcases the value of a strong vision, integrated design, collaborative working with the public sector and monitoring and adapting the building over time.

4.5 Conclusion

This chapter has presented the compelling evidence for built environment interventions that will promote health through increasing biodiversity, resource efficiency and progress to zero carbon. Hierarchies and frameworks were introduced that can inform design or policy measures at different scales. The importance of monitoring was described, alongside the introduction of specific data sources that show how planetary health trends are changing over time. The Bullitt Center case study combined city-scale policies and sustainable development incentives with building-scale design measures and technologies, showcasing an internationally recognised deep green workplace.

There is overlap between the THRIVES health impact scales of planetary and ecosystem health. The next chapter will pick up similar themes to those covered earlier, but it will turn to a spatial scale that is closer to urban interventions and policies by focusing on the immediate ecosystem in and around a city.

References

Birkeland, J. (2008). *Positive development: From vicious circles to virtuous cycles through built environment design*. Earthscan.

Birkeland, J. (2017). Net-positive design and development. *Landscape Review, 17*, 83–87.

Birkeland, J. (2018). Challenging policy barriers in sustainable urban design. *Bulletin of Geography: Socio-Economic Series, 40*, 41–56.

Birkeland, J., & Knight-Lenihan, S. (2016). Biodiversity offsetting and net positive design. *Journal of Urban Design, 21*, 50–66.

Breed, M. F., Cross, A. T., Wallace, K., Bradby, K., Flies, E., Goodwin, N., Jones, M., Orlando, L., Skelly, C., Weinstein, P., & Aronson, J. (2020). Ecosystem restoration: A public health intervention. *EcoHealth, 18*, 269–271.

Bull, J. W., Gordon, A., Watson, J. E. M., & Maron, M. (2016). Seeking convergence on the key concepts in 'no net loss' policy. *Journal of Applied Ecology, 53*, 1686–1693.

Bullitt Center. (2021). The Bullitt Center composting toilet system a white paper on lessons learned. Retrieved November 26, 2021, from https://bullittcenter.org/wp-content/uploads/2021/03/The-Bullitt-Center-Composting-Toilet-System-FINAL.pdf

Bullitt Foundation. (n.d.-a). History. Retrieved March 07, 2021, from https://www.bullitt.org/about/history/

Bullitt Foundation. (n.d.-b). Deep green buildings. Retrieved March 07, 2021, from https://www.bullitt.org/programs/deep-green-buildings/

Burpee, H., Beck, D. A. C., & Meschke, J. S. (2014). *Health impacts of green buildings*. American Institute of Architects.

Burpee, H., Gilbride, M., Douglas, K., Beck, D., & Meschke, J. S. (2015). *Health impacts of a living building*. University of Washington.

Buse, C. G., Oestreicher, J. S., Ellis, N. R., Patrick, R., Brisbois, B., Jenkins, A. P., McKellar, K., Kingsley, J., Gislason, M., Galway, L., McFarlane, A., Walker, J., Frumkin, H., & Parkes, M. (2018). Public health guide to field developments linking ecosystems, environments and health in the Anthropocene. *Journal of Epidemiology and Community Health, 72*, 420–425.

Butler, C. D. (2016). Sounding the alarm: Health in the anthropocene. *International Journal of Environmental Research and Public Health, 13*, 665.

Carrus, G., Scopelliti, M., Lafortezza, R., Colangelo, G., Ferrini, F., Salbitano, F., Agrimi, M., Portoghesi, L., Semenzato, P., & Sanesi, G. (2015). Go greener, feel better? The positive effects of biodiversity on the well-being of individuals visiting urban and peri-urban green areas. *Landscape and Urban Planning, 134*, 221–228.

Cedeño-Laurent, J. G., Williams, A., MacNaughton, P., Cao, X., Eitland, E., & Spengler, J. D. (2018b). Building evidence for health: Green buildings, current science, and future challenges. *Annual Review of Public Health, 39*, 291–308.

Clifford, B., Canelas, P., Ferm, J., & Livingstone, N. (2020). *Research into the quality standard of homes delivered through change of use permitted development rights*. Ministry of Housing, Communities and Local Government.

Clifford, B., Ferm, J., Livingstone, N., & Canelas, P. (2018). *Assessing the impacts of extending permitted development rights to office-to-residential change of use in England*. Royal Institution of Chartered Surveyors.

Coleman, S., Touchie, M. F., Robinson, J. B., & Peters, T. (2018). Rethinking performance gaps: A Regenerative sustainability approach to built environment performance assessment. *Sustainability, 10*, 1829.

Convention on Biological Diversity. (2006). Article 2: Use of terms. Retrieved March 11, 2021, from https://www.cbd.int/convention/articles/?a=cbd-02

Costello, A., Abbas, M., Allen, A., Ball, S., Bell, S., Bellamy, R., Friel, S., Groce, N., Johnson, A., Kett, M., Lee, M., Levy, C., Maslin, M., McCoy, D., McGuire, B., Montgomery, H., Napier, D., Pagel, C., Patel, J., … Patterson, C. (2009). Managing the health effects of climate change. *The Lancet, 373*, 1693–1733.

Cowan, S., Davies, B., Diaz, D., Enelow, N., & Halsey, K. (2014). *Optimizing urban ecosystem services: The Bullitt center case study*. Ecotrust.

de Coninck, H., Revi, A., Babiker, M., Bertoldi, P., Buckeridge, M., Cartwright, A., … Waterfield, T. (Eds.). (2018). Global warming of 1.5°C. An IPCC Special Report on the impacts of global warming of 1.5°C above pre-industrial levels and related global greenhouse gas emission pathways, in the context of strengthening the global response to the threat of climate change, sustainable development, and efforts to eradicate poverty (p. 132). Intergovernmental Panel on Climate Change.

Deutch, J. (2020). Is net zero carbon 2050 possible? *Joule, 4*, 2237–2240.

Ellen MacArthur Foundation. (2015). *Growth within: A circular economy vision for a competitive Europe*. Ellen MacArthur Foundation.

Fuller, R. A., Irvine, K. N., Devine-Wright, P., Warren, P. H., & Gaston, K. J. (2007). Psychological benefits of greenspace increase with biodiversity. *Biology Letters, 3*, 390–394.

Fusco Girard, L., & Nocca, F. (2019). Moving towards the circular economy/city model: Which tools for operationalizing this model? *Sustainability, 11*, 6253.

Gardner, T. A., Von Hase, A., Brownlie, S., Ekstrom, J. M. M., Pilgrim, J. D., Savy, C. E., Theo Stephens, R. T., Treweek, J., Ussher, G. T., Ward, G., & Kate, K. T. (2013). Biodiversity offsets and the challenge of achieving no net loss. *Conservation Biology, 27*, 1254–1264.

Garrard, G. E., Williams, N. S. G., Mata, L., Thomas, J., & Bekessy, S. A. (2018). Biodiversity sensitive urban design. *Conservation Letters, 11*, e12411.

Gilbride, M., Loveland, J., Burpee, H., Kriegh, J., & Meek, C. (2016). Occupant-behavior-driven energy savings at the Bullitt Center in Seattle, Washington. In *From components to systems, from buildings to communities* (p. 15). Presented at the ACEEE Summer Study on Energy Efficiency in Buildings, American Council for an Energy Efficient Economy, Pacific Grove, California.

Holden, M. (2007). Revisiting the local impact of community indicators projects: Sustainable Seattle as prophet in its own land. *Applied Research in Quality of Life, 1*, 253–277.

Holden, M. (2008). Social learning in planning: Seattle's sustainable development codebooks. *Progress in Planning, 69*, 1–40.

Institution of Mechanical Engineers. (2020). The energy hierarchy: Supporting policy making for "Net Zero." Institute of Mechanical Engineers.

IPBES. (2019). *Summary for policymakers of the global assessment report on biodiversity and ecosystem services of the Intergovernmental Science-Policy Platform on Biodiversity and Ecosystem Services*. IPBES Secretariat, Bonn, Germany.

Karvonen, A. (2010). Metronatural^{TM}: Inventing and reworking urban nature in Seattle. Prog. Plan., Metronatural^{TM}: Inventing and Reworking Urban Nature in Seattle. *Progress in Planning, 74*, 153–202.

Lawrance, E., Thompson, R., Fontana, G., & Jennings, N. (2021). *The impact of climate change on mental health and emotional wellbeing: Current evidence and implications for policy and practice*. Grantham Institute.

Lehmann, S. (2014). Introduction Low carbon cities: More than just buildings. In S. Lehmann (Ed.), *Low carbon cities: Transforming urban systems* (pp. 1–55). Routledge.

Leising, E., Quist, J., & Bocken, N. (2018). Circular economy in the building sector: Three cases and a collaboration tool. *Journal of Cleaner Production, 176*, 976–989.

London Legacy Development Corporation. (2013). Legacy communities scheme: Biodiversity action plan 2014–2019. Retrieved March 5, 2021 from https://www.queenelizabetholympicpark.co.uk/-/media/lldc/sustainability-and-biodiversity/legacy-communities-scheme-biodiversity-action-plan-2014-2019.ashx?la=en

MacArthur, D. E. (2015). *The surprising thing I learned sailing solo around the world*. TED. Retrieved March 18, 2021 from https://www.ted.com/talks/dame_ellen_macarthur_the_surprising_thing_i_learned_sailing_solo_around_the_world

Mayor of London. (2020a). Energy assessment guidance: Greater London Authority guidance on preparing energy assessments as part of planning applications. Greater London Authority.

Nassauer, J. I. (1995). Messy ecosystems, orderly frames. *Landscape Journal, 14*, 161–170.

Nayar, J. (2015). The Bullitt center: Raising the bar with the living building challenge [WWW Document]. Interface Human Spaces. Retrieved March 19, 2021, from https://blog.interface.com/raising-the-bar-bullitt-center/

Peña, R., Meek, C., & Davis, D. (2017). The Bullitt center: A comparative analysis between simulated and operational performance. *Technology|Architecture + Design, 1*, 163–173.

Peña, R. B. (2014). *Living proof: The Bullitt Center*. High Performance Building Case Study. University of Washington Center for Integrated Design.

Pineo, H., Glonti, K., Rutter, H., Zimmermann, N., Wilkinson, P., & Davies, M. (2018a). Urban health indicator tools of the physical environment: A systematic review. *Journal of Urban Health, 95*, 613–646.

Pineo, H., & Moore, G. (2021). Built environment stakeholders' experiences of implementing healthy urban development: An exploratory study. *Cities Health, 0*, 1–15. https://doi.org/10.1080/23748834.2021.1876376

Pizarro, R. (2009). Urban Form and Climate Change: Towards appropriate development patterns to mitigate and adapt to climate change. In S. Davoudi, J. Crawford, & A. Mehmood (Eds.), *Planning for climate change: Strategies for mitigation and adaptation for spatial planners* (pp. 33–45). Earthscan.

RIBA, Hay, R., Bradbury, S., Dixon, D., Martindale, K., Samuel, F., & Tait, A. (2017). *Building knowledge: Pathways to post occupancy evaluation: Value of architects*. University of Reading.

Rice, J. L. (2010). Climate, carbon, and territory: Greenhouse gas mitigation in Seattle, Washington. *Annals of the Association of American Geographers, 100*, 929–937.

Ritchie, H., & Roser, M. (n.d.). Emissions by sector [WWW Document]. *Our World Data*. Retrieved April 16, 2021, from https://ourworldindata.org/emissions-by-sector

Rockström, J., Steffen, W., Noone, K., Persson, Å., Chapin, F. S. I., Lambin, E., Lenton, T. M., Scheffer, M., Folke, C., Schellnhuber, H., Nykvist, B., De Wit, C. A., Hughes, T., van der Leeuw, S., Rodhe, H., Sörlin, S., Snyder, P. K., Costanza, R., Svedin, U., … Foley, J. (2009). Planetary boundaries: Exploring the safe operating space for humanity. *Ecology and Society, 14*, 32.

Romanelli, C., Cooper, H. D., Campbell-Lendrum, D., Maiero, M., Karesh, W. B., Hunter, D., & Golden, C. D. (2015). *Connecting global priorities: Biodiversity and human health, a state of knowledge review*. World Health Organization and Secretariat for the Convention on Biological Diversity.

Seattle Department of Construction & Inspections. (n.d.). Living building & 2030 challenge pilots. Retrieved March 22, 2021, from https://www.seattle.gov/sdci/permits/green-building/living-building-and-2030-challenge-pilots

Sellers, S., Ebi, K. L., & Hess, J. (2019). Climate change, human health, and social stability: Addressing interlinkages. *Environmental Health Perspectives, 127*, 045002.

Shanahan, D. F., Fuller, R. A., Bush, R., Lin, B. B., & Gaston, K. J. (2015). The health benefits of urban nature: How much do we need? *Bioscience, 65*, 476–485.

Steg, L., Bolderdijk, J. W., Keizer, K., & Perlaviciute, G. (2014). An integrated framework for encouraging pro-environmental behaviour: The role of values, situational factors and goals. *Journal of Environmental Psychology, 38*, 104–115.

Stone, S. B., Myers, S. S., & Golden, C. D. (2018). Cross-cutting principles for planetary health education. *Lancet Planet Health, 2*, e192–e193.

Trasande, L., Zoeller, R. T., Hass, U., Kortenkamp, A., Grandjean, P., Myers, J. P., DiGangi, J., Bellanger, M., Hauser, R., Legler, J., Skakkebaek, N. E., & Heindel, J. J. (2015). Estimating Burden and Disease Costs of Exposure to Endocrine Disrupting Chemicals in the European Union. *The Journal of Clinical Endocrinology and Metabolism, 100*, 1245–1255.

United Nations Environment Programme. (2021). *Making peace with nature: A scientific blueprint to tackle the climate, biodiversity and pollution emergencies*. UNEP, Nairobi, Kenya.

United Nations General Assembly. (2015). Resolution adopted by the general assembly on 25 September 2015: Transforming our world: The 2030 Agenda for Sustainable Development. United Nations.

Urban Land Institute. (2015). *ULI case studies*. Bullitt Center. Retrieved September 07, 2020 from https://casestudies.uli.org/wp-content/uploads/2015/02/TheBullittCenter1.pdf

Van Buren, N., Demmers, M., Van der Heijden, R., & Witlox, F. (2016). Towards a circular economy: The role of Dutch logistics industries and governments. *Sustainability, 8*, 647.

Watts, N., Amann, M., Arnell, N., Ayeb-Karlsson, S., Beagley, J., Belesova, K., Boykoff, M., Byass, P., Cai, W., Campbell-Lendrum, D., Capstick, S., Chambers, J., Coleman, S., Dalin, C., Daly, M., Dasandi, N., Dasgupta, S., Davies, M., Di Napoli, C., … Costello, A. (2021). The 2020 report of The Lancet Countdown on health and climate change: Responding to converging crises. *The Lancet, 397*, 129–170.

Whitmee, S., Haines, A., Beyrer, C., Boltz, F., Capon, A. G., de Souza Dias, B. F., Ezeh, A., Frumkin, H., Gong, P., Head, P., Horton, R., Mace, G. M., Marten, R., Myers, S. S., Nishtar, S., Osofsky, S. A., Pattanayak, S. K., Pongsiri, M. J., Romanelli, C., … Yach, D. (2015). Safeguarding human health in the Anthropocene epoch: Report of The Rockefeller Foundation–Lancet Commission on planetary health. *The Lancet, 386*, 1973–2028.

Wilkinson, S., Remøy, H. T., & Langston, C. (2014). *Sustainable building adaptation: Innovations in decision-making, Innovation in the built environment*. Wiley-Blackwell.
Williams, J. (2016). *Circular cities: Strategies, challenges and knowledge gaps*. Circular Cities Hub.
Williams, J. (2019). Circular cities: Challenges to implementing looping actions. *Sustainability, 11*, 423.
Younger, M., Morrow-Almeida, H. R., Vindigni, S. M., & Dannenberg, A. L. (2008). The built environment, climate change, and health. *American Journal of Preventive Medicine, 35*, 517–526.

Chapter 5
Ecosystem Health

5.1 Introduction

Even though many people living in cities may feel far away from nature, cities have an enormous impact on the health of ecosystems locally and globally, and in turn, ecosystems impact human health at multiple scales. The benefits that ecosystems provide for humans are called ecosystem services and they are fundamental to the achievement of society's most fundamental collective goals, such as eliminating poverty (Millennium Ecosystem Assessment, 2005). A selection of ecosystem services is shown in Table 5.1 with descriptions of their 'providers' and spatial scales at which the services operate, adapted from Kremen (2005). The table also provides example health impacts from Romanelli et al. (2015) and Barnes et al. (2019) and gives clear evidence for the selection of urban planning and design goals in THRIVES, including sustaining air, water and soil quality and greenspace, and improving sanitation, waste and mobility infrastructure. These factors are included in the 'ecosystem health' scale of THRIVES because of their impact on local/regional ecosystems and their amenability to policy or design interventions at the city and district scales of decision-making, although action to support ecosystem health can be taken at smaller urban scales. Biodiversity is a goal at the 'planetary health' scale of THRIVES because it operates at local to global scales.

Planetary and ecosystem health are deeply interconnected fields of study that offer many important concepts for practitioners of healthy urbanism. Knowledge of the interactions between humans, ecosystems and non-human organisms is rooted with diverse indigenous cultures and dates back to Hippocrates in Western knowledge traditions (Buse et al., 2018). The urgent threats to human health of environmental degradation have led to a proliferation of research in these fields, that Chris Buse and colleagues posit has led to seven distinct field developments from occupational and environmental health in the mid-nineteenth century, to political ecology of health, environmental justice, ecohealth and One Health in the late twentieth

H. Pineo, *Healthy Urbanism*, Planning, Environment, Cities,
https://doi.org/10.1007/978-981-16-9647-3_5

Table 5.1 A selection of ecosystem services, service providers/trophic level, the scale at which they operate and their health impact

Ecosystem service	Ecosystem service providers/ trophic level	Scale of operation	Examples of health impacts from Romanelli et al. (2015) and Barnes et al. (2019)
Aesthetic and cultural	All biodiversity	Local–global	Psychological and wellbeing benefits
Ecosystem goods	Diverse species	Local–global	Multiple, sources of medicine
UV protection	Biogeochemical cycles, micro-organisms, plants	Global	Skin cancers and cataracts, other effects through air pollution and climate change
Purification of air	Micro-organisms, plants	Regional–global	Various impacts to pulmonary, cardiac, vascular and neurological systems
Flood mitigation	Vegetation	Local–regional	Disease transmission via contaminated drinking water and injury, death or displacement from flooding
Drought mitigation	Vegetation	Local–regional	Crop failure and water contamination
Climate stability	Vegetation	Local–global	Extreme weather and disasters, leading to injury, death or displacement; diseases spread via water, food and vectors
Pollination	Insects, birds, mammals	Local	Food quantity, quality and variety and nutritional content
Pest control	Invertebrate parasitoids and predators and vertebrate predators	Local	Food supply, vector-borne diseases
Purification of water	Vegetation, soil micro-organisms, aquatic micro-organisms, aquatic invertebrates	Local–regional	Disease transmission, food supply
Detoxification and decomposition of wastes	Leaf litter and soil invertebrates; soil micro-organisms; aquatic micro-organisms	Local–regional	Emissions to the atmosphere, soil and water, vector-borne disease transmission, food supply
Soil generation and soil fertility	Leaf litter and soil invertebrates; soil micro-organisms; nitrogen-fixing plants; plant and animal production of waste products	Local	Food supply
Seed dispersal	Ants, birds, mammals	Local	Food supply

Source: Adapted from Kremen (2005)

century and ecological public health and planetary health in the early twenty-first century. These field developments are underpinned by diverse theories and methods, resulting in different perspectives about similar problems. Yet there are important similarities across these fields that relate to healthy urbanism: recognition of the complex interdependencies of human and environmental factors, the important role of equity and the need to include diverse types of knowledge to understand and create healthy places.

This chapter touches upon environmental issues that many people in high-income countries currently take for granted, such as access to clean water and functioning sanitation infrastructure. Yet the environmental justice movement has shown that low-income, minority ethnic and indigenous communities are disproportionately exposed to environmental pollution in places like the USA, Australia and Canada. Crescentia Dakubo's (2011) book on ecosystem health argues that 'ecosystem degradation must be examined from the basis of the structural inequalities surrounding the use of and control of ecosystem services, and how such inequalities drive various land use practices that degrade ecosystems' (p. 4). She examines the complex causes of ecosystem degradation in diverse global settings, noting that indigenous communities are particularly affected as they often rely on the land for subsistence. Dakubo describes a case in Canada where the Grassy Narrows First Nation community suffered from mercury poisoning over an eight-year period, caused by pollution from a paper mill 320 km upstream. The community's water and food sources were contaminated, leading to severe birth defects, eye problems and other health impacts. The solutions to this problem required cross-sectoral collaboration with the community, public and private organisations, alongside action to redress unequal power dynamics and marginalisation of indigenous knowledge. Thus, Dakubo contends that ecosystem approaches 'must be infused with political ecology theorizing to illuminate the hidden underlying forces that produce poor health in such marginalised communities' and poststructuralism 'to interrogate Western constructions of Indigenous peoples, their health practices' and the devaluation of their knowledge systems (p. 143). Healthy urbanism requires inclusive policy and design practices that elevate diverse needs and knowledge types that have typically been ignored.

5.2 Achieving Ecosystem Health Framework Goals

In the face of rapid urbanisation and climate change impacts, new approaches to infrastructure development are urgently required to support local ecosystems. The human health risks of poor water, waste and sanitation infrastructure are most strongly (although not exclusively) felt in low- and middle-income countries, particularly in informal settlements. To meet such challenges in terms of urban water infrastructure, Sarah Bell (2015) challenges the wisdom of applying conventional approaches to infrastructure engineering that have sought to dominate natural resources, such as by burying and channelling rivers under cities, harming local ecosystems and human health. Bell argues that a transition to ecosystem sensitive engineering is required, although it may be slower in low- and middle-income countries where immediate health risks may take priority. Bell advocates for new approaches that understand water as both scarce and multi-purposed. Instead of centralised expert-led solutions, water should be seen as a 'partner in the urban environment' requiring solutions to be adopted at many urban scales from households to city planning (ibid., p. 23). Water sensitive urban design, sustainable (urban) drainage systems and ecological sanitation are examples of such approaches (later discussed under water quality) that could also be called nature-based solutions. The concept of nature-based solutions applies more broadly than water infrastructure. Such solutions are advocated for supporting ecosystem services and human health by international bodies, such as the World Bank, International Union for Conservation of Nature and the European Union (Pauleit et al., 2017).

Nature-based solutions have been defined as 'actions which are inspired by, supported by, or copied from nature' that aim 'to address a variety of environmental, social and economic challenges in sustainable ways' (ECDG, 2015, p. 5). There are many views about what the concept of nature-based solutions includes and how it can support urban health. A review by van den Bosch and Ode Sang (2017) found that scientific literature on this concept can be grouped in three topic areas which represent frameworks and solutions: green infrastructure, ecosystem-based adaptation and ecosystem services. Looking specifically at the scholarly literature on the health and wellbeing benefits of nature-based solutions, their review found that publications were largely conceptual and have been interpreted through diverse lenses, including biophilic cities (Beatley, 2016), ergonomics (Richardson et al., 2017) and pro-environmental behaviour (Annerstedt van den Bosch & Depledge, 2015). Nature-based solutions function by using 'the features and complex system processes of nature, such as its ability to store carbon and regulate water flow, in order to achieve desired outcomes, such as reduced disaster risk, improved human well-being and socially inclusive green growth' (ECDG, 2015, p. 5). For example, green infrastructure at different urban scales (e.g. green walls and roofs, parks, street trees, etc.) can be used to store carbon and to reduce air pollution, the urban heat island effect and the negative perception of noise—all supporting diverse societal goals including health (van den Bosch & Ode Sang, 2017). This chapter will describe examples of nature-based solutions and other strategies for healthy urbanism that support ecosystem health.

5.2.1 Air Quality

The estimated seven million global deaths every year from indoor and outdoor air pollution have been described by the WHO (2020) as the 'tip of the iceberg'. A substantial burden of health, social and economic costs from air pollution arise from sickness, hospitalisation, reduced life expectancy and lower productivity—all driven by the fact that globally 'about 9 of 10 persons are exposed to air pollution at levels above the WHO air quality guidelines' (ibid., p. 10). Air pollution is a leading risk factor for poor health, as shown in the ranking in Fig. 5.1 of risk factors by disease burden, measured in disability-adjusted life years (DALYs; see Chap. 3), in

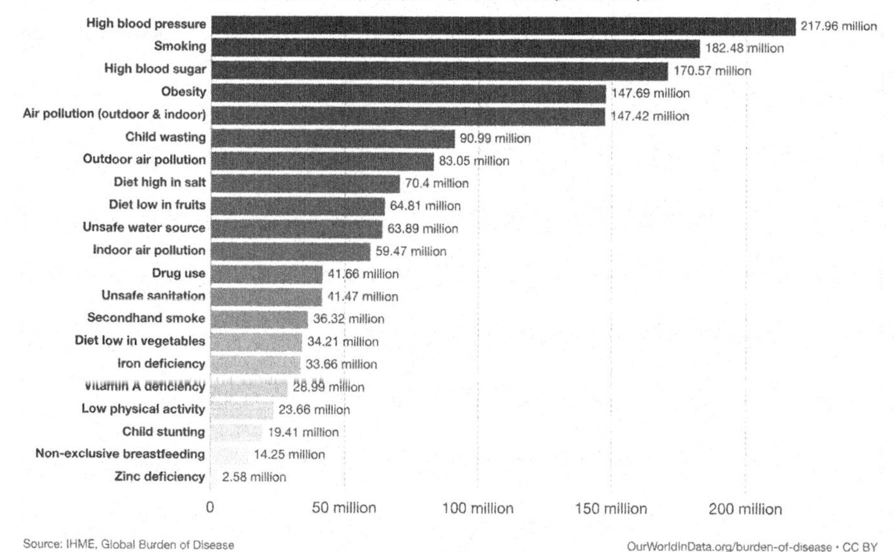

Fig. 5.1 Burden of disease by risk factor in 2017. (Source: Ritchie and Roser (2017) using data from IHME, Global Burden of Disease (Stanaway et al., 2018))

which indoor and outdoor air pollution rank fifth after obesity (Ritchie & Roser, 2017). Air pollution also contributes to the risk of transgressing several of the planetary boundaries described in Chap. 4, such as atmospheric aerosol loading and climate change, reducing the capacity of the Earth to support human life. Thus, the health impacts of air pollution are greater than the tallies of deaths and disease that can be attributed to air pollution and should include the impacts associated with planetary processes, such as climate change.

Pollutants can travel many miles and they can cross borders, putting constraints on the potential for city mayors and planners to protect local populations. In most cases, less than 40% of a city's particulate matter (specifically $PM_{2.5}$) concentrations originate from that city, underscoring the need for regional and cross-sectoral cooperation to reduce pollution (WHO, 2020). The anthropogenic sources of air pollution (also called atmospheric or ambient pollution) arise from activities including transport, agriculture, buildings, energy production and industrial processes (Fig. 5.2). Sources of pollution can also be natural, such as sand, dust, trees, volcanic eruptions and wildfires. Furthermore, pollution sources may be stationary, as in power plants, or mobile, as in ships and vehicles. Multiple sources contribute to the air pollution levels in any given city, and the primary pollutants from these sources can mix together to create secondary pollutants. Indoor sources of air pollution are generated from heating and cooking systems, tobacco smoke, inadequate ventilation leading to mould growth, furnishings and building materials, adhesives and

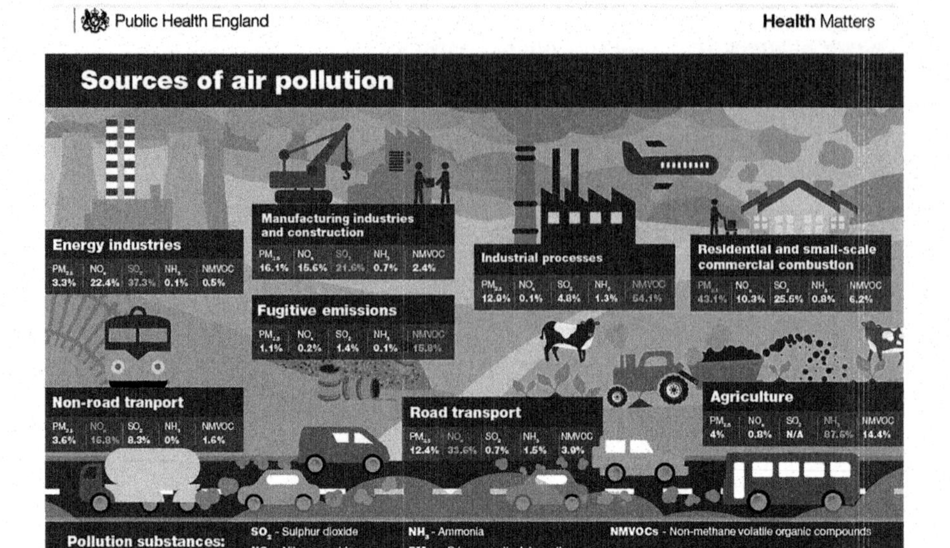

Fig. 5.2 Anthropogenic sources of air pollution. (Source: Public Health England, Health Matters, n.d.)

coatings, cleaning products and other sources (WHO, 2018b). Globally, around three billion people cook using open fires or stoves fuelled by kerosene, biomass and coal, which leads to 3.8 million premature deaths annually (ibid.). Given these diverse sources, reducing air pollution in cities requires action from multiple sectors and tiers of government.

There are no safe levels of air pollution, and the existing thresholds produced by public health and environmental agencies are regularly exceeded in cities around the world (WHO, 2013). Table 5.2 lists common pollution sources with guideline thresholds produced by the WHO in 2005 and updated in 2021. Governments take these thresholds under advisement and set their own, which are often higher than the WHO levels. For example, the European Union and US Environmental Protection Agency have set thresholds for annual mean concentrations of $PM_{2.5}$ at 25 micrograms per cubic metre ($\mu g/m^3$) and 12 $\mu g/m^3$, respectively, both of which are higher than the 2005 WHO threshold of 10 $\mu g/m^3$ and the 2021 threshold of 5 $\mu g/m^3$. Given that there is no safe level of air pollution, it is concerning that current air pollution concentrations in many countries do not meet the WHO thresholds or those set locally. Growing certainty in the evidence base linking air pollution with specific health effects and rising public awareness of these dangers may change political will to tighten regulations. Accountability for air pollution is increasing as shown by cases such as the landmark court ruling in the UK that listed air pollution as a child's cause of death (see Box 5.1).

Table 5.2 Selected types of air pollution with associated sources and guideline thresholds from WHO (2021a)

Pollutant	Significant sources	WHO guideline thresholds
Carbon monoxide (CO)	Petrol combustion in road vehicles, forest fires and biomass burning	See indoor guidelines[a]
Sulphur dioxide (SO_2)	Fuel combustion (coal, oil, diesel), shipping, metal smelting	40 μg/m^3 24-hour mean
Nitrogen oxides (NO_x)	Fuel combustion, primarily in road traffic and electricity generation, forest fires and biomass burning	NO_2: 10 μg/m^3 annual mean 25 μg/m^3 24-hour mean
Particulates (e.g. $PM_{2.5}$ and PM_{10})	Fuel combustion, forest fires & biomass burning, trees and other vegetation, dust storms	$PM_{2.5}$: 5 μg/m^3 annual mean, 15 μg/m^3 24-hour mean PM_{10}: 15 μg/m^3 annual mean 45 μg/m^3 24-hour mean
Volatile organic compounds (VOCs)	Fuel combustion, leakage or evaporation from gases, solvents, forest fires and biomass burning, trees and other vegetation	See indoor guidelines[a]

[a] Guideline thresholds for indoor air pollutants, including benzene, carbon monoxide, formaldehyde, naphthalene, nitrogen dioxide, polycyclic aromatic hydrocarbons, radon, trichloroethylene and tetrachloroethylene, can be found in WHO (2018b)

Box 5.1 Increasing accountability for air pollution and health: A landmark court ruling in the UK

A court ruling in London in December 2020 exposed the country's failure to meet particulate matter and nitrogen dioxide levels set in EU and domestic law (Dyer, 2020). The coroner ruled that these failures possibly contributed to the death of nine-year-old Ella Adoo Kissi-Debrah, who was hospitalised 27 times over three years before a fatal asthma attack in February 2013 (Vaughan, 2020). Her mother, Rosamund Adoo Kissi-Debrah, was never given information about the health risks of the air pollution in their home which was located 25 metres away from one of London's busiest roads. Following a prior inquest in 2014 that did not acknowledge air pollution as a cause of death, the 2020 ruling means that the official 'medical cause of death' is now stated 'as (1a) acute respiratory failure, (1b) severe asthma, and (1c) air pollution exposure' (Barlow, 2021). This finding demonstrates how growing scientific knowledge about the health effects of environmental pollution may have a direct effect on the accountability of responsible authorities. In 2021, the coroner sent a Report to Prevent Future Deaths to government and professional societies raising concerns about UK pollution levels and calling for action related to: (1) adjustments to the UK's pollution thresholds in line with WHO levels, (2) public education campaigns and (3) communication of risk from healthcare professionals to patients (ibid.).

Air pollution affects people differently and there are inequities in exposure to air pollution within cities and globally. Low-income and minority ethnic communities often have higher exposures than other population groups due to their neighbourhoods being in closer proximity to environmental hazards such as waste facilities and busy roads. Unborn babies, children, elderly people and those with existing conditions are disproportionately affected by air pollution. The health risks for children are significant. The WHO (2018c) reports that 93% of all children live in areas with air pollution concentrations that exceed their guidelines and this exposure contributed to 543,000 deaths from respiratory tract infections in children under five years in 2016. Children are more susceptible to the adverse effects of air pollution for environmental, social and physiological reasons. For example, they breathe faster than adults and take in more air pollutants, their bodies and lungs are still developing and are more vulnerable to inflammation and, in most cases, they are powerless to leave a home polluted from tobacco smoke or fuel combustion (ibid.). Furthermore, exposure to pollution before a baby is conceived, whilst they are in the womb and in early childhood can damage their body in ways that increase long-term vulnerability. Bringing in Krieger's ecosocial theory (see Chap. 3), the constructs of embodiment, pathways of embodiment and cumulative interplay help explain how biological and social factors are both responsible for the health effects of air pollution across the life course and across generations. There are short- and long-term effects of air pollution that can be produced following brief and chronic exposures to different pollutants. The evidence is strong that respiratory and cardiovascular diseases are caused by air pollution (WHO, 2013). Studies have also found links between chronic exposures to $PM_{2.5}$ and adverse birth outcomes, neurodevelopment deficits, diabetes (ibid.) and dementia (Peters et al., 2015). There is emerging evidence linking air pollution to obesity (Lam et al., 2021) and mental health conditions, including depression and anxiety (Braithwaite et al., 2019). In addition, poor air quality in buildings is linked to absenteeism, reduced productivity and sick building syndrome (Kelly & Fussell, 2019).

Cities have an important role to play in reducing air pollution sources and decreasing exposure to air pollution, particularly through design and engineering measures. Cutting pollution at its source requires a range of actions across society and international collaboration. At the city level, there are overlaps with the recommended design and policy measures from the 'zero carbon' section of Chap. 4, such as transitioning to low-carbon energy and transport systems, creating compact cities and using green infrastructure to absorb pollutants. Each specific measure may suffer from delays, lack of compliance or other setbacks, thus a comprehensive set of measures will be more successful than a single policy option. Box 5.2 describes the efforts in Beijing, China, to reduce air pollution, including the complex interconnections among public attention to pollution, policy levers, health effects, international guidelines and other factors.

Reducing exposure to pollution can be achieved through design and engineering measures from the whole city to the building scale. Homes should not be located near sources of pollution, such as factories and busy roads. Where this cannot be avoided, porous (e.g. trees and vegetation) and solid (e.g. parked cars, noise barriers

Box 5.2 Achieving transformative air pollution reduction in Beijing
Air pollution control in China over the past decade has been impressively rapid compared to the progress made in other settings during the twentieth century. Although pollution levels are still relatively high, decisive action in cities like Beijing has resulted in significant reductions. Extreme pollution events in 2012 and 2013 drew international attention and public outcry, resulting in national policy changes that filtered down to regions and municipalities (Wang et al., 2018). The National Ambient Air Quality Standards (NAAQS) were changed in 2012 to include a concentration limit for $PM_{2.5}$ using the WHO Interim target-1 (IT-1) of 35 $\mu g/m^3$ annual mean level. Pollution in China was often significantly higher than the WHO limits causing premature deaths and significant environmental impacts to climate and agriculture. Data gathered in January 2013 showed that during this extreme period, the hourly maximum $PM_{2.5}$ concentration in Beijing reached 791 $\mu g/m^3$ (Andersson et al., 2015). In 2013, China adopted the Air Pollution Prevention and Control Action Plan with the most stringent targets applied in the Beijing-Tianjin-Hebei region. The action plan required cross-sector action and strict enforcement with monitoring across the goals, which include: 'industrial restructuring, clean energy generation, coal and oil quality management, small coal-fired boiler control, industrial emissions (dust and Volatile Organic Compounds—VOCs), municipal dust control, vehicle pollution control, air pollution control investment, building energy-saving and heat metering management, and atmospheric environment management' (Wang et al., 2018, p. 4). Following these changes, Beijing's $PM_{2.5}$ decreased by 33% (from 2013 to 2017) and the city met its target of annual average concentrations of 60 $\mu g/m^3$ from the action plan (ibid.). However, there were concerns about the long-term effectiveness of the control measures and the social inequities that resulted from industrial restructuring, among other bans and regulations. Analysis by Huang et al. (2019) found that household income and education were negatively correlated with ambient air quality. Therefore, they recommended that the government increased educational campaigns about air filtration for socio-economically deprived groups.

and low boundary walls) barriers can be used to reduce exposure (Gallagher et al., 2015). Trees planted between a major road and residential buildings in Beijing, China, possibly reduce air and noise pollution for residents (Fig. 5.3). From an equity and inclusion perspective, design teams should avoid placing more vulnerable building occupants near higher pollution exposures. Development economics often dictate that less desirable locations (e.g. overlooking busy roads) are reserved for affordable or social housing units, yet these are likely to be occupied by the most vulnerable residents who have greater susceptibility to adverse effects and fewer resources to manage the effects.

Fig. 5.3 Trees and vegetation between a major road and residential buildings in Beijing, China, potentially reducing residents' exposures to air and noise pollution. (Source: Author, 2019)

The three main strategies to reduce exposure to indoor air pollution are (1) cutting indoor and outdoor emissions at their sources, (2) providing adequate ventilation and (3) using air cleaning technologies when required (Kelly & Fussell, 2019). Each of these strategies can be achieved through numerous design measures and technologies that will vary internationally; see WHO Housing and Health Guidelines (2018b). It is important to note that existing building codes and regulations may have gaps that result in unacceptable levels of indoor air pollution (see WHO, 2021b), thus design teams may need to consider additional measures. Building standards can be used to facilitate the use of materials with lower levels of VOCs, formaldehyde, allergens, particular matter and other pollutants (Allen et al., 2015). In areas of high outdoor air pollution, natural ventilation strategies may not be sufficient to provide indoor air quality. Air purification can be achieved through mechanical filtration and this is the most common technique, yet it can fail through poor design or without sufficient management and maintenance. See Kelly and Fussell (2019) for a review of other technologies and their health benefits in different settings.

5.2.2 *Water Quality*

Water is essential for human health and hygiene and the ecosystems that humans depend upon. Urbanisation increases water demand and one in four global cities are already water stressed (McDonald et al., 2014), a challenge likely to increase due to climate change (Schultz, 2019). Unsafe water sources are ranked tenth in global risk factors by disease burden in Fig. 5.1 (Ritchie & Roser, 2017). SDG 6 contains targets for safe drinking water, sanitation and hygiene; however, there are still considerable inequities in access. International data from 2015 showed that 844 million people lacked a basic drinking water service, 263 million travelled over 30 minutes round trip to collect water from an improved source, and 159 million people collected surface water for drinking (WHO and UNICEF, 2017). Inequitable access is problematic across and within countries, between urban/rural populations and the least and most deprived populations. For example, in Angola, there is a 65% gap between the wealthiest and poorest population quintiles for access to basic drinking water (ibid.).

In cities, water is needed for drinking, cooking, cleaning and sanitation systems, but it is also a source of food, recreation and aesthetic benefit. Contaminated water can spread infectious and non-infectious diseases, such as cholera, diarrhoea illnesses, dysentery, hepatitis A, typhoid and lead poisoning (WHO, 2018b). A systematic review of the health benefits of so-called blue infrastructure (e.g. lakes, rivers, and the sea) found an overall positive association between greater levels of exposure and mental health, wellbeing and physical activity (Gascon et al., 2017). Another review by Britton et al. (2020) looked at 'blue care' or health interventions specifically in blue infrastructure/spaces and found that these had direct benefits for mental health and psycho-social wellbeing (also see 'nature-based interventions' under section 'Greenspace'). The goal of water quality in THRIVES includes all of these potential purposes that relate to human health and the requirement for functioning aquatic, riparian and coastal ecosystems.

There are various scales at which urban planners and design teams may seek to improve water quality for different purposes and these will vary based on the local context and priorities. Sarah Bell (2015) contrasts the experiences of city dwellers in wealthy and poor cities, characterising water as either their servant or their master, determining how much effort they must exert to meet basic health and hygiene requirements. Bell's approach to water sensitive urban design (WSUD) eschews conventional engineering approaches that seek to dominate water as a resource, instead she advocates for conceptualising 'water as a partner in urban design' (ibid., p. 20). Urban planners should ensure that water is not contaminated as it passes through a city, rinsing pollutants off of urban surfaces and discharging them into aquatic ecosystems. Instead, planners can use sustainable urban drainage systems (SUDS) that mimic the function and form of natural hydrological systems. In such systems, storm water slowly filters through a city, gathering on green roofs, in ponds and wetlands and in cisterns where it can be used in gardens. Restoring urban rivers from concrete channels and culverts supports better drainage, biodiversity and human health benefits, exemplified by the restoration of the Cheonggyecheon River in Seoul, Korea, which required the removal of a major highway.

Water-sensitive urban design can apply to urban projects of multiple scales. A smaller-scale example can be seen in the Bullitt Center case study from Chap. 4. A greywater infiltration system is used to process the building's greywater from sinks and showers. Biodegradable soap is used in the washing facilities and the water drains to a storage tank. Up to 500 gallons of this water is then pumped through the constructed wetland (on the third floor of the building) where it passes through porous gravel and soil several times before being safely released into the bio-swales at street level (Fig. 4.5). There are 20 feet of gravel below the bio-swales which provide further filtration for the building's greywater and stormwater runoff in the area. This system means that 61% of the building's water is returned to the ecosystem through ground infiltration or evaporation (Bullitt Center, n.d.). A large-scale example of water-sensitive urban design is seen in China's 'Sponge City' programme, in which cities like Ningbo have adopted multiple strategies to reduce flood risk, conserve water and increase water quality (see Box 5.3).

Box 5.3 China's Sponge Cities: The case of Ningbo

Ancient and modern approaches to sustainable urban water management in Ningbo may be equally relevant as part of China's Sponge City programme (Tang et al., 2018). Ningbo is a port city located in a floodplain that was heavily engineered during the Tang Dynasty (833 AD) to secure freshwater supplies and reduce flooding. Faced with rapid population growth, rising sea levels and increased precipitation due to climate change, the city is targeting the following five issues with the Sponge City initiative: 'pollution remediation, flood mitigation, reducing saturated soils by improving the drainage system, securing the freshwater supply, and reducing unnecessary water consumption' (ibid., p. 3). Water-sensitive urban design techniques from the city's past may be a part of future solutions to the flooding risks that grey infrastructure are unlikely to solve (see Fig. 5.4). In ancient times, the residents of Ningbo, a so-called water town, built canals that were lined with mulberry bushes to help stabilise the soil and provide food for silkworms, and over time these constructed channels were indistinguishable from those that formed naturally. The general public were involved in the construction and maintenance of water gates and farmers chose flood-resilient crops. These practices were eventually replaced with concrete and steel 'grey' infrastructure systems that were centrally managed. Tang and colleagues suggest that some of Ningbo's ancient practices can be renewed through modern blue-green infrastructure projects, such as the Sponge City programme, in which Ningbo is constructing woodlands, rain gardens, wetlands and streams in newly developed areas to sustainably manage water.

Fig. 5.4 Modern and historical visions of the Chinese 'water town' of Ningbo. Left: Remnants of an ancient canal at Mashan Mountain wetland with many modifications over time. Right: The rooftop view from Liyuan North Road, Haishu District, in Ningbo shows the city's extensive waterways. (Source: Author, 2019)

5.2.3 Soil Quality

In many parts of the world, soil is contaminated from anthropogenic causes or geological sources at very low levels that do not pose health risks. However, soil contamination can be very harmful for health. Site assessments should always be conducted to determine the level of risk for proposed land uses and appropriate mitigation strategies. Past and current industrial processes have left inorganic and organic contaminants in many urban areas and these can harm health when they are ingested or touched, as well as through contamination of food and groundwater supplies (Yolande et al., 2014). Brownfield land refers to previously developed land (sometimes industrial uses), and as cities grow, there is increasing pressure to redevelop on this land to meet housing needs and increase density to support sustainable development. Local regulations are likely to dictate how and when contaminated land can be redeveloped and they may also require the construction industry to reduce its own soil polluting practices.

Whilst there are well-established processes for measuring chemical toxicity and exposures from soil, the ability to accurately estimate health effects is constrained by uncertainties such as lack of knowledge about chemicals, synergistic effects, latency of effects and low understanding about potential psychological impacts (Yolande et al., 2014). These uncertainties can create challenges for decision-makers and may result in cases where health risks were not fully anticipated prior to development. An urban development on a prior gasworks site in the Southall community in London was the subject of significant concern among neighbouring residents who experienced health effects which they attributed to the project (see Box 5.4). This example demonstrates that built environment professionals who seek to create healthy developments will need to carefully consider the findings of health impact assessments and ideally involve local residents to identify risks. There may be health effects even when exposures are below regulatory limits. Working with

Box 5.4 Understanding residents' concerns about soil decontamination in London

Residents near a major brownfield development in Southall, London, reported a strong petrol-like odour and health effects during the soil decontamination process. Newspaper articles and videos in *The Guardian* told the story from the perspective of residents (Griffin, 2019, 2020a, 2020b). Multiple government authorities had a role in the issue, including the Environment Agency, the local authority and Public Health England (PHE), which contributed to lack of clarity for residents and local politicians about who was accountable for any potential health effects. A report was produced for the community group Clean Air for Southall and Hayes (CASH), suggesting that although PHE found the pollution levels associated with site remediation to be within regulatory limits, residents may have been more susceptible to these exposures due to their chronic exposures to ambient air pollution and psychosocial stressors caused by deprivation (Centric Lab, 2020). Community groups from several affected sites across London and Brighton have created an alliance called Gasworks Communities United (GCU) to campaign for independent assessment of remediation works, better mitigation for local communities and further research into health risks (Griffin, 2020a).

public health colleagues and the community in the risk assessment process may result in the identification of additional monitoring or mitigation measures to reduce residents' exposures and concerns (Ghiassee et al., 2014).

5.2.4 Greenspace

The benefits for physical and mental health and wellbeing from being in and around nature are of great significance in cities. Greenspace (or 'green infrastructure') benefits include: keeping cities cool, supporting flood resilience, removing air pollution, buffering noise pollution, reducing risk of poor mental health, increasing recovery times (and other healthcare setting benefits), providing place for mental restoration and children's play, enhancing social connections, reducing crime and increasing productivity. Well-maintained parks and greenspaces are visited by people of all ages and income levels. They can help to reduce inequalities and may promote physical activity (Twohig-Bennett & Jones, 2018; Allen & Balfour, 2014). Achieving many of these benefits requires that people can access urban greenspaces; however, studies have shown that green infrastructure is not distributed evenly across cities and that low-income and minority ethnic neighbourhoods tend to have reduced access. As shown in Fig. 5.5, green infrastructure can be applied at many scales within a city including green roofs and walls, street trees, linear parks (e.g. the High Line in New York), pocket parks, squares, courtyards and larger parks and woodlands. The cumulative impact of green infrastructure at all scales can be substantial for nature and people.

Fig. 5.5 Urban green infrastructure at multiple scales. Top left: Trees and vegetation on the High Line in New York City. Bottom left: A large urban park overlooks London. Top right: A green roof adjacent to the Crystal Siemens building in London. Bottom right: A river flowing in a culvert below ground is marked with wooden posts on a green buffer between a road and houses in London. (Source: Author)

The mental health benefits of greenspace have been evaluated in many studies across the world showing its value for mitigating stress (see Box 5.5), anxiety and depression, reducing mental fatigue and improving the ability to cope with stressful situations (Kondo et al., 2015). The use of gardens and access to nature is a form of treatment that can be prescribed or used by healthcare professionals to support patients with mental health illnesses and other conditions. Sahlin et al. (2015) studied the practice of 'nature-based rehabilitation' (NBR) for Swedish employees on long-term sick leave due to stress-related mental illness. They found that participants of NBR had decreased scores for self-assessed burnout, depression and anxiety, and increased wellbeing scores, compared to their scores before the rehabilitation. Pretty and Barton (2020) studied the concept of 'nature-based interventions' (NBIs) whereby a 'dose of nature' can be used to support health and wellbeing as part of social prescribing (see Box 6.6 in Chap. 6) and other non-clinical community interventions (p. 4). Nature-based interventions include forest and woodland schools, wilderness and adventure therapies, ecotherapies and green care, social farms and gardens and walking for health. Pretty and Barton also studied mind-body interventions (MBIs) and practices (which include tai-chi, yoga, qigong and mindfulness),

Box 5.5 Stress and the urban environment

Life's stressors from work or school can be exacerbated by the pollution and constant activity of city living. Stress impacts the body through psychological and physiological responses that are interrelated and can affect health in multiple ways. Prolonged stress is associated with serious health conditions, including cardiovascular disease, diabetes type II, depression, schizophrenia and anxiety. The importance of considering stress in urban design is underscored by Lawrence Frank et al.'s (2019) framework of built environment impacts on health which situates stress as one of three interconnected biological responses (alongside BMI/obesity and systemic inflammation) to behavioural and environmental exposures (e.g. physical activity and noise pollution) which lead to mental and physical chronic diseases. They highlight air and noise pollution, traffic safety and crime as the causes of stress.

Restorative natural environments and biophilic design may reduce some of the stress caused by modern living. Rather than focusing on the activities within natural spaces (e.g. nature-based therapeutic activities), these schools of thought link the simple act of being in or near to nature (or nature-inspired design) as solutions to stress. As with water (see Chap. 2), parks and gardens have been used as places to heal for millennia, with examples found in the Roman and Persian Empires and the Middle Ages (Stigsdotter et al., 2011). Multiple theories have been proposed linking natural environments to health (see review by Stigsdotter et al., 2011). One such theory that has been very influential in design fields is Attention Restoration Theory (ART) (Kaplan & Kaplan, 1989; Kaplan, 1995).

In considering ART and natural environments, Kaplan (1995) argues that prolonged mental tasks, such as problem solving or studying, lead to directed attention fatigue. This fatigue can be restored in part by sleep, but this is not sufficient. Instead, he calls for 'an alternative mode of attending that would render the use of directed attention temporarily unnecessary' (ibid., p. 172). In other words, people need to be in an environment that allows mental space for reflection and restoration of the capacity for directed attention; and natural environments are an ideal solution. These spaces need to have several qualities, including: (1) allowing the feeling of 'being away' from one's normal thoughts through seeing or being in a different environment, (2) constituting a space that is large or rich enough to engage one's mind, known as 'extent', (3) supporting 'compatibility' between visitors' desired actions and what the space allows, and (4) being 'fascinating' by having features that engage the mind, either allowing for reflection (i.e. 'soft fascination' through watching rustling leaves or passing clouds) or fully encompassing attention (i.e. 'hard fascination' through watching a sports game). A key point is that supporting attention restoration in an urban environment does not require an expansive park. Smaller greenspaces, such as the manicured style of Japanese gardens, designed with the above features can suffice.

noting that MBIs and NBIs support the practice of attentiveness and immersion that affect life satisfaction and happiness. In their study of four NBI and MBI programmes in England, they found increased life satisfaction and happiness scores amongst participants and they noted the value of these interventions for health equity. In order for health professionals and the general public to use nature-based rehabilitation and interventions, these spaces need to be available and maintained in urban areas.

Equitable *access* to greenspaces is not the only important factor for health and wellbeing. Mick Lennon et al. (2017) offer a framework for planning and designing urban greenspaces that encourages consideration of the relational nature of *quality*. Using Gibson's (1979) ecological approach to perceptions, they argue that the diverse functions and features of greenspaces are experienced in different ways by different users. Lennon et al. adopt Gibson's term 'affordances' to describe the 'opportunities or constraints that exist within an environment relative to the characteristics of the organism perceiving them' (2017, p. 783). Applying an affordance dimension to greenspace means that 'all aspects of the emerging experience of such spaces are produced in a relative and integrative fashion, such that no dimension exists as an a priori attribute' (ibid.). Figure 5.6 provides images for the types of

Fig. 5.6 Urban greenspaces support diverse users' needs. Top left: Sheltered space for play, resting and social interaction in Seattle, USA. Top right: Den building in urban woodlands in London, UK. Bottom left: Adult exercise equipment overlooking children's play equipment for use by parents and carers, in Beijing, China. Bottom right: Accessible toilets for all genders with clean drinking water and baby changing facilities in London, UK. (Source: Author)

spaces that Lennon et al. describe as supporting different users, such as benches for resting, places for parents and carers to observe children whilst they play safely, den building for children and hang-out spaces for teenagers. Spaces, uses and visitors are likely to change throughout the day and across seasons. Planners and designers can use their six-point 'Affordances Star' relational framework to plan and enhance the quality of greenspaces for different users to promote health. The star dimensions include spaces, actions, times, persons, scales and objects. The framework can be used as a heuristic to ensure that these dimensions and the interactions between them are considered. Lennon et al. describe how use of the 'Affordances Star' should be participative and must recognise the diverse needs and perceptions of different greenspace users, across ages, ethnicities, physical abilities and genders.
Implementation of health-enhancing green infrastructure can be achieved by different professionals and members of the public. A green infrastructure strategy or plan may form part of a city's spatial planning policy. Plans should consider how different plant species and locations would relate to exacerbating allergens, absorbing pollutants, providing shade, supporting biodiversity and having associated water requirements. Greenspace design should be inclusive to ensure provision of appropriate facilities (Corkery, 2015). Planners should consider how barriers to access for disadvantaged communities could be overcome (based on engagement), including perceived safety, poor maintenance and lack of transport, among others (Institute of Health Equity, 2014). Property developers and design teams should consider the aforementioned points alongside more local impacts such as the benefits of views to nature and any potential adverse effects from poor maintenance. Public health and parks officials should consider how the provision of greenspace may be complemented by organised activities and classes, possibly as part of social prescribing (Hunter et al., 2015). Residents and schools may initiate activities in local parks that promote health, such as bird watching, running groups and forest schools.

Knowledge from diverse stakeholders should be used to explore the benefits and disbenefits of green infrastructure for different policy objectives, as shown in Table 5.3 (van den Bosch & Ode Sang, 2017; Kondo et al., 2015). For instance,

Table 5.3 Selected benefits and disbenefits of urban green infrastructure

Benefits	Disbenefits
• Air pollution reduction, carbon sinks and buffers to air and noise pollution • Some greenspaces are associated with safety and lower rates of crime • Increased value to adjacent real estate • Human mental and physical health benefits • Places for social interaction to build social capital • Reduced urban temperatures (which may reduce crime) • Flood risk reduction • Wildlife habitat	• Generation of pollutants and allergens from some plant species • Some greenspaces are associated with fear of crime and actual crime • Displacement of low-income residents following greenspace renovations • Reduced wellbeing and personal safety in locations of perceived/actual crime

greenspaces have the potential to both increase and decrease air pollution and crime, but contextually appropriate design and landscaping solutions can make benefits more likely than disbenefits. For example, perceived safety is likely to be improved through well-maintained vegetation that allows greenspaces to be over-looked and where visitors' views within the space are not blocked, such as through shorter trees and vegetation (Kondo et al., 2015). Tree and plant species can be selected to reduce exposure to allergens in cities. The issue of environmental gentrification, where greenspace renovations displace low-income or vulnerable residents (Cole et al., 2017, 2021), could be avoided through mechanisms such as rent controls.

5.2.5 *Sanitation*

Significant disparities exist internationally and within countries in access to basic sanitation systems. According to WHO and UNICEF data (2017), 39% of the global population use a safely managed sanitation service (Table 5.4), while 2.3 billion people do not have basic sanitation services. They operate a ladder of service provision for drinking water and sanitation which is reflected in SDG targets, as shown in Table 5.4 for sanitation only. Improved facilities are defined as including 'flush/

Table 5.4 The WHO/UNICEF Joint Monitoring Programme for Water Supply, Sanitation Hygiene (JMP) ladder for sanitation with associated SDG targets

Sanitation ladder	Description	SDG targets
Safely managed	Use of improved facilities that are not shared with other households and where excreta are safely disposed of in situ or transported and treated offsite.	SDG 6.2 by 2030, *achieve access to adequate and equitable sanitation and hygiene for all* and end open defecation, paying special attention to the needs of women and girls and those in vulnerable situations
Basic	Use of improved facilities that are not shared with other households.	SDG 1.4 by 2030, ensure all men and women, in particular the poor and vulnerable, have equal rights to economic resources, as well as access to basic services
Limited	Use of improved facilities shared between two or more households.	N/A
Unimproved	Use of pit latrines without a slab or platform, hanging latrines or bucket latrines.	N/A
Open defecation	Disposal of human faeces in fields, forests, bushes, open bodies of water, beaches or other open spaces, or with solid waste.	SDG 6.2 by 2030, achieve access to adequate and equitable sanitation and hygiene for all and end open defecation, paying special attention to the needs of women and girls and those in vulnerable situations

Source: WHO and UNICEF (2017, pp. 2–8)

pour flush to piped sewer systems, septic tanks or pit latrines; ventilated improved pit latrines, composting toilets or pit latrines with slabs' (ibid., p. 8). The UN highlights gender inequities related to water and sanitation because women's health and safety are disproportionately affected by poor access to these services. In many places, women and girls spend more time collecting water and they are also at greater risk of sexual harassment/assault while gathering water or using shared sanitation facilities.

Lack of access to basic and safely managed sanitation services is primarily a problem in low- and middle-income countries, resulting in health impacts through infectious and non-infectious diseases and injuries. Sanitation systems in high-income countries can still pose health risks, for example, when sewage is leaked into rivers and lakes due to misconnected pipes or flooding events. As with drinking water, it should not be assumed that conventional grey infrastructure solutions are optimal for sanitation requirements. Bell (2015) argues that 'a more ecologically sensible and socially just approach to water in cities needs to acknowledge these multiple flows of water and their relationship to social life in devising systems that are not based on domination and control of people and nature', which could involve a combination of nature-based and grey infrastructure solutions (p. 8). Sanitation infrastructure can be applied at multiple scales, such as the Bullitt Center's composing toilets and the retrofitting of basic services in informal settlements in Belo Horizonte, Brazil (described later in this chapter). A nature-based solution for the treatment of wastewater (including sewage) was adopted in informal settlements in Indonesia with the support of local communities as part of the Revitalising Informal Settlements and their Environments (RISE) project (Fig. 5.7). The RISE project adopted a Water-Sensitive Cities approach that involved working with nature and local communities to design decentralised technologies for wastewater treatment, such as the 'treatment train' (Fig. 5.8) 'which safely collects, cleans, and discharges blackwater within informal settlements using natural systems and bio-filtration processes' (ADB and RISE, 2021). This technology is resilient to shocks, supports biodiversity, is low cost and has many other benefits for people and local ecosystems.

5.2.6 *Waste*

Urban waste management is a global issue that affects health at multiple scales, as discussed in section 'Resource Efficiency' in Chap. 4. Uncollected solid waste can have significant local health impacts such as diarrhoea, respiratory disease and dengue fever (Hoornweg & Bhada-Tata, 2012). Accumulated waste can cause soil, air and water pollution and increase vectors for disease, such as mosquitoes and rodents (Alirol et al., 2011). Health impacts from waste are more likely to affect low-income residents. The Belo Horizonte case study later in this chapter shows that residents of informal settlements often do not benefit from municipal waste collection services partly because the roads in these areas would not support large trucks. European

Fig. 5.7 Images of the completed water-sensitive cities infrastructure in Makassar, Indonesia, showing the subsurface wetland and biofilters. (Source: RISE)

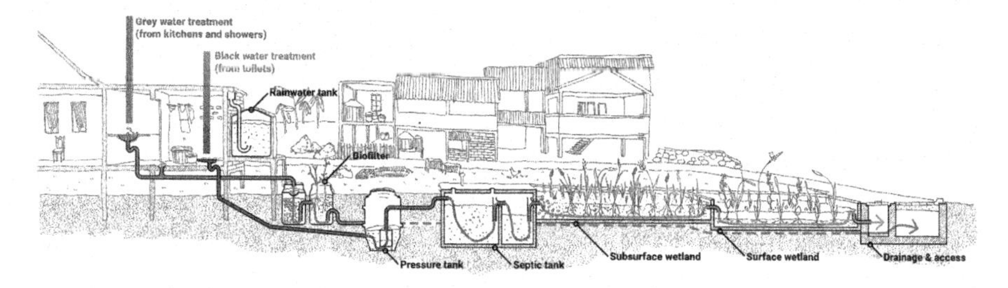

Fig. 5.8 The 'treatment train' collects, cleans and discharges blackwater using natural systems and bio-filtration processes. (Source: RISE)

studies have found that neighbourhood deprivation is associated with proximity to landfills, which can cause cancers, birth defects, respiratory diseases and annoyance (WHO, 2016). Rich countries export their waste to poorer countries, harming local ecosystems and health, and waste accumulates in the world's oceans, harming marine life. China was the biggest importer of foreign waste (China banned waste imports in late 2017) and its residents will suffer the health effects of persistent toxic substances that were released into the air, water and soil for decades (Liu & Wang, 2019). Waste also contributes to greenhouse gas emissions, but sustainable waste management can reduce these impacts. This section adds briefly to the more comprehensive discussion of waste in Chap. 4.

Waste planning is a complex process that is often covered by national regulations and systems. City planners who are not waste specialists can still support sustainable waste management practices. Reducing consumption of single-use products is a key component of waste management and this can be facilitated by food vendors and supermarkets that sell unwrapped produce, bulk items (e.g. nuts and grains) and refill stations for cleaning products. Planners may be able to incentivise shops to support these practices through the planning system and they can allocate space for outdoor markets in local plans. Planners and designers need to provide adequate space for waste disposal and sorting in homes (e.g. space for recycling, compost and waste bins in the kitchen) and near the street (e.g. individual or communal bin collection). Waste bins are often left on roadsides and pavements on collection days, creating barriers for pedestrians and cyclists. Whilst the problem of waste storage is difficult to solve through retrofit, in newly planned communities, refuse collection points may be centralised and underground, such as in new Swedish developments.

5.2.7 Mobility Infrastructure

Urban transport and mobility infrastructure include the physical structures (e.g. roads, pavements and trains) and other mechanisms (e.g. transport fares and levies) that enable people and goods to move from one place to another. This infrastructure can contribute positively and negatively to health and equity in myriad ways as shown in Mark Nieuwenhuijsen's (2016) framework for urban and transport planning and health (Fig. 5.9). The construction and operation of roads, train lines and vehicles create pollution and deaths and injuries from collisions. The transport sector is responsible for 16.2% of global greenhouse gas emissions (Ritchie & Roser, n.d.). Public transport systems and other forms of active travel (or active transport), such as walking and cycling, are far less polluting and dangerous than personal vehicles and they enable people to access work and school, social interaction and physical activity (discussed in Chap. 6). Mobility infrastructure is thus an important public health measure that can be applied across a city (e.g. through low speed limits and public transport) or targeted at specific groups (e.g. offering reduced travel fares for vulnerable groups). This section shows how policy and design decisions at multiple urban scales will affect residents' ability and motivation to use active transport, highlighting some of the key considerations. Comprehensive guidance is available for mobility infrastructure and active design more generally (see Nieuwenhuijsen & Khreis, 2019; City of New York, 2010; Edwards & Tsouros, 2008).

The health impacts of transport are not evenly distributed across society, with younger people and those in low-and middle-income countries experiencing a higher burden. Globally, road traffic injury is the leading cause of death for 5- to 29-year-olds and the eighth leading cause of death for all age groups (WHO, 2018a). From an inclusion and equity perspective, mobility infrastructure is one of the key

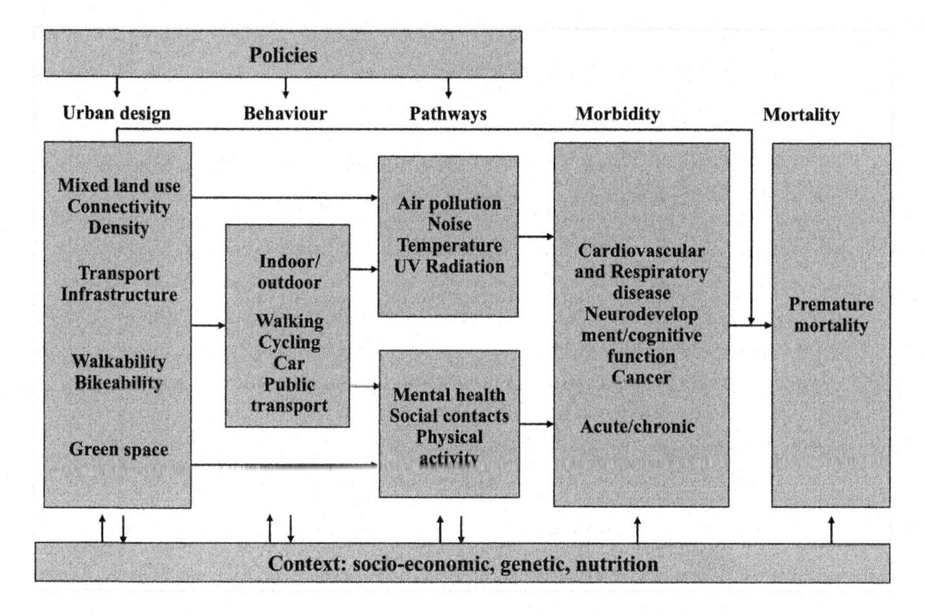

Fig. 5.9 Framework showing the health impacts of urban and transport planning. (Source: Reproduced from Nieuwenhuijsen, 2016)

urban health topics that requires careful consideration of the diverse needs of different population groups. American research teams have shown that public transit providers do not always make adequate provisions for women's safety needs (Loukaitou-Sideris & Fink, 2008) and that the bicycle infrastructure needs of low-income minority communities were absent from transport and Crime Prevention Through Environmental Design (CPTED) guidance in Boston (Lusk et al., 2019). Investment and design decisions for mobility infrastructure should adopt equity and inclusion lenses and they should be informed or co-produced with a diverse mix of residents. Chapter 8 describes methods for incorporating community knowledge into urban design and planning, including a case study of the Los Angeles (USA) Metro Equity Platform which sought to redress disparities in transport access in terms of race and socio-economic status.

Transitioning to sustainable urban mobility systems involves integrated urban planning and infrastructure provision processes at multiple scales, as seen in Fig. 5.9. Urban form and layout can support active transport through net residential density, intersection density and public transport density (Sallis et al., 2016), which relate to the requirements of proximity (neighbourhood density and mix of uses) and connectivity (ease of movement between destinations) (Saelens et al., 2016). At the street scale, pedestrians require well-maintained and adequately lit pavements, safe crossing points, natural surveillance to improve perceptions of safety from crime and places to rest (among other factors). At the urban scale, cities can provide convenient and affordable public transport, low speed limits and interconnected segregated cycle lanes. Many of these design and policy measures would also

Box 5.6 Seattle's Vision Zero initiative

Vision Zero initiatives have been adopted around the world to eliminate traffic-related injuries and deaths. The City of Seattle's Vision Zero initiative was launched by the Mayor in 2015 to end traffic deaths and serious injuries by 2030 (City of Seattle, 2015). The Vision Zero plan includes a range of mobility infrastructure measures to improve safety including: reducing speed limits in the city centre to 25 mph, with 20 mph in residential zones near parks and schools; improving unsafe intersections (e.g. by giving pedestrians a head-start); increasing protected bicycle lanes; installing traffic cameras in school zones and providing additional traffic enforcement measures. Pedestrian crossings have been retrofitted with Leading Pedestrian Intervals at nearly 300 intersections across the city giving pedestrians the priority to cross before vehicles turn. This has reduced collisions by 48% and serious injury and death from collisions by 34% (City of Seattle, n.d.).

support greater social connectivity for people who are increasingly socially isolated, such as the elderly and people with multiple chronic conditions. Box 5.6 describes Seattle's Vision Zero initiative where positive health impacts have been achieved through mobility infrastructure provision for health.

5.3 Monitoring Indicators for Ecosystem Health

As introduced in Chap. 4, the indicators in Table 5.5 were gathered from Pineo et al.'s (2018a) systematic review. These should be seen as a starting point for urban health indicators related to ecosystem health scale impacts.

5.4 Inclusive Processes for Health in the Vila Viva Programme

Solving the health risks of informal settlements and increasing climate change resilience involves in situ 'slum-upgrading' initiatives rather than demolition and relocation (Satterthwaite et al., 2018; Corburn & Sverdlik, 2017). Brazil has received funding from international agencies since the 1980s to promote upgrade initiatives. This case study focuses on a specific programme called Vila Viva in the city of Belo Horizonte, the state capital of Minas Gerais. Beginning in 1993 with the creation of the Municipal Housing Policy, the Belo Horizonte City Government has developed programmes aimed at the recovery and upgrading of the city's precarious settlements. The interventions rely on federal resources, obtained through financing from

Table 5.5 Example monitoring indicators for built environment impact on ecosystem and human health identified by Pineo et al. (2018a)

Goal	Monitoring indicators
Air quality	Number of exceedances of environmental standards for PM_{10}, $PM_{2.5}$, NO_2, ozone, or other pollutants
	Average annual concentrations of pollutants (e.g. PM_{10}, $PM_{2.5}$, NO_2, ozone)
	Proportion of people who are either very or mostly satisfied with air quality in their neighbourhood
	% of area covered by trees
Water quality	% of residents in households recycling waste water
	Rate of clogged storm drain reports per 1000 Residents
Soil quality	Number of days with soil moisture deficit
	Soil erosion
Greenspace	Proportion of households that are within 1/4 mile of a community garden
	% respondents satisfied with quality of parks and greenspaces
	% of residents living within 0.25 mile radial distance of a park
	Total hectares of greenspace per 1000 residents
	% of buildings with green roofs
Sanitation infrastructure	Presence of public toilets in the midst of human dwellings
	Presence of overflowing septic tanks and aqua-privy systems in the community
Waste infrastructure	Non-recyclable waste generated by households: expressed as an average volume (kilograms) per household
	Proportion of people who recycle all or most recyclable waste
	Mounds of uncollected garbage within community
	Paper and plastic litter within community
Mobility infrastructure	Walk trip occurrence for exercise
	Transit stop density (count/km^2 in buffer)
	Share of trips being made by cycling per demographic group (e.g. age)
	Consumer spending on transportation by income level
	% of roadways in settled areas with sidewalks

the National Bank for Social Development (BNDES) and Federal Savings Bank (Caixa Econômica Federal). This example highlights the participatory and inclusive process that informed the city's policies associated with Vila Viva. This is also an interesting case because urban health indicators from the Observatory for Urban Health in Belo Horizonte (OSUBH) were used to inform and evaluate policies and investments. Personal correspondence with the director of OSUBH, Professor Waleska Caiaffa, informed this section and it is important to note that Caiaffa and her colleagues (among others) view the term 'slum upgrade' as problematic, preferring instead revitalisation, rehabilitation or urban transformation (W Caiaffa 2021, personal communication, 14 September).

Latin American countries are highly urbanised with high rates of poverty, violence and social exclusion. The population in Latin America and the Caribbean more than tripled from 1950 to 2019 (UN, 2019b) and the percentage of people living in cities grew from 41% to 88%, but the urban growth rate has declined in recent years (UN, 2019a). This rapid urbanisation meant that many cities grew in an unplanned way without sufficient provision of housing and infrastructure. Informal settlements (or slums) are a distinctive feature of cities in this region. In 2018, a fifth of the population (21%) lived in these settlements, compared to 36% in 1990. In this region, slums are also called favelas, chabolas, villas miseria and barriadas, and they are often labelled as dangerous places to live (Beato & Velásquez, 2021). Around the world, informal settlements are characterised by lack of sanitation, water and waste infrastructure, poor-quality and insecure housing, overcrowding, high unemployment and crime, limited access to food and healthcare and other challenges associated with land tenure irregularities (WHO and UN Habitat, 2010). As such, slum dwellers suffer from the triple threat of disease (infectious diseases, chronic conditions and injury) more than other urban dwellers. For these reasons, slum dwellers have been particularly vulnerable to COVID-19 (see Chap. 9).

In Brazil, the creation of slums was driven not only by rapid population growth but also by socio-spatial polarisation (i.e. the gap between wealthy and disadvantaged populations) and a centralised model of governance that prevented public participation (Lopez et al., 2010). Stemming from city policies developed in the 1990s and revised over time, the Vila Viva project in Belo Horizonte aimed to use participatory governance models to achieve social, economic and environmental benefits through urban revitalisation. The timeline of policy and investment that supported Vila Viva is as follows (Dias et al., 2015, 2019; Lopez et al., 2010):

- 1993: Social participation was implemented in local government, including a Participatory Budget process, whereby the public decided how to allocate resources.
- 1994: The Municipal Housing Policy (PMH in Portuguese) included plans for structural interventions in slums, including legalising land holdings and improvements to infrastructure and housing.
- 1996: The city's Master Plan and urban planning legislation encompassed informal settlements as Special Social Interest Zones (ZEIS in Portuguese), with specified directives and schedules to undertake studies of each area.
- 2005: Start of the integrated structural interventions foreseen in the PMH (from 1994), called Vila Viva Program from then onwards, implemented through the Urbanization Company of Belo Horizonte (URBEL in Portuguese, the managing body of the PMH) with public participation.

General geographic and demographic data are provided for Belo Horizonte in Table 5.6 as reported by Friche et al. (2015) and Dias et al. (2019). Belo Horizonte is the sixth-largest city in Brazil, with per capita gross domestic product and Human Development Index figures that hide the level of inequality within the city. The homicide mortality rates in Belo Horizonte's slums are three to seven times higher than the city-wide average.

The Vila Viva process involved several steps beginning with assessment of the slum for an elaboration of the Specific Global Plan (PGE in Portuguese) and

Table 5.6 Key social, economic and environmental indicators for Belo Horizonte, Brazil

Indicators	Value
Population	2,375,151
Number of households	600,000
Per capita gross domestic product	BRL 13,636.00
Human Development Index	0.839
Number of Special Social Interest Zones (ZEIS)	216
Number of ZEIS categorised as slums	186
Residents living in slums	385,395
Percentage of the population living in slums	16.2%
Number of households in slums	130,670
Area of the city categorised as slums	16.4 km^2
Percentage of the city's total area categorised as slums	5%
Average homicide mortality rate per 100,000 inhabitants*	25.9

Source: Friche et al. (2015) and Dias et al. (2019)
* indicates the final indicator (average homicide) is taken from Dias et al. (2019), while the rest of the data are from Friche et al. (2015)

implementation of the built environment works and social and educational activities, followed by participatory evaluation (Dias et al., 2019; Lopez et al., 2010). The interventions were preceded by the elaboration of a Specific Global Plan, an instrument for a detailed assessment of the slums, creating a reference baseline to guide the government's interventions and community demands. The plans identify paths for the social, environmental and legal recovery of the settlements and set priorities for implementing the actions and public works. Production of the plans involved deep community participation through URBEL technicians and Reference Groups consisting of community leaders and other local residents that participate in the project's elaboration, monitoring and follow-up and financial outlays. The representatives received training to plan, manage and implement projects to ensure a level of autonomy from the city.

The Specific Global Plans (SGPs) are perceived as indispensable instruments for the decision-making process by the government and local communities, aiding with fundraising and being a requirement to access resources from the Municipal Participatory Budget. The plans covered actions related to urban and environmental recovery, strategies for legalising land holdings and socio-organisational development proposals (e.g. for healthcare, education, employment and community organisation). Assessments were carried out in relation to housing, road quality and access, water quality, sanitation systems, geology and geotechnical hazards, social issues and land ownership. Carvalho et al. (2012) state that the SGPs are different to other public policies because they approached urbanisation, land regularisation and social policies in an integrated manner with significant public participation, the latter also highlighted by Caiaffa (W Caiaffa 2021, personal communication, 14 September). The main demands of the public were related to 'availability and/or increase of coverage of services such as water supply, sewerage network, and collection of solid waste including installation of waste bins; and also elimination of undesirable animals and rats', in addition to road interventions and slope containment (Silveira et al., 2019, p. 1168).

There have been multiple evaluations and studies of the Vila Viva programme, including its impact on crime and homicide (Beato & Velásquez, 2021; Dias et al., 2019), health and other social effects (Friche et al., 2015; Carvalho et al., 2012) and the planning process (Silveira et al., 2019; Lopez et al., 2010). A summary of key evaluation findings across these studies is provided in Table 5.7. There are some contradictory findings. For example, an early case study produced by Lopez et al. (2010) states that the 'community's participation in all phases of the programme is perhaps the main factor responsible for the success obtained so far' (p. 6). While a later study by Silveira et al. (2019) said that although community participation was a key component of the Vila Viva programme, it 'was effective only at the stage of ascertaining the demands of community; its effectiveness was low in the other

Table 5.7 Summary of evaluation findings about the Vila Viva programme

Summary finding	Description
Reduced homicide rates	Homicide rates decreased in four out of five upgraded slums, compared to five non-upgraded slums, during the study period (2005–2011) and the effect was higher when considering the cumulative time of exposure after the slum upgrade (Dias et al., 2019).
Increased connections to city infrastructure	In the pilot slum of Aglomerado da Serra, >2500 residents received connection to the sewerage network and by urbanising 30 km of alleys and 20 routes the city implemented 96% coverage of networked drinking water, electricity and basic sanitation (Beato & Velásquez, 2021). There were six new parks, four children's centres/schools, two social assistance centres and two sports centres (W Caiaffa 2021, personal communication, 14 September).
Reduced number of homes at risk	In 2009, an URBEL consultant reported that the number of 'high-risk homes' fell by 74.5% since 2005 (Lopez et al., 2010).
Improved road access, leading to greater public service access	Improvements to streets, alleys and avenues (including street widening) led to increased access to public services (see city infrastructure above), including refuse collection and the passage of ambulances which could previously not access the slums, and motor vehicle traffic increased (Lopez et al., 2010).
Improved housing/ neighbourhood conditions	The upgrades reduced the population density and exposure to water-borne diseases through improved drainage. Children and adults gained access to new leisure and community facilities (Lopez et al., 2010).
Resettlement of some residents outside of the original slum	Regarding resettlement in the pilot slum of Aglomerado da Serra previous research found some negative impacts, with approximately 26% of the 1500 financially compensated households unable to find local accommodation (Carvalho et al., 2012). The latest figures show that 2535 families were removed and relocated and 936 housing units were produced. Of the relocated families, 74% moved back into the area, 17% acquired housing in other Belo Horizonte neighbourhoods, 7% moved to the wider metropolitan region and 2% moved elsewhere in Minas Gerais state (W Caiaffa 2021, personal communication, 14 September).

phases' (p. 1168). A Social Network Analysis of community participation, specifically in the pilot slum called Aglomerado da Serra, found that the perspectives of some members of the public were undervalued and this contributed to increased control by drug traffickers in the slum (Beato & Velásquez, 2021). Some residents felt that resettlement and removal of a portion of slum residents was considered a negative impact (Carvalho et al., 2012). None of the participatory planning activities had resulted in requests for resettlement or removal of residents, yet this was deemed necessary to complete the works (Silveira et al., 2019).

The participatory budgeting used in Belo Horizonte may be an important mechanism that produces positive health and wellbeing impacts in the city beyond the Vila Viva programme. In general terms, participatory budgeting involves a multi-stage process where members of the public and government officials deliberate over the allocation of public money on local services. A systematic review of evaluations of participatory budgeting projects described four stages through which the process and its outcomes could support health: (1) participation in public finance decision-making; (2) collaborating to exercise political rights, develop civic skills and build social cohesion; (3) prioritising improvements to public services that support health, such as housing or healthcare; and 4) allocating resources according to need, resulting in more efficient and transparent public spending (Campbell et al., 2018). This international review demonstrated that Brazil has been at the forefront of participatory budgeting with a clear social justice rationale, supported by substantial evaluations. Given the complex range of interventions that participatory budgeting supports, and the many pathways through which it could support health and wellbeing, the existing evidence is relatively weak regarding the public health value of the process itself. Nevertheless, participatory budgeting is a part of a suite of governance measures that can be used internationally to increase public involvement in policy and decision making.

5.5 Conclusion

Achieving ecosystem health through urban development requires collaborative processes to match appropriate technologies with local social, economic and environmental requirements. This chapter has showcased innovative and ancient solutions that have been applied in China, Indonesia, Brazil and other settings. The next chapter shifts to the 'local health' scale of health impact in the THRIVES framework, describing neighbourhood design measures that support health and wellbeing primarily through social interaction and physical activity.

References

Alirol, E., Getaz, L., Stoll, B., Chappuis, F., & Loutan, L. (2011). Urbanisation and infectious diseases in a globalised world. *The Lancet Infectious Diseases, 11*, 131–141.

Allen, J., & Balfour, R. (2014). *Natural solutions for tackling health inequalities*. UCL Institute of Health Equity.

Allen, J. G., MacNaughton, P., Laurent, J. G. C., Flanigan, S. S., Eitland, E. S., & Spengler, J. D. (2015). Green buildings and health. *Current Environmental Health Reports, 2*, 250–258.

Andersson, A., Deng, J., Du, K., Zheng, M., Yan, C., Sköld, M., & Gustafsson, Ö. (2015). Regionally-varying combustion sources of the January 2013 severe haze events over Eastern China. *Environmental Science & Technology, 49*, 2038–2043.

Annerstedt van den Bosch, M., & Depledge, M. H. (2015). Healthy people with nature in mind. *BMC Public Health, 15*, 1232.

Asian Development Bank, Revitalising Informal Settlements and their Environments (RISE). (2021). *Water-sensitive informal settlement upgrading: Overall principles and approach*.

Barlow, P. (2021). Regulation 28: Report to prevent future deaths. Ref: 2021-0113. Retrieved April 23, 2021 from https://www.judiciary.uk/publications/ella-kissi-debrah/

Barnes, P. W., Williamson, C. E., Lucas, R. M., Robinson, S. A., Madronich, S., Paul, N. D., Bornman, J. F., Bais, A. F., Sulzberger, B., Wilson, S. R., Andrady, A. L., McKenzie, R. L., Neale, P. J., Austin, A. T., Bernhard, G. H., Solomon, K. R., Neale, R. E., Young, P. J., Norval, M., … Zepp, R. G. (2019). Ozone depletion, ultraviolet radiation, climate change and prospects for a sustainable future. *Nature Sustainability, 2*, 569–579.

Beatley, T. (2016). *Handbook of biophilic city planning and design*. Island Press.

Beato, C., & Velásquez, C. (2021). Participatory Slum upgrading and urban peacebuilding challenges in Favela settlements: The Vila Viva Program at Aglomerado da Serra (Belo Horizonte, Brazil). *The Journal of Illicit Economies and Development, 2*, 155–170.

Bell, S. (2015). Renegotiating urban water. *Progress in Planning, 96*, 1–28.

Braithwaite, I., Zhang, S., Kirkbride, J. B., Osborn, D. P. J., & Hayes, J. F. (2019). Air pollution (particulate matter) exposure and associations with depression, anxiety, bipolar, psychosis and suicide risk: A systematic review and meta-analysis. *Environmental Health Perspectives, 127*, 126002.

Britton, E., Kindermann, G., Domegan, C., & Carlin, C. (2020). Blue care: A systematic review of blue space interventions for health and wellbeing. *Health Promotion International, 35*, 50–69.

Bullitt Center. (n.d.). Greywater system. Retrieved April 14, 2021, from https://bullittcenter.org/building/building-features/wastewater-use/

Buse, C. G., Oestreicher, J. S., Ellis, N. R., Patrick, R., Brisbois, B., Jenkins, A. P., McKellar, K., Kingsley, J., Gislason, M., Galway, L., McFarlane, A., Walker, J., Frumkin, H., & Parkes, M. (2018). Public health guide to field developments linking ecosystems, environments and health in the Anthropocene. *Journal of Epidemiology and Community Health, 72*, 420–425.

Campbell, M., Escobar, O., Fenton, C., & Craig, P. (2018). The impact of participatory budgeting on health and wellbeing: A scoping review of evaluations. *BMC Public Health, 18*, 822.

Carvalho, A. M., Rezende, B. de M., Santos, D. G. O., Miranda, I. G., Merladet, F. A. D., Coelho, L. X. P., de Oliveira, R. A. P., & Isaías, T. L. S. (2012). Vila Viva, a project of urban, social and political organization of Aglomerado da Serra: Analysis of effects.

Centric Lab. (2020). *Air pollution and health in Southall*. A report for C.A.S.H.

City of New York. (2010). *Active design guidelines: Promoting physical activity and health in design*.

City of Seattle. (2015). Seattle launches Vision Zero plan to end traffic deaths and injuries by 2030 [WWW Document]. *Seattle.gov*. Retrieved February 08, 2017, from http://murray.seattle.gov/seattle-launches-vision-zero-plan-to-end-traffic-deaths-and-injuries-by-2030/

City of Seattle. (n.d.). Vision Zero—transportation. Retrieved April 16, 2021, from https://www.seattle.gov/transportation/projects-and-programs/safety-first/vision-zero

Cole, H. V. S., Anguelovski, I., Connolly, J. J. T., García-Lamarca, M., Perez-del-Pulgar, C., Shokry, G., & Triguero-Mas, M. (2021). Adapting the environmental risk transition theory for urban health inequities: An observational study examining complex environmental riskscapes in seven neighborhoods in Global North cities. *Social Science & Medicine, 277*, 113907.

Cole, H. V. S., Lamarca, M. G., Connolly, J. J. T., & Anguelovski, I. (2017). Are green cities healthy and equitable? Unpacking the relationship between health, green space and gentrification. *Journal of Epidemiology and Community Health, 71*, 1118–1121.

Corburn, J., & Sverdlik, A. (2017). Slum upgrading and health equity. *International Journal of Environmental Research and Public Health, 14*, 342.

Corkery, L. (2015). Urban greenspaces and human well-being. In H. Barton, S. Thompson, M. Grant, & S. Burgess (Eds.), *The Routledge Handbook of planning for health and well-being: Shaping a Sustainable and healthy future* (pp. 239–253). Taylor and Francis.

Dakubo, C. Y. (2011). *Ecosystems and human health: A critical approach to ecohealth research and practice*. Springer.

Dias, M. A. d. S., Friche, A. A. d. L., Mingoti, S. A., Costa, D. A. d. S., Andrade, A. C. d. S., Freire, F. M., de Oliveira, V. B., & Caiaffa, W. T. (2019). Mortality from homicides in slums in the city of Belo Horizonte, Brazil: An evaluation of the impact of a re-urbanization project. *International Journal of Environmental Research and Public Health, 16*, 154.

Dias, M. A. d. S., Friche, A. A. d. L., de Oliveira, V. B., & Caiaffa, W. T. (2015). The Belo Horizonte Observatory for Urban Health: Its history and current challenges. *Cadernos de Saúde Pública, 31*, 277–285.

Dyer, C. (2020). Air pollution from road traffic contributed to girl's death from asthma, coroner concludes. *BMJ, 371*, m4902.

ECDG. (2015). Towards an EU research and innovation policy agenda for nature-based solutions & re-naturing cities: Final report of the Horizon 2020 expert group on 'Nature based solutions and re naturing cities': (full version). Publications Office, Brussels.

Edwards, P., & Tsouros, A. D. (2008). *A healthy city is an active city: A physical activity planning guide*. WHO Regional Office for Europe, Denmark.

Frank, L. D., Iroz-Elardo, N., MacLeod, K. E., & Hong, A. (2019). Pathways from built environment to health: A conceptual framework linking behavior and exposure-based impacts. *Journal of Transport and Health, 12*, 319–335.

Friche, A. A. d. L., Dias, M. A. d. S., Reis, P. B. d., Dias, C. S., & Caiaffa, W. T. (2015). Urban upgrading and its impact on health: A "quasi-experimental" mixed-methods study protocol for the BH-Viva Project. *Cadernos de Saúde Pública, 31*, 51–64.

Gallagher, J., Baldauf, R., Fuller, C. H., Kumar, P., Gill, L. W., & McNabola, A. (2015). Passive methods for improving air quality in the built environment: A review of porous and solid barriers. *Atmospheric Environment, 120*, 61–70.

Gascon, M., Zijlema, W., Vert, C., White, M. P., & Nieuwenhuijsen, M. J. (2017). Outdoor blue spaces, human health and well-being: A systematic review of quantitative studies. *International Journal of Hygiene and Environmental Health, 220*, 1207–1221.

Ghiassee, C., Urquhart, G., Duarte-Davidson, R., Wilding, J., Landeg-Cox, C., & Gittins, A. (2014). Key concepts and framework for investigation. In N. Bradley, H. Harrison, G. Hodgson, R. Kamanyire, A. Kibble, & V. Murray (Eds.), *Essentials of environmental public health science* (pp. 20–56). Oxford University Press.

Gibson, J. J. (1979). *The ecological approach to visual perception*. Houghton Mifflin.

Griffin, J. (2019). Families hit out at London gasworks redevelopment. *The Observer*. May 5, 2019. Retrieved from https://www.theguardian.com/environment/2019/may/04/brownfield-site-new-homes-building-wrecking-health-southall

Griffin, J. (2020a). Residents demand new clean air rules for former gasworks sites in England. *The Guardian*. September 22, 2020. Retrieved from https://www.theguardian.com/environment/2020/sep/22/residents-demand-new-clean-air-rules-for-former-gasworks-sites-in-england

Griffin, J. (2020b). Londoners claim toxic air from gasworks damaging their health. *The Guardian*. August 27, 2020. Retrieved from https://www.theguardian.com/environment/2020/aug/27/londoners-claim-toxic-air-from-gasworks-damaging-their-health

Hoornweg, D., & Bhada-Tata, P. (2012). *What a waste: A global review of waste management*. Urban Development Series; Knowledge Papers no. 15. World Bank, Washington, DC.

Huang, G., Zhou, W., Qian, Y., & Fisher, B. (2019). Breathing the same air? Socioeconomic disparities in PM2.5 exposure and the potential benefits from air filtration. *Science of the Total Environment, 657*, 619–626.

Hunter, R. F., Christian, H., Veitch, J., Astell-Burt, T., Hipp, J. A., & Schipperijn, J. (2015). The impact of interventions to promote physical activity in urban green space: A systematic review and recommendations for future research. *Social Science & Medicine, 124*, 246–256.

Institute of Health Equity. (2014). *Local action on health inequalities: Improving access to green spaces*. Health Equity Briefing 8. Public Health England, London.

Kaplan, R., & Kaplan, S. (1989). *The experience of nature: A psychological perspective*. Cambridge University Press.

Kaplan, S. (1995). The restorative benefits of nature: Toward an integrative framework. *Journal of Environmental Psychology, 15*, 169–182.

Kelly, F. J., & Fussell, J. C. (2019). Improving indoor air quality, health and performance within environments where people live, travel, learn and work. *Atmospheric Environment, 200*, 90–109.

Kondo, M. C., South, E. C., & Branas, C. C. (2015). Nature-based strategies for improving urban health and safety. *Journal of Urban Health, 92*, 800–814.

Kremen, C. (2005). Managing ecosystem services: What do we need to know about their ecology? *Ecology Letters, 8*, 468–479.

Lam, T. M., Vaartjes, I., Grobbee, D. E., Karssenberg, D., & Lakerveld, J. (2021). Associations between the built environment and obesity: An umbrella review. *International Journal of Health Geographics, 20*, 7.

Lennon, M., Douglas, O., & Scott, M. (2017). Urban green space for health and well-being: Developing an 'affordances' framework for planning and design. *Journal of Urban Design, 22*, 778–795.

Liu, D., & Wang, S. (2019). The global issue of foreign waste. *Lancet Planetary Health, 3*, e120.

Lopez, E., Carvalho, P., & Pedrosa Nahasunder, M. I. (2010). Belo Horizonte, Brazil, Villa Viva Programme—Aglomerado da Serra. UCLG Committee on Social Inclusion, Participatory Democracy and Human Rights.

Loukaitou-Sideris, A., & Fink, C. (2008). Addressing women's fear of victimization in transportation settings: A survey of U.S. transit agencies. *Urban Affairs Review, 44*, 554–587.

Lusk, A. C., Willett, W. C., Morris, V., Byner, C., & Li, Y. (2019). Bicycle facilities safest from crime and crashes: Perceptions of residents familiar with higher crime/lower income neighborhoods in Boston. *International Journal of Environmental Research and Public Health, 16*, 484.

McDonald, R. I., Weber, K., Padowski, J., Flörke, M., Schneider, C., Green, P. A., Gleeson, T., Eckman, S., Lehner, B., Balk, D., Boucher, T., Gril, G., & Montgomery, M. (2014). Water on an urban planet: Urbanization and the reach of urban water infrastructure. *Global Environmental Change, 27*, 96–105.

Millennium Ecosystem Assessment. (2005). *Ecosystems and human well-being: Synthesis*. Island Press, Washington, DC.

Nieuwenhuijsen, M., & Khreis, H. (Eds.). (2019). *Integrating human health into urban and transport planning: A framework*. Springer International Publishing.

Nieuwenhuijsen, M. J. (2016). Urban and transport planning, environmental exposures and health-new concepts, methods and tools to improve health in cities. *Environmental Health, 15*, S38.

Pauleit, S., Zölch, T., Hansen, R., Randrup, T. B., & Konijnendijk van den Bosch, C. (2017). Nature-Based Solutions and Climate Change—Four Shades of Green. In N. Kabisch, H. Korn, J. Stadler, & A. Bonn (Eds.), *Nature-based solutions to climate change adaptation in urban areas: Linkages between science, policy and practice, theory and practice of urban sustainability transitions* (pp. 29–49). Springer International Publishing.

Peters, R., Peters, J., Booth, A., & Mudway, I. (2015). Is air pollution associated with increased risk of cognitive decline? A systematic review. *Age and Ageing, 44*, 755–760.

Pineo, H., Glonti, K., Rutter, H., Zimmermann, N., Wilkinson, P., & Davies, M. (2018a). Urban health indicator tools of the physical environment: A systematic review. *Journal of Urban Health, 95*, 613–646.

Pretty, J., & Barton, J. (2020). Nature-based interventions and mind–body interventions: Saving public health costs whilst increasing life satisfaction and happiness. *International Journal of Environmental Research and Public Health, 17*, 7769.

Public Health England. (n.d.). *Health matters: Air pollution* [WWW Document]. GOV. UK. Retrieved April 13, 2021, from https://www.gov.uk/government/publications/health-matters-air-pollution/health-matters-air-pollution

Richardson, M., Maspero, M., Golightly, D., Sheffield, D., Staples, V., & Lumber, R. (2017). Nature: A new paradigm for well-being and ergonomics. *Ergonomics, 60*, 292–305.

Ritchie, H., & Roser, M. (2017). Air pollution [WWW Document]. *Our World Data*. Retrieved April 13, 2021, from https://ourworldindata.org/air-pollution

Ritchie, H., & Roser, M. (n.d.). Emissions by sector [WWW Document]. *Our World Data*. Retrieved April 16, 2021, from https://ourworldindata.org/emissions-by-sector

Romanelli, C., Cooper, H. D., Campbell-Lendrum, D., Maiero, M., Karesh, W. B., Hunter, D., & Golden, C. D. (2015). *Connecting global priorities: Biodiversity and human health, a state of knowledge review*. World Health Organization and Secretariat for the Convention on Biological Diversity.

Saelens, B. E., Sallis, J. F., & Frank, L. D. (2016). Environmental correlates of walking and cycling: Findings from the transportation, urban design, and planning literatures. *Annals of Behavioral Medicine, 25*, 80–91.

Sahlin, E., Ahlborg, G., Tenenbaum, A., & Grahn, P. (2015). Using nature-based rehabilitation to restart a stalled process of rehabilitation in individuals with stress-related mental illness. *International Journal of Environmental Research and Public Health, 12*, 1928–1951.

Sallis, J. F., Cerin, E., Conway, T. L., Adams, M. A., Frank, L. D., Pratt, M., Salvo, D., Schipperijn, J., Smith, G., Cain, K. L., Davey, R., Kerr, J., Lai, P.-C., Mitáš, J., Reis, R., Sarmiento, O. L., Schofield, G., Troelsen, J., Van Dyck, D., … Owen, N. (2016). Physical activity in relation to urban environments in 14 cities worldwide: A cross-sectional study. *The Lancet, 387*, 2207–2217.

Satterthwaite, D., Archer, D., Colenbrander, S., Dodman, D., Hardoy, J., & Patel, S. (2018). *Responding to climate change in cities and in their informal settlements and economies*. International Institute for Environment and Development.

Schultz, J. M. (2019). Disasters. In S. Galea, C. K. Ettman, & D. Vlahov (Eds.), *Urban health* (pp. 156–165). Oxford University Press.

Silveira, D. C., Carmo, R. F., & Luz, Z. M. P. (2019). Planning in four areas of the Vila Viva Program in the city of Belo Horizonte, Brazil: A documentary analysis. *Ciência & Saúde Coletiva, 24*, 1165–1174.

Stanaway, J. D., Afshin, A., Gakidou, E., Lim, S. S., Abate, D., Abate, K. H., Abbafati, C., Abbasi, N., Abbastabar, H., Abd-Allah, F., Abdela, J., Abdelalim, A., Abdollahpour, I., Abdulkader, R. S., Abebe, M., Abebe, Z., Abera, S. F., Abil, O. Z., Abraha, H. N., … Murray, C. J. L. (2018). Global, regional, and national comparative risk assessment of 84 behavioural, environmental and occupational, and metabolic risks or clusters of risks for 195 countries and territories, 1990–2017: A systematic analysis for the Global Burden of Disease Study 2017. *The Lancet, 392*, 1923–1994.

Stigsdotter, U. K., Palsdottir, A. M., Burls, A., Chermaz, A., Ferrini, F., & Grahn, P. (2011). Nature-based therapeutic interventions. In K. Nilsson, M. Sangster, C. Gallis, T. Hartig, S. de Vries, K. Seeland, & J. Schipperijn (Eds.), *Forests, trees and human health* (pp. 309–342). Springer.

Tang, Y.-T., Chan, F. K. S., O'Donnell, E. C., Griffiths, J., Lau, L., Higgitt, D. L., & Thorne, C. R. (2018). Aligning ancient and modern approaches to sustainable urban water management in China: Ningbo as a "Blue-Green City" in the "Sponge City" campaign. *Journal of Flood Risk Management, 11*, e12451.

Twohig-Bennett, C., & Jones, A. (2018). The health benefits of the great outdoors: A systematic review and meta-analysis of greenspace exposure and health outcomes. *Environmental Research, 166*, 628–637.

United Nations, Department of Economic and Social Affairs, Population Division. (2019a). World urbanization prospects: The 2018 revision, File 3: Urban Population at Mid-Year by Region, Subregion, Country and Area, 1950–2050 (thousands).

United Nations, Department of Economic and Social Affairs, Population Division. (2019b). World population prospects 2019: Highlights (ST/ESA/SER.A/423).

van den Bosch, M., & Ode Sang, Å. (2017). Urban natural environments as nature-based solutions for improved public health—A systematic review of reviews. *Environmental Research, 158*, 373–384.

Vaughan, A. (2020). Landmark ruling says air pollution contributed to death of 9-year-old. *New Scientist*. December 16, 2020. Retrieved from https://www.newscientist.com/article/2263165-landmark-ruling-says-air-pollution-contributed-to-death-of-9-year-old/

Wang, L., Zhang, F., Pilot, E., Yu, J., Nie, C., Holdaway, J., Yang, L., Li, Y., Wang, W., Vardoulakis, S., & Krafft, T. (2018). Taking action on air pollution control in the Beijing-Tianjin-Hebei (BTH) region: Progress, challenges and opportunities. *International Journal of Environmental Research and Public Health, 15*, 306.

WHO. (2013). *Review of evidence on health aspects of air pollution—REVIHAAP project: Technical report*. WHO Regional Office for Europe.

WHO. (2016). *Waste and human health: Evidence and needs*. WHO Meeting Report, 5–6 November 2015, Bonn, Germany. WHO Regional Office for Europe, Copenhagen, Denmark.

WHO. (2018a). *Global status report on road safety*. WHO, Geneva, Switzerland.

WHO. (2018b). *WHO housing and health guidelines*. WHO, Geneva, Switzerland.

WHO. (2018c). *Air pollution and child health: Prescribing clean air*. WHO, Geneva, Switzerland.

WHO. (2020). *Personal interventions and risk communication on air pollution*. WHO, Geneva, Switzerland.

WHO. (2021a). *WHO global air quality guidelines: Particulate matter (PM2.5 and PM10), ozone, nitrogen dioxide, sulfur dioxide and carbon monoxide*. WHO, Geneva, Switzerland.

WHO. (2021b). *Policies, regulations and legislation promoting healthy housing: A review*. WHO, Geneva, Switzerland.

WHO, United Nations Children's Fund. (2017). *Progress on drinking water, sanitation and hygiene. 2017 update and SDG baseline*. WHO and United Nations Children's Fund, Geneva, Switzerland.

WHO, United Nations Human Settlements Programme (Eds.). (2010). *Hidden cities: Unmasking and overcoming health inequities in urban settings*. WHO; UN-HABITAT, Kobe, Japan.

Yolande, M., Kerry, F., Paul, H., Louise, U., Sian, M., & George, K. (2014). Contaminated land and public health. In N. Bradley, H. Harrison, G. Hodgson, R. Kamanyire, A. Kibble, & V. Murray (Eds.), *Essentials of environmental public health science* (pp. 115–145). Oxford University Press.

Chapter 6
Local Health: Neighbourhood Scale

6.1 Introduction

The activities of daily life—travelling to work, collecting children from school and picking up food for dinner—have a much bigger impact on health and wellbeing than most people realise. Depending on where one lives in the city, these activities will feel completely different, and this is likely to vary by a neighbourhood's deprivation status. Residents of more affluent neighbourhoods may experience pleasant journeys through well-connected tree-lined streets, stopping to chat with neighbours, while residents from a poorer area may experience a harder landscape—noisier and greyer—where walking and lingering in public places is not enjoyable and may be considered dangerous, either from road traffic or from crime. These features of neighbourhood life matter for health because they influence social connection, feelings of belonging, access to other parts of the city and physical activity.

Inclusive urban design is one of the key themes of this chapter which focuses on how people experience cities at the level of streets and public spaces. It has become increasingly apparent that these spaces have been traditionally designed without adequate consideration of the diversity of needs and uses in public space. A common example is found in pavements. Able-bodied designers and engineers with no caring responsibilities may have little understanding of the importance of ramps and unobstructed pavements because they can easily navigate these obstacles. Whereas people with mobility impairments or pushing a pram with toddlers in tow will instantly recognise the daily struggle of navigating the city when surface levels change, pavements suddenly disappear or street furniture blocks the path. Improving population health requires design teams to create neighbourhoods that support everybody to conduct their daily lives safely and conveniently without risk of injury or personal harm. Two functions of the neighbourhood environment are frequently mentioned in this chapter due to their important role in health and wellbeing—physical activity and social interaction.

H. Pineo, *Healthy Urbanism*, Planning, Environment, Cities,
https://doi.org/10.1007/978-981-16-9647-3_6

6.1.1 *Physical Activity Through Neighbourhood Design*

The previous chapter described the role of land use, residential density and mobility infrastructure in supporting active travel. This chapter will focus more on the streetscape, public space and services that support physical activity. Being active is important for people of all ages and abilities to protect against disease, including non-communicable and infectious diseases, and mental illness (Fontaine, 2000; Lee et al., 2012; Chastin et al., 2021), while its role in maintaining a healthy weight is less clear (see Box 6.1). Physical inactivity or sedentary behaviour is considered an epidemic in many countries with the WHO (2018a) reporting that globally a quarter of adults and three-quarters of adolescents (aged 11–17 years) do not meet their activity recommendations. Economic development leads to changes in transport, rapid urbanisation and other shifts that have resulted in inactivity levels reaching 70% of the population in some countries (ibid.). The health and wider economic costs of physical inactivity are substantial. For example in the UK, Public Health England reports that physical inactivity contributes to one in six deaths and costs

Box 6.1 Reducing obesity through the built environment
Obesity is a global health and economic problem linked to the changing ways that people live and work. Being obese is a key risk factor for non-communicable diseases, leading to reduced quality of life and premature death. In recognition of the health risks and changing behaviours in society, health agencies and charities have started talking about obesity as 'the new smoking'. This framing of obesity is underscored by data in the UK where the number of obese people outnumber smokers two to one (Cancer Research UK, 2019). The energy balance between calorie intake (diet) and expenditure (physical activity) is the prime focus for obesity research and policy; however, other causal factors may be important, including psychosocial stress and sleep (Ross et al., 2016). Design and planning decisions related to urban form and fast food availability have the strongest evidence to reduce obesity; however, it should be noted that there are shortcomings with the evidence base due to the complex social, environmental, economic and cultural factors affecting obesity, meaning other factors may also be important. An umbrella review of associations between the built environment and obesity showed that fast food exposure, urban sprawl and land use mix are important factors (Lam et al., 2021). Another review found that walkable environments, presence of recreation facilities and nearby shops and services help to increase physical activity, while land use mix/sprawl, aesthetics, overall food environment and availability of parks/playgrounds are supportive of healthy weight (Dixon et al., 2020). If an improvement to the built environment specifically aims to reduce obesity, it may be stronger if it is coupled with behavioural interventions such as park running programmes (Wilkie et al., 2018).

£7.4 billion to business and wider society every year (Petrokofsky & Davis, 2016). There are many co-benefits of increasing physical activity such as job growth, increased productivity, reduced crime and better social cohesion.

Physical activity is about expending energy through bodily movement and it can be achieved through many daily activities, such as walking and cleaning, alongside sports and recreation. Built environment professionals should consider ways to integrate activity in building design, walkable neighbourhoods and provision of varied sports facilities. Establishing good physical activity behaviours in childhood, such as walking or cycling to school, helps to ensure that people will remain active throughout their lives. Being in the habit of walking and cycling for transport can be a substantial part of achieving one's physical activity requirements later in life. Research in England found that at least a quarter of respondents under age 45 met their activity recommendations only from active travel (Brainard et al., 2019). The Bullitt Center case study in Chap. 4 showed that staircase design can support physical activity within buildings. This chapter will describe the importance of services, perceived and actual personal safety and public spaces in supporting physical activity as a part of day-to-day life. The WHO (2018a) highlights the importance of approaching this topic with an equity and inclusion lens because 'girls, women, older adults, people of low socioeconomic position, people with disabilities and chronic diseases, marginalized populations, indigenous people and the inhabitants of rural communities often have less access to safe, accessible, affordable and appropriate spaces and places in which to be physically active' (p. 15). The section on personal safety (both perceived and actual) will explore these inequities in more depth.

6.1.2 Social Interaction Through Neighbourhood Design

The second key function of the neighbourhood environment for supporting health and wellbeing is social interaction, which can refer to chance encounters between neighbours, planned visits with friends or family and connections that are either in-person or remote (e.g. phone calls). Regular interaction leads to social relationships within a community that are important for resilience and health. For instance, following a burst water pipe or power outage, residents may look after older neighbours by taking them drinking water or fuel. Chapter 1 described the tragic case of Hurricane Sandy in New York, where more than 100 people died from the storm, many of whom were elderly and not known to the emergency services as being at risk. This breakdown in community resilience is linked to the worrying rise in many cities of social isolation and loneliness. Social isolation refers to a low number of social relations and a denuded social network, while loneliness is a subjective negative perception of the quality of social relations (Yang & Victor, 2011). The health and quality of life impacts of loneliness and isolation include heart disease, depression, dementia and suicide (ibid.).

Table 6.1 Summary of three place-related wellbeing constructs for older people developed by Burton et al. (2011)

Construct	Description
Functional	Perceptions of independence, safety (from traffic and non-motorised transport), noise and air quality. Incidence of falling outside.
Social	Perceptions of community spirit and safety from crime (before and after dark). Extent of social interaction.
Emotional	Self-rated quality of life and enjoyment of trips in the neighbourhood. Perceived attractiveness and satisfaction with the neighbourhood.

The ageing population is one driver of rising social isolation in cities, but any age group can be affected, such as young mothers or disabled people, and neighbourhood design can be used to reduce this isolation. The work of Elizabeth Burton and Lynne Mitchell has been influential in informing the design of environments that support health and wellbeing for older people, and for social interaction more generally. Mitchell et al. (2003) synthesised literature on dementia-friendly environments and developed design principles to ensure that outdoor environments are familiar, legible, distinctive, accessible, comfortable and safe. Burton et al. (2011) used focus groups with older people and scientific studies to develop place-related wellbeing constructs for older people that fall into three categories: functional, social and emotional. Table 6.1 describes each of these constructs and the associated wellbeing measures and perceptions of environmental characteristics that influence each. The perception of being safe from traffic and crime is important for the functional and social constructs. Design teams should refer to these constructs to create places that support social interaction and wellbeing, particularly for older people.

Designing for social interaction across all age groups overlaps heavily with the measures used to increase physical activity in cities. Social isolation and connection are affected by the following: residential density; mixed land use; street layout and design; transition between public/private space; environmental cues for crime and safety; greenspace; public transport; and local facilities for leisure and recreation (including cafés, pubs, religious facilities, etc.). Community severance (barriers between communities, e.g. a major road) reduces social interaction and access to key services and can have multiple health impacts (Mindell & Karlsen, 2012). These topics will be discussed in more detail in the next section.

This chapter explores how urban form and design at the district and neighbourhood scale affects health, focusing on access to services, safety, culture, public space and food. The planning and design goals that align with 'Local Health' in the THRIVES framework are aligned with neighbourhood and building scales of decision-making—explained across this chapter and the next. In reality, the policies and decisions to achieve these goals are made by public and private sector actors working at multiple scales, from national policy-makers to housebuilders. Nevertheless, it is conceptually useful to distinguish primary scales of health impact and decision-making, which are not always perfectly aligned.

6.2 Achieving Local Health Framework Goals in Neighbourhoods

The importance of the neighbourhood unit for health was encompassed in twentieth-century planning and design strategies. Corburn (2013) describes how Clarence Perry's Neighbourhood Unit influenced the American Public Health Association's reports on healthy housing and neighbourhoods in 1938 and 1948, respectively. Shaping neighbourhoods requires consideration of how they connect up and down in spatial scales to the wider city and to individual blocks and buildings. Following the lockdowns of many cities during COVID-19, the concept of restructuring cities into a series of connected liveable and sustainable hubs grew in prominence, often under the banner of the so-called 15-minute city (see Chap. 2), but this could be seen as a new label for more widely practised urban design principles. Black and Sonbli (2019) have distilled principles of good urban design which can be applied at multiple scales: the city/region, the neighbourhood/district ('the scale that humans can relate to comfortably') and the site/block ('the "human scale"...the strongest social unit of the city') (pp. 22–23). They group the principles of urban design into three themes: arrangements, networks and features. The latter is about the way that places look and feel, as summarised in Box 6.2. Through the successful use of these principles, urban design can create places that are sustainable, 'inclusive, equitable, resilient and healthy' (ibid., p. 27). The features should be considered throughout the following sub-sections of this chapter which constitute each of the design and planning goals in THRIVES at the neighbourhood scale.

Box 6.2 Ten features of good urban design

The following ten features of good urban design have been drawn from Black and Sonbli's (2019, pp. 27–35) book, *The Urban Design Process*, which builds on their research and prior literature, including the work of Jane Jacobs, Jan Gehl, Kevin Lynch and Christopher Alexander.

1. *Places for people*: Places of all scales and types are designed for the needs of 'each and every citizen equally' and they promote activity and relationships (p. 27).
2. *Character*: Places have their own identity, they are 'distinctive and memorable' because they reflect the history and culture of the community (p. 28).
3. *Mixed uses*: Places are 'enjoyable and useful' when they have a mixture of densities, typologies and uses, further supporting identity and community (p. 28).
4. *Continuity and enclosure*: Public and private spaces should be well-defined through 'animated edges and a sense of enclosure' which is achieved through 'continuity of building lines, frontages, and considered, consistent setbacks', creating a buffer between public and private spaces (p. 29).

(continued)

Box 6.2 continued

5. *Quality of the public realm*: Public spaces should be 'attractive, visible and well-used' with clear routes connecting public spaces that are 'safe, legible, and designed for all', allowing the public realm to be the 'lifeblood of the city' (p. 30).
6. *Legibility and transparency*: Places are easy to understand and move around when there are 'recognisable paths (routes), strategic nodes, and well-considered landmarks to assist people' and avoid feelings of 'confusion or anxiety' (p. 31).
7. *Adaptability*: Places are flexible and able to respond to 'new social, market or environmental demands', including through the use of technology (p. 32).
8. *Diversity/complexity*: Places support the diversity of the human condition by providing choice through 'a range of architecture, a variety of landscape and well-considered complexity that strikes a balance between the mundane and the confusing' (p. 33).
9. *Nature and landscape*: Places that integrate nature 'can impact positively on quality of life, both physical and mental' (p. 34).
10. *Human-scaled*: The core principle, underpinning all of the above, is that 'all design should be human-scaled' with people 'remaining central to all aspects' (pp. 34–35).

6.2.1 Services

Services is a catch-all term in THRIVES for the infrastructure (e.g. institutions and facilities) that serve society in terms of places for work, education, healthcare, sports, banking, retail and other functions. The focus here is on the services (sometimes called facilities or amenities) that affect health primarily through social and economic mechanisms and physical activity, such as childcare, education, employment, healthcare, sports and play (Table 6.2). These services support physical and mental health as outlined in Table 6.2 (this is not a comprehensive list, cultural and food-related services are discussed later in this chapter). They can also affect health across the life course, creating the foundations for health or ill-health throughout people's lives in terms of knowledge, confidence, skills and cognitive development.

The goal of infrastructure planning is to coordinate the delivery and location of services to meet the needs of existing and new populations in an integrated and timely fashion, yet this is a major challenge for planners around the world due to the complex nature of the task. Considering the multiple pathways between services and health outlined in Table 6.2 and elsewhere in THRIVES, built environment professionals should aim to promote health through infrastructure planning beyond funding healthcare services. The location and provision of services influences whether people can conveniently access work, school and other places that they require on a daily basis, as discussed earlier in relation to compact and complete

Table 6.2 List of selected urban services and their relevance for health and wellbeing with sources of further information

Service	Links to health and wellbeing	Reference for further information
Childcare (~0–5 years)	High-quality public or private childcare can support parents (especially mothers) to work and benefit child development.	Behbehani et al. (2019), Costa et al. (2019)
Education	Supporting health via (1) leading to employment (e.g. income and sick leave), (2) psycho-social environment (e.g. social standing and support) and (3) health-related knowledge (e.g. diet and exercise).	Hahn and Truman (2015)
Employment	Generally beneficial for health and reduces risk of depression. Other positive and negative effects on health relate to job quality. High-quality jobs support financial security, purpose in life, social interaction and so forth. Low quality jobs may have harmful exposures (e.g. noise pollution), job precarity, risk of injury and stress.	van der Noordt et al. (2014), Utzet et al. (2020)
Healthcare	Access to affordable healthcare is a basic human right. Universal health coverage (UHC) means that everybody can access quality health services without financial hardship. UHC impacts population health, productivity, education and poverty.	WHO (2016b)
Sports	Participation in group and individual sports leads to increased self-esteem and social interaction, whilst reducing depression. The physical activity achieved in sports leads to multiple physical health benefits (depending on the sport), such as better cardiovascular function.	Eime et al. (2013), Oja et al. (2015), Oliveira et al. (2017)
Play	Outdoor play for children and adolescents is important for physical activity; motor, visual, and cognitive development; socio-emotional learning and mental health.	Lambert et al. (2019)

neighbourhoods. To serve dense urban populations, some services are located in close proximity to where people live, such as primary schools, post offices and sports pitches, and there are typically threshold population levels that are used to determine their location. Large cultural and civic institutions are likely to be centrally located, such as museums, theatres and city halls, creating the vibrant urban centres that attract many people to live in a city. The nature of urban services in terms of type, location, provision and quality will depend on many factors, such as whether they are coordinated by government activities (e.g. through spatial planning), how they are funded (e.g. tax or private investment) and political processes. Regrettably, the nature of many infrastructure delivery systems results in significant spatial inequities in service provision and quality within cities. Although this is sometimes called a 'postcode lottery', it is not a matter of chance. Low-income and minority ethnic groups are systematically disadvantaged in terms of service delivery and this directly harms their health and wellbeing (WHO, 2016a). Another equity and inclusion component to services relates to the provision and protection of places for specific communities, such as faith groups and LGBTIQ+ communities. In response to research demonstrating a significant decrease in LGBT+ venues in London, Planning Out (2019) offers guidance on using the planning system to protect existing venues (from closure or redevelopment) and run events.

Focusing on the service of education and the role of schools in wider community health serves as an example to highlight the interconnected nature of service planning. Buildings can be used for multiple purposes and the co-location of services can benefit residents in terms of convenient access. Education is a public service that significantly influences health and often varies in quality in different parts of a city. Hahn and Truman (2015) conceptualise education as a process and a product that affects health throughout the life course. Through the *process* of education, children gain 'knowledge, skills of reasoning, values, socio-emotional awareness and control, and social interaction' and the *product* of education can be seen as the 'array of knowledge, skills, and capacities' gained through formal and informal learning that are 'an attribute of a person' (ibid., p. 658). This conceptualisation of education leads to health through three main factors: employment (e.g. income and sick leave), psycho-social environment (e.g. social standing and support) and health-related knowledge (e.g. diet and exercise). In their review of the evidence linking health and education, Hahn and Truman found that high educational attainment is associated with higher income, better self-assessed health, reduced rates of several diseases and psychological conditions and better health behaviours. The provision of early years' education can have lasting positive health impacts throughout one's life (see Table 1.2). Schools also perform important health-related functions within communities, such as meals for low-income children. There are complex interconnections between the quality of schools and neighbourhood socio-economic status that have a direct influence on children's health and wellbeing.

Beyond the provision of education within schools, public and private school buildings can be used for other purposes or multiple public services can be located near to each other to increase efficiency and convenient access. For example, school facilities can be jointly used by the community for sports, clubs, farmer's markets and more. Jeffrey Vincent (2014) explored the mechanisms through which schools could be part of sustainable and healthy communities in California, USA. Most examples were achieved through partnership between various local authorities (e.g. school districts, housing authorities and parks and recreation departments) and they required investment. In one example, Los Angeles was home to one of the largest school-based health clinics in the country. A simpler case was from Berkeley, where the school district had a policy to leave gates open for unstructured community use of facilities, such as play equipment. Vincent argues that joint use models are an efficient use of public funds and land, yet he identified concerns and obstacles that require partners to build trust and relationships. Other public and privately run buildings are suitable for multiple uses, such as town halls and places of worship. The opportunity to co-locate multiple health and community services offers many benefits for local communities that are further discussed in the Barton Park case study. Chapter 8 describes diverse international models for public service and infrastructure delivery, particularly focusing on funds produced through the planning system.

6.2.2 (Perceived) Safety

Residents' perceptions of safety are important for health and wellbeing. Feeling safe may even outweigh actual safety risks in certain contexts and population groups by inhibiting people from living their daily life as they would like, for example, where the fear of crime is disproportionate to actual crime. In the UK and USA, rising rates of childhood obesity have coincided with reductions in physical activity and increased restrictions on children's ability to roam free in their neighbourhood environment without supervision. Parents' shifting perceptions of safety are partly driving this change (see Box 6.5). For older people, the perceived safety of the neighbourhood environment can stop them from leaving their homes, increasing their social isolation and decreasing their physical activity. Women's daily lives are influenced by perceived safety in terms of violence and crime, which can be influenced by urban design, such as avoiding narrow footpaths where visibility is constrained (Fig. 6.1). Creating healthy cities means understanding the safety-related needs of different groups within society and adjusting the built environment to provide greater security so that everybody can participate in day-to-day life.

The built environment can be designed to increase perceptions of safety in terms of traffic, crime and personal safety, but perceptions and design requirements will vary among people of different ages, physical abilities, ethnic backgrounds and other characteristics. A growing body of research is revealing different perceptions of safety for specific groups and contexts. Safety from crime, and particularly

Fig. 6.1 Examples of footpaths where women and girls may feel unsafe because visibility is constrained. Street lighting supports perceived safety. (Source: Author, 2021)

perceived safety, has been repeatedly associated with increased physical activity and walking in older people (Barnett et al., 2017) and sidewalk maintenance is a key element for wider neighbourhood safety perceptions among the elderly (Won et al., 2016). Nicole Fiscella et al. (2021) investigated physical activity engagement among children (6–17 years) with Autism Spectrum Disorder (ASD) and found that feeling safe in the neighbourhood was a significant predictor of their activity levels. They suggested that children with ASD may interact differently with the built environment than other children and that hyper- or hyposensitivity to sensory stimuli may be an important component of greenspace design to enable the use of such spaces for exercise. Lisa Groshong et al. (2020) examined safety perceptions among low-income and minority ethnic participants in Kansas City, Missouri (USA) positioning this work in the environmental justice literature due to the inequities in park quality and quantity across cities. They found that both social and structural elements contributed to perceived safety among participants. Structural elements included constraints such as traffic, litter and poorly maintained spaces, and facilitators such as well-maintained trees and grass. They recommended partnership working across planners, parks and recreation professionals, community groups and the police (among others) to develop locally appropriate parks solutions. One example was the 'Parks After Dark' project in Los Angeles which supports adolescents to use parks on summer evenings as a safe space.

It has become increasingly problematic that changing perceptions of safety have led to reduced opportunities for children's outdoor 'risky' play, resulting in negative impacts for physical activity and health. This dilemma is highlighted by Mariana Brussoni et al.'s (2015) systematic review which defines risky play as 'thrilling and exciting play that can include the possibility of physical injury' such as playing at heights, at high speeds and in unsupervised places (p. 6425). This type of play has been associated with improved self-esteem, motor skills, independence, risk management and other health and social benefits for children. Brussoni et al. describe how the majority of playground injuries in places like Canada and the USA tend to be very minor (not requiring medical attention), yet there have been many safety changes to playground equipment (e.g. installation of rubber ground surfaces) which have not reduced injuries, but they have made playgrounds less enticing, thereby causing an unintended negative consequence for health. Perhaps some of the most striking research in this area shows the substantial generational changes in unsupervised outdoor play in the streets around children's homes. Brussoni et al. cited numerous studies in England and the USA (e.g. Gaster, 1991; Clements, 2004; Natural England, 2009) documenting how adults report substantial differences with their own childhood outdoor unsupervised play compared to younger generations. These shifts relate to parents' perceptions of danger from traffic and abduction (so-called stranger danger) among other factors (see Box 6.3). Brussoni et al.'s review found that unsupervised outdoor play leads to increased physical activity and social health, thus there is a need to find acceptable solutions for outdoor risky play. Urban play space best practice resources provide guidance for creating places to play (not only in playgrounds but integrated into other public spaces) that will meet the needs of different ages, genders and abilities (Gill, 2021). Specific built environment

Box 6.3 The 'social trap' of perceived danger for children
Safety is a complex construct that can be best understood using systems thinking. Focusing specifically on children's health, parents' perceived safety of the outdoor environment is an important factor. Parents' perceptions of neighbourhood traffic safety have been positively associated with physical activity in adolescents (Esteban-Cornejo et al., 2016) and low perceived road traffic safety has been associated with reductions in children's and adults' physical activity (Powell-Wiley et al., 2013). A review by Carver et al. (2008) highlights the diversity of perspectives about safety between parents and children of different ages and genders. For example, they describe a Scottish study (Burman et al., 2000) showing that 58% of surveyed girls aged 13 to 16 years old were worried about being sexually assaulted. Carver et al. also discuss an Australian 1998 report by the Criminology Research Council that 'highlights a "risk-victimisation paradox" whereby parents are overly anxious about their children's safety and exaggerate the risk of "stranger danger"' (Carver et al. p. 219), which can affect children's perceptions of risk. From a systems thinking perspective, a reinforcing feedback loop can be identified whereby fear of traffic injury or stranger danger leads to restricted independent time outside, which increases other parents' perceptions of danger. This phenomenon has been termed a 'social trap' that could be solved by increasing social interaction in neighbourhoods (ibid.). Noonan (2021) investigated the social trap of UK parents' safety fears about travelling to school. He found that more affluent families were more likely to be driven to school than less affluent families. The 2019 National Travel Survey in England showed that although 44% of children aged 5 to 16 walk to school, 35% were driven in cars/vans (DfT, 2020), missing a potentially key opportunity to increase daily physical activity among children.

factors that are associated with children's and adolescent's outdoor play include lower traffic volumes (ages 6–11), garden/yard access (ages 3–10) and increased neighbourhood greenness (ages 2–15) (Lambert et al., 2019). During the COVID-19 pandemic, cities around the world looked for solutions to create safe residential streets, such as Los Angeles 'Slow Streets' and London's Low Traffic Neighbourhoods (LTNs) (Fig. 6.2). In London, the LTNs were rapidly implemented as temporary measures that would be monitored and evaluated before becoming permanent (also see Chap. 9).

Practical evidence-based measures to improve safety and perceived safety in the built environment can be applied at multiple scales and they should be informed by inclusive design processes. According to the health-related research on this topic, increasing perceived safety is achieved by the following: segregating motor vehicles from cycling and pedestrian infrastructure; ensuring continuity of segregated routes, calming traffic through street design, for example, road humps and narrowing, and reduced speed limits, ideally to 20 mph; providing passive surveillance in public

Fig. 6.2 Temporary Low Traffic Neighbourhood (LTN) infrastructure in Hackney, London (UK), installed during the COVID-19 pandemic allows passage for emergency services, whilst restricting entrance for other motor vehicles. (Source: Author, 2021)

spaces, maintaining walkways and the wider public realm; and providing street lighting. Carmona (2021) provides a comprehensive discussion of design approaches to improve actual and perceived safety from multiple threats in urban spaces, including the widely adopted Crime Prevention Through Environmental Design (CPTED) method. Carmona discusses a number of tensions between measures to deter crime or undesirable behaviour with health-related outcomes. For instance, anti-skateboarding design would reduce opportunities for adolescents to gather for social interaction and physical activity. When approaching urbanism through a health lens, recommended methods to reduce crime should be carefully scrutinised for unintended consequences, especially considering under-represented population groups.

6.2.3 Culture

In comparison with other topics in this book, culture receives less attention in the healthy place-making literature, yet it is undoubtedly important for health and well-being. Cultural factors influence health-related behaviours such as diet, physical activity and social interaction. The UN Educational, Scientific and Cultural

Organisation defines culture as 'the set of distinctive spiritual, material, intellectual and emotional features of society or a social group, and that it encompasses, in addition to art and literature, lifestyles, ways of living together, value systems, traditions and beliefs' (UNESCO, 2001). The WHO has taken steps to increase policy-makers and public health professionals' understanding of the importance of culture in health and wellbeing through a policy briefing by David Napier et al. (2017). The briefing calls for actions including: consideration of culture in health impact assessment (taking into account the diverse health needs and impacts within a society); redressing biases in the evidence base that do not account for cultural diversity; taking people-centred approaches to policy-making; and adopting multi-disciplinary and whole-of-society approaches to health governance. This report adds helpful weight to the increasing calls for built environment professionals to consider cultural diversity in planning and design. This imperative is reinforced by the authors' point that 'culture is not a static set of beliefs and practices, but rather an ever-emerging array of collective values, ethics, assumptions and ideals' (p.viii) and that it should not be conflated with race or ethnicity. Culture should be considered as part of the THRIVES core principle of inclusion in healthy urbanism. However, it is also listed as a policy and design goal to reflect the need for culture to be supported through specific urban places, such as those related to faith, art, music, food and other practices.

Supporting culture through place-making could involve community art installations, creating public space for cultural activities, finding creative financing mechanisms or delivering buildings for diverse cultural purposes. An inclusive design and planning process, such as community-based asset mapping, can be used to understand residents' needs for urban places and culture. Parks and other public spaces are important urban settings for cultural activities which could be directly related to health, such as the nature-based and mind-body interventions studied by Pretty and Barton (2020), described in Chap. 5. Efforts should be made to understand and reduce any conflicts arising from diverse uses of public space for cultural purposes, such as noise from loud music. There is also research examining the role of other urban spaces in supporting health through cultural activities. For example, places of worship (or faith-based organisations) have been found to be successful settings for health interventions, such as diabetes education for Black Americans (Newlin et al., 2012).

One example of neighbourhood scale urban development which sought to support the diverse cultures of existing residents is the Mariposa project in Denver, Colorado (USA). Mariposa is a master-planned development of mixed-income housing, senior housing, retail and public space (e.g. playgrounds and gardens) located on the site of a former public housing project. Celebrating cultural diversity was a key part of the project's vision that was integrated throughout the design goals. For instance, Goal I of the master-plan was to 'provide amenities and site features that meet the needs of families and residents of different ages and cultures. Promote community interaction and active participation' (Mithun, 2010). Existing residents' needs were gathered through iterative and in-depth community participation. The development amenities listed on the Mariposa website clearly relate to diverse cultural offerings, including healthy affordable food, a non-profit youth arts education programme and an urban food initiative with a weekly farmer's market, among others.

Alysson Burbeck (2020) explored the role of murals depicting the Chicano/a residents in Denver place-making activities, looking back at the history of Emanuel Martinez and community members who produced public murals in the 1970s, and tracing this through to more recent evolutions. The original murals were a strong expression of local culture that has been continued through the Mariposa redevelopment, with a contemporary mural by JOLT referencing the strong women and butterflies of Martinez's murals alongside representation of new Somali residents in Mariposa. However, Burbeck also described another side to Denver murals which did not represent local culture and were instead deployed by 'city officials, urban planners, and corporate developers [who] have effectively co-opted street art practices in order to commodify the urban landscape' (ibid., p. 48). Her research is situated in a wider body of literature on gentrification and art that relates directly to health. Urban development should reflect local cultures and identities but if this results in displacement of those communities, it becomes a social justice and health problem. The Mariposa development was successful in providing urban improvements that are supportive of local cultures without displacing residents, however this was challenging for the project and it remains a key challenge for similar projects going forward (Joseph, 2019).

6.2.4 *Public Space*

Public space is often described as the space between buildings where civic life happens in the city. The importance of these places for health and wellbeing cannot be underestimated, but it can be harder to quantify than other urban components, such as pollution. Urban design scholars have written extensively on the value of public space for individual and communal benefits related to health including identity, belonging, social capital (see Box 6.4), sense of security and other factors. A comprehensive overview of designing quality public spaces is provided by Matthew Carmona's (2021) book *Public Spaces Urban Places* where he defines and distinguishes between public space, the public realm and public life. The public realm 'encompasses the sites and settings of formal and informal public life', including public spaces that are material or virtual (ibid., p. 326). He notes the complexity of the concept of public space with varying conceptualisations addressing issues of ownership, access and use. Public spaces are not necessarily in the ownership of public bodies but may be privately owned and freely accessible to the public, such as shopping malls and cafés, which are so-called third places. Carmona draws on his previous research to outline many different types of public spaces and their uses, which may cause positive or negative health effects. 'Negative' public spaces, for instance, include those for motorised transport, such as underpasses and motorways, which may cause noise and air pollution in addition to feeling or being unsafe.

Box 6.4 Social capital and public space

Social capital is one of the key proposed health-related benefits of public spaces, particularly parks. Ellen Bassett (2018) reviewed the evidence on social capital and public space, looking back to the work of Frederick Law Olmsted who sought to create social cohesion and interaction across societal groups in the design of parks, such as Central Park in New York City. Bassett defines social capital as 'the features of social organization, such as civic participation, norms of reciprocity, and trust in others, which facilitate cooperation for mutual benefit' (ibid., p. 63). Studies have identified health and equity benefits of social cohesion in countries around the world, some focusing on the particular importance of social capital for women's health. These benefits are achieved from public spaces through their role in supporting community connections. Looking specifically at small public spaces, Bassett's review found that health is supported by spaces that are good quality, in close proximity to residents and well-maintained. These features are regularly described as being important for quality public spaces.

There is no blueprint for designing high-quality public spaces, Carmona argues, instead designers should aim to meet four idealised yet rarely achieved qualities of the public realm, which he summarises from Tiesdell and Oc (1998) as follows:

1. 'Universal access (open to all).
2. Neutral territory (free from coercive forces).
3. Inclusive and pluralist (accepting and accommodating difference).
4. Symbolic and representative of the collective and of sociability (rather than individual privacy)' (Carmona, 2021, p. 327).

This set of qualities can be readily aligned to the core principles of healthy urbanism, particularly equity and inclusion. The 'Public Space' goal in the THRIVES framework can therefore be interpreted as an imperative to provide public spaces with the qualities outlined earlier that are equitably distributed throughout a city. To support equity, urban planners and designers should use inclusive and participatory processes (see Chap. 8) to map the existing provision and quality of public space across cities and within neighbourhoods and plan to remedy gaps. To support inclusion, design professionals should work with a range of residents to understand how well existing or proposed spaces support their needs.

It is important for public space design to start with exploration of the people who will use the space and its proposed function(s). The output of an inclusive design process should not exclude people because of their characteristics such as age, race/ethnicity, gender, disability and others. Unfortunately, many public spaces do not meet this requirement, instead they exclude or make life inconvenient for large portions of society. Sometimes this exclusion is done intentionally, such as the use of pointed paving stones to deter unhoused people from sleeping around building entrances. Agyeman (2013) summarises work by Madanipour (2010) regarding the history of exclusion in public spaces: 'women, foreigners, and slaves had no place in the Greek Agora, and intergroup interactions in the mediaeval Square were dominated by social rules and hierarchies' (Agyeman, 2013, p. 141). Exclusion from public spaces today is the result of decades, if not centuries, of design practices based on the needs of able-bodied adult men. Many resources are becoming available to help professionals rethink their assumptions about design and planning through the lens of inclusion. For instance, the *Inclusive Healthy Places* resource by Gehl Institute (2018) provides a conceptual framework for inclusion and health in public space with four guiding principles related to understanding context, inclusive processes, health equity and social resilience. In addition to changing professional practices, there is also a need to increase diversity within built environment professions. The remainder of this section highlights public space design needs of specific population groups, highlighting existing resources and practical examples.

Women and girls have been under-represented in urban design and planning with consequential negative health impacts. This problem is referenced throughout this book, but it merits further attention in this section on public space, where lack of inclusive design is acutely felt. Urban spaces can exclude women and girls at different life stages, such as the following: feeling or being unsafe in spaces that are not overlooked or there is no clear way of escaping attack (Fig. 6.1), being unable to rest on a bench or access a toilet during pregnancy or menstruation, working hard to navigate the city with a stroller during motherhood (Kern, 2020). In its *Handbook for Gender-Inclusive Urban Planning and Design*, the World Bank (2020) states that generally 'cities work better for heterosexual, able-bodied, cisgender men than they do for women, girls, sexual and gender minorities, and people with disabilities' (p. 8). The authors explain how European historical practices of urban planning and design reinforced patriarchal gender norms where men left the home for work and women stayed behind for caregiving. These norms were then perpetuated through colonialism and displaced local practices, such as matrilineal inheritance and collective land ownership, in countries such as Kenya and Australia. Feminist scholars began challenging these norms in the 1970s and urban planners and designers in cities such as Vienna, Austria and Toronto, Canada spearheaded the idea of 'gender mainstreaming', which is 'a process that systematically integrates gender perspectives into legislation, public policies, programs, and projects' (ibid., p. 17). Designing public spaces that are supportive of different genders requires greater community participation in design and planning processes, increased recognition of gender-related design needs in educational curricula and greater diversity on design and planning teams. The resources in Box 6.5 support gender mainstreaming in urban design and planning.

Public spaces need to meet the design needs of different age groups to support physical activity and social interaction across the life course. As discussed earlier, public health studies about healthy urban environments for children have established positive health benefits for children from provision of road safety and active travel measures and playground equipment (Audrey & Batista-Ferrer, 2015). The WHO Global Network of Age-Friendly Cities and Communities has developed

Box 6.5 Guidance on designing public spaces for women and girls
An online resource by UN Habitat called *Her City* aims to support cities with involving girls in urban planning. The online tool guides users through nine 'flexible building blocks' covering: stakeholder engagement, city wide assessment, site specific assessment, analysing challenges, designing ideas, recommendations for action, action plans, sharing results and finally, implementation and follow-up. https://hercity.unhabitat.org/ (accessed 19 May 2021)

The UK-based Make Space for Girls charity campaigns for facilities and public spaces (including parks and play equipment) for girls. They have produced case studies of European examples that highlight the following design interventions: 'better lighting; wider entrances to play areas; smaller, subdivided sports areas, or adding a second more open court; seating areas which are arranged in groups rather than lines; circular paths around the perimeter of the park; more swings; good quality toilets'. http://makespaceforgirls.co.uk/ (accessed 19 May 2021)

skills capacity and knowledge resources on this topic (WHO, 2007, 2018b). The network calls for fundamental shifts in the way older people's needs are considered, touching all aspects of urban policy from public spaces to housing and transportation. The elderly are increasingly active but they need safe mobility infrastructure, including benches for resting and public toilets (which are also important for children and women). Alidoust and Bosman's (2015) work on neighbourhood environments and the elderly highlights the potential importance of what Jan Gehl (1986) called 'soft edges'. Gehl proposed that 'soft edges' in the form of semi-private spaces between the public street and the private home, such as forecourts, front yards and porches, could be important for social interaction in residential environments. The size and potential uses of these spaces vary in different places, where some fronts of houses are dominated by parked cars and rubbish bins, whilst others may have vegetation and seating. Designing enough space for residents to have a semi-private space outside their home could support health and social capital by enabling residents to personalise their space and use it for play or social interaction in a way that they would not use a public space.

Designing public spaces for diverse races and ethnicities is important for health and wellbeing, yet this is one of the least developed areas in urban design and planning curricula. Agyeman (2013) states that built environment professionals should not assume that there are 'homogenous life experiences based upon racial or ethnic similarities' that translate into design preferences for public spaces (p. 143). He reviews multiple US studies on the diverse public space preferences and barriers reported among Asians, Latinos, African Americans, Whites and others, noting that fear or experience of discrimination and interracial tensions have adversely impacted minority residents' use of public spaces. Understanding and acting upon these diverse experiences and injustices require cultural competency. Agyeman and

Erickson (2012) describe cultural competency as 'the ability to work effectively with difference, in cross-cultural situations' (p. 362). One step better on the development continuum of this skill is cultural proficiency, meaning 'practice that *proactively* engages diversity and promotes intercultural relations' (ibid.). The five systemic elements of cultural competency are relevant beyond the topic of public space to the wider design and policy work of built environment (and other) professionals, they include:

> (1) valuing diversity, (2) the capacity for cultural self-assessment, (3) consciousness of the 'dynamics' of cultural interaction, (4) the institutionalization of cultural knowledge, and (5) the development of adaptations to service delivery based on understanding diversity inter- and intraculturally. (ibid., p. 362)

A rapid shift in knowledge and skills is required to put the principles of cultural competency into action. A highly useful resource for practical and up-to-date guidance is the US-based Project for Public Spaces. For instance, their *Playbook for Inclusive Placemaking* is available in four online articles, the first of which covers community process drawing upon Agyeman and Erickson's work on cultural competency (Peinhardt & Storring, 2019).

6.2.5 Food

Many people would recognise the importance of a nutritious diet that is low in fat, sugar and salt in maintaining overall health and wellbeing, but few would link this topic to urban design and planning. The epidemiological evidence about the health impact of access to food in the built environment is not straightforward; however, there is moderate evidence from an international review that neighbourhood food environments influence diet (Caspi et al., 2012) and the evidence base is growing. In the UK and the USA, reviews have highlighted dietary inequalities whereby fast-food outlets and other unhealthy food vendors are more prevalent in low-income and minority ethnic neighbourhoods, while healthy food is harder to access (Black et al., 2014; Kraft et al., 2020). Recent reviews have also highlighted that the food environment around schools affects children's diets (Townshend & Lake, 2017; Kraft et al., 2020). This body of research has termed geographical areas with reduced access to healthy food as 'food deserts' (Beaulac et al., 2009) while areas with higher concentrations of fast-food outlets have been called 'fat swamps' or 'food swamps' (Saunders et al., 2015). Alongside this focus on the environment and availability of foods with varying nutritional value, researchers have highlighted the complexity of shopping and food intake, noting that individual agency should not be discounted (Thompson et al., 2013), that planning policies alone cannot address the existing overabundance of unhealthy foods in some neighbourhoods (Rudge et al., 2013) and that culture is an important component of nutrition that should not be ignored (Napier et al., 2017).

Urban planning initiatives that support healthy diet could be grouped into two activities: (1) those which seek to increase healthy food availability and consumption and (2) those which try to restrict unhealthy food provision, particularly near schools and low-income communities. The former includes supporting community gardens, urban farms, farmer's markets and supermarkets that are accessible to all residents. Such support may be in the form of changing zoning legislation or using development impact fees to fund gardening projects. There is a growing body of evidence that community gardens, or gardening initiatives in specific settings such as schools, may reduce overweight/obesity by simultaneously increasing physical activity and increasing healthy diet and food supply (Heise et al., 2017; Savoie-Roskos et al., 2017; Ohly et al., 2016). It is important to engage with the local community about their food experiences and requirements to ensure that any design and planning measures will be successful.

In considering how urbanism may support healthy diets, it is essential to recognise the socio-cultural, economic and environmental factors that influence diet and how these vary across urban population groups. In his evaluation of the American trend for local food production and consumption, Agyeman (2013) explores the multi-scalar and complex social justice and sustainability impacts that are often underappreciated in urban food discussions. Local food movements have grown substantially in the twenty-first century and they are popular in North America, Europe and Australia where they are primarily driven by White middle-class consumers who want their money to support local farmers and the environment. Agyeman argues that these 'yuppie chow' movements have been dominated by White cultural histories and privileges that have ignored race, culture and justice issues. He provides the example of Boston's Zoning Commission which took a decision in 2011 to allow two vacant lots to be used for urban agriculture in a neighbourhood with predominantly African American residents who were not involved in this decision-making process. Whilst Boston's urban farming project has been widely supported by planners, Agyeman explains why it is problematic:

> These initiatives often involve the establishment of urban farms or community gardens and encourage community residents to grow their own produce and provide food security for themselves. Yet, sometimes, those being targeted have specific cultural histories that can see farming and growing food as reminders of past oppression. This can be the case especially with African American communities whose ancestors have had a tormented past of slavery and share-cropping in the United States and whose residents may want little to do with growing their own food. An act seen as positive and empowering from a local food perspective reflecting white cultural histories can be perceived as an unwanted reminder of past injustices from a black cultural perspective. (2013, p. 70)

In response to this type of local food challenge and others, Agyeman invokes the need for 'thinking beyond the local'—which relates to the THRIVES key message to think beyond the boundaries. There are three insights from Agyeman that apply equally well to urban food initiatives which seek to promote health. First, local food production/consumption is not in itself sufficient to create social justice and ecologically sustainable practices, but instead these principles need to be implicit aims of food programmes. Second, culture heavily influences food and programmes should take account of the multiple histories and cultural experiences present in a place using socially inclusive and democratic processes. Third, the ultimate goal of food initiatives should be shared responsibility for good food provision that integrates issues of sustainability and inclusion. Agyeman offers North American food policy councils (FPCs) as a potential way forward. FPCs operate between governments and the community and through collaborative processes they organise local food initiatives that support the community, the economy and the environment. Example FPC initiatives include those which aim to increase community food security for low-income populations, including: establishing farmer's markets in low-income areas, community-based mapping of food deserts, linking public transport with healthy food outlets and supporting new grocery stores to open in food deserts.

Planners and public health officials may be looking for ways to limit certain food and beverage outlets, such as the sale of alcohol (which contributes to chronic diseases, violence and crime) or the sale of unhealthy foods. Imposing such restrictions may fall outside of the control of urban planning policy and this will vary across planning systems. A review of English planning authorities' policies to reduce unhealthy food access by Keeble et al. (2019) found that half of English authorities (164/325) had a policy targeting food takeaways, with 56 explicitly focused on health. The most common health-related policy aimed to reduce hot food takeaways via exclusion zones around places for children and families, whilst the second most common policy aimed to limit density of takeaway food outlets in retail areas. Fast-food exclusion policies are relatively new and further evaluative research is required to understand the impact of these interventions. In relation to restricting alcohol outlets, policy-makers in Baltimore, Maryland (USA), revised their zoning legislation in 2017, called *TransForm Baltimore*, to reduce the oversupply of alcohol outlets. One part of this policy prohibits retailers of alcoholic beverages (for off-premise consumption) from being located within 300 feet of each other (with the exception of five districts). Hippensteel et al. (2019) evaluated the potential impact of this policy and highlighted its likely effectiveness if accompanied by appropriate enforcement. They found that most of these retailers were located in communities with higher Black, single-parent, low-income and unemployed populations demonstrating a health inequity that could be reversed. The authors warned that the policy could result in increased oversaturation of these alcohol retailers in vulnerable communities if the enforcement activities or community objections were stronger elsewhere potentially causing spill-over effects if the premises were then able to open in areas with less opposition.

6.3 Monitoring Indicators for Local Health at the Neighbourhood Scale

There are many available indicators for the health impact of neighbourhood design. The indicators in Table 6.3 were gathered from Pineo et al.'s (2018a) systematic review and these should be seen as a starting point for further exploration of appropriate indicators, depending on local priorities and context.

Table 6.3 Selected monitoring indicators for local health impacts at neighbourhood scale

Goal	Monitoring indicators
Services	Parks and recreation facilities in the neighbourhood (public or private) are accessible for all residents
	% of residents that have access to public exercise facilities in their neighbourhood
	Distance from postal code centroid to nearest school (metres, measured on the street network)
	Proximity to early childhood education centres of any type
	% of respondents satisfied with children's playgrounds and play areas
	Jobs/mile^2
(Perceived) safety	% of people who feel safe walking alone in the local area during the day/night
	% of respondents who agree locally, antisocial behaviour is a problem
	% of respondents who think street litter is a problem
	% of people who feel that streets and public spaces are well-lit at night
	% of roads with speed limits of 20 mph or less
Culture	Number of performing arts companies, museums, concert venues, sports stadiums and movie theatres per 10,000 people
	% of the adult population with enough opportunity to participate in arts and related activities in local area
	% of the adult population who agree that it is a good thing for a society to be made up of people from different cultures
Public space	% of people using outdoor space for exercise/health reasons
	% of occupants who feel a sense of shared values and trust among neighbours
	% of adults who report ever experiencing discrimination, been prevented from doing something or been hassled or made to feel inferior because of their race, ethnicity (or other characteristics)
	Proportion of people who think that public spaces are either mostly or completely welcoming to a range of cultural (and other) groups
	Number of sitting spots on benches per household (include seating in front yards)
Food	Average walking time to healthy food retailers
	Supermarket density (count/km^2)
	Takeout restaurant density (count/km^2)
	Facilities have functioning, safe and freely available water fountains

6.4 Thinking Beyond the Boundaries at Barton Park

Accommodating high housing demand usually requires large-scale development projects that create hundreds, if not, thousands of new homes alongside public services and amenities. New development at the scale of an entire neighbourhood or community is challenging because it takes many years to plan and build, it is subject to failure or unforeseen changes through democratic decision-making processes, and it is logistically difficult to align the financing and provision of infrastructure with housing. However, these projects can deliver many health benefits in terms of affordable housing, infrastructure and services for local communities. Large projects deliver economies of scale that enable the use of innovative technologies/

materials and decentralised low carbon energy systems. New supply chains and jobs can be created that can have lasting benefits for communities. The NHS England Healthy New Towns programme (introduced in Chap. 2) focused on creating health benefits through this scale of new development in ten demonstration sites around the country. This case study looks at the process and current progress on one of the demonstration sites, called Barton Park, located three miles from Oxford city centre. The focus here is on Barton Park's impact beyond the boundaries of the new development and the provision of health-promoting services and amenities for new and existing residents.

The context for this case study is one of significant demand for new housing, particularly affordable housing, as set out in Oxford City Council's strategic spatial plans in the *Core Strategy 2026* (adopted in March 2011). The city's population was expected to grow from 151,000 to 176,100 people between 2007 and 2026. To manage this growth, the housing stock was expected to increase by 400 homes per year (2000 in the first five years of the plan) over the same time period. Barton Park was allocated as a strategic site for housing and the 2012 *Barton Area Action Plan* provides details on the city's ambitions for the greenfield development (885 homes across 38 hectares) located adjacent to the existing neighbourhood of Barton (~1500 homes). The Council plans to use the new development to improve Barton which is one of the most deprived neighbourhoods in the city and among the 20% most deprived areas in the country. The development model involved creating a joint venture between Oxford City Council and Grosvenor Developments Ltd to deliver 885 new homes, 2500 square metres of retail, a community hub (with a primary school, pavilion and sports pitches) and 12.29 hectares of public open space (Fig. 6.3).

Barton Park has a number of features that would support health across multiple spatial and temporal scales. The project's health impact assessment (Arup, 2017) used the NHS Healthy Urban Development Unit's (HUDU) guidance for rapid HIAs, which sets out 11 health determinants for assessment. Table 6.4 describes selected positive health impact findings of the Barton Park HIA against the NHS HUDU (2019) categories, such as the inclusion of community social spaces, sustainable urban drainage supporting wildlife and access to nature, and achieving the Code for Sustainable Homes (CSH) level 4 (which includes health and wellbeing credits for Lifetime Homes, sound insulation, daylight and private outdoor space).[1] Some negative impacts may result from the roughly ten-year construction process, related to dust, noise or lack of perceived safety. One source of data for the HIA and the development's steering group was a report compiled using primary and secondary data about the health and wellbeing of Barton's existing residents (Rapkins, 2017). This report provides a baseline that can be used to monitor and evaluate changes in the community during and after construction on the new development. The use of interviews alongside surveys and routinely collected data provided a more nuanced picture of the population's health and wellbeing status. For example,

[1] The Arup HIA for Barton Park used an earlier version of the NHS HUDU Rapid HIA tool.

Fig. 6.3 Infrastructure and services at Barton Park were delivered in time for the first residents, including the community hub with sports pitches and playground (top left), primary school (bottom left), bus services and shelters (top right) and sustainable drainage blue/green infrastructure (bottom right). (Source: Author, 2021)

Table 6.4 Selected findings from Barton Park health impact assessment (Arup, 2017) organised by the 11 health determinants from NHS Healthy Urban Development Unit's Rapid HIA Tool

Category	Selected findings
Housing design and affordability	40% of housing provision will be affordable. Supports accessibility through adherence to the Lifetime Homes Standard and Building for Life Criteria.
Access to health and social care services and other social infrastructure	New bus routes will connect to Northway, Barton and the city centre via key facilities, including the John Radcliffe Hospital. A new primary school and community hub are provided at the centre of the development with sports pitches and community pavilion.
Access to open space and nature	A network of greenspaces runs throughout the site, with trails connecting the new and old residential areas. A new park includes a playground and wetland as part of sustainable drainage. The existing allotments allow for food growing.
Air quality, noise and neighbourhood amenity	The Construction Environmental Management Plan (CEMP) will be used to monitor dust, noise and other impacts. The park, greenspaces, community hub and sports area provide amenities.

(continued)

Table 6.4 (continued)

Category	Selected findings
Accessibility and active travel	The development supports active travel modes (partly to reduce local air pollution from vehicles) through ample cycle storage, bus routes, electric vehicle charging points and pedestrian-friendly street design.
Crime reduction and community safety	Secure by Design principles were integrated into the Design Code, such as defensible space, natural surveillance and active frontages.
Access to healthy food	The community centre plans to have a food store. The development supports Good Food Oxford, an initiative to connect residents with allotments, food stores and farmers' markets.
Access to work and training	There will be jobs associated with construction, the commercial centre (with food store, primary school, community hub, sports pitches and pavilion) and the new bus route.
Social cohesion and inclusive design	The design aims to encourage social connection on pedestrian-friendly streets and open spaces. Amenities are shared with existing and new residents. The community hub has flexible space for use by different groups.
Minimising the use of resources	Soil will be reused on site. Buildings are designed with ample space for storing compostable waste and recycling indoors. Rainwater harvesting and sustainable drainage are incorporated in site design.
Climate change	The site layout was tested to ensure that photovoltaic solar panels can deliver a 20% reduction in energy use for homes. Achieving Code for Sustainable Homes Level 4 and BREEAM 'Very Good' for non-residential buildings. Greenspace, wetland and bat boxes support biodiversity.

survey scores for wellbeing were equivalent to the English average and scores for loneliness were low, yet interview data showed that there were clusters of isolated residents, particularly among the elderly and middle-aged men (ibid.). HIAs are now routinely used on large developments in the UK; however, it is less common to find a health and wellbeing report such as the one conducted for Barton Park. It is good practice to have this baseline data to inform the development and ongoing monitoring.

In addition to the research data used for planning the healthy new town, Barton Park has also been praised for its approach to community engagement, new models of healthcare, integrated care with the voluntary sector and delivering key services before residents moved in to new accommodation (NHS England, 2019). These achievements of the partnership relate to understanding and addressing the health needs of residents beyond the boundary of the development. Services such as new schools and healthcare centres may be funded using the value created from new development, and this model was adopted in Barton. In England and Wales, Section 106 of the 1990 Town and Country Planning Act sets out the arrangements for charging developers to mitigate the impact of their proposed development. These are commonly called 'Section 106 agreements' and they are negotiated between the developer and the local planning authority as a condition of the planning consent. Gallent et al. (2020) provide further detail about how these arrangements work among other infrastructure delivery mechanisms in the UK and elsewhere, described in Chap. 8.

A Section 106 agreement was used at Barton Park to partly fund redevelopment of the Barton Neighbourhood Centre. The funds were used to extend the health and wellbeing services at the Centre and to triple the size of the clinical space, creating extra capacity for Barton's new and existing residents (NHS England, 2019). Through this support, the existing social prescribing activities (see Box 6.6) were bolstered and residents with long-term conditions were offered specially commissioned classes, including Strength and Balance, Dance for Health and Mindfulness. The participants were able to stay after classes and socialise in the Centre's café. An evaluation showed that 332 patients used the service over a 2.5-year period at an estimated cost of £79,779 (Chanan et al., 2019). The physical and social benefits were deemed to reduce demand on primary and secondary health care services at a savings of £79,426, thus making the social prescribing project cost neutral with

Box 6.6 Social prescribing and the built environment

Social prescribing involves the use of non-medical sources of support or interventions to support patients with social, emotional or practical needs—it is also called community referral (Chatterjee et al., 2018). The support or interventions offered through social prescribing can be grouped into four purposes: '(i) for advice and knowledge (e.g., on benefits or housing); (ii) for skills development (e.g., computing, food and cooking); (iii) for activities in social groups (e.g., befriending and self-help groups, dance, art, or crafts); and (iv) for activities with therapeutic design, especially nature-based (e.g., walking for health or woodland therapies) and formal counselling' (Pretty & Barton, 2020, p. 3).

The Bromley by Bow Centre in East London was the first to offer social prescribing over 25 years ago and it has served as an inspiration to many other areas, including the Barton Neighbourhood Centre (Chanan et al., 2019). Bromley by Bow was also an example of the co-location of health services with other community facilities, which include childcare, job support services and a community garden. A systematic review of social prescribing programmes found multiple benefits, including improved self-esteem, confidence, sense of control, empowerment and wellbeing; there were also reductions in anxiety or depression, negative mood and use of healthcare services (Chatterjee et al., 2018). The benefits of social prescribing can create considerable cost savings for healthcare services, with one estimate stating: 'For each £1 invested in SP, benefits are £1.43 (3 months); £2.30 (1 year); £1.98 (5 years); and reduced costs of the order of £250–500 per person per year' (Pretty & Barton, 2020, p. 4).

For social prescribing to be successful, healthcare providers need organised activities and a place for those activities, which can be supported by built environment professionals. A flexible community centre, park or other public building could serve as the location for classes and activities.

Fig. 6.4 Views of a small greenspace enclosed by housing at Barton Park, including the approach from Bluebell Mews (bottom left), and a diversity of housing types. (Source: Author, 2021)

added benefits to residents (ibid.). Since most developments will not be comprehensive in terms of the services that they provide, investment in existing services, as done in Barton, can be an important way to support health for future occupants and neighbouring communities.

Overall, Barton Park has a lot to offer as an example of healthy urban development on the edge of a small city. The urban design could be commended for its provision of ample public spaces for different purposes and the project has good sustainability credentials. Figure 6.4 shows images of one of the squares enclosed by houses, featuring a communal eating table, benches, vegetation and lawn. A diversity of housing typologies is visible, from detached homes to an apartment block. A critique of the design in terms of health and wellbeing would note that cars and garages take up considerable space in the public realm. There may be times of the day, such as mornings and evenings, when cars are frequently parking or leaving, reducing the sense of safety and serenity that could otherwise be provided in this greenspace and the residential streets. This could be contrasted with the oft-cited exemplar of healthy development, Vauban in Freiburg, Germany, where cars can unload on residential streets but must park elsewhere. Disabled residents or those with mobility difficulties may find it uncomfortable or impossible to join the communal table. There are very few buffers between public and private spaces. These observations may be the result of project constraints, such as economic viability (see Chap. 8), but it is worth considering how these observations could relate to future design proposals.

6.5 Conclusion

This chapter marked a clear shift from the planetary and ecosystem health chapters, which focused more on environmental exposures (e.g. different types of pollution), to the ways in which neighbourhood design can increase or diminish opportunities for physical activity and social interaction—two key behaviours for health. There are debates in both public health and urban design literature about the extent to which individual agency is overlooked in the rhetoric of healthy and sustainable design methods. As Carmona (2021) explains, the relationship between design and social dimensions of place value (e.g. health) is not deterministic, but instead the environment creates potentials or possibilities. People will make choices about how to use a place, but social and environmental justice literature shows that these choices are constrained for many groups within society. This chapter has emphasised that inclusive design processes must be used to understand people's needs in a specific place.

It is useful for healthy urban design and planning practitioners to understand that although the role of individual agency and the local population's needs should be integral to design and planning, there are some common design features that regularly support health. In other words, practitioners are not starting from scratch on every project. Carmona's (2019) review of studies about the value of quality built environments provides strong evidence that the following place features are supportive of multiple social, economic and environmental benefits:

> Greenness in the built environment (notably the presence of trees and grass, water, and open space—the latter if of good quality); a mix of uses (notably the diversity of land uses within a neighbourhood); low levels of traffic; the walkability and bikeability of places (derived from their strategic street-based connectivity and the quality of the local public realm); the use of more compact (less sprawling and fragmented) patterns of development; and ready convenient connection to a good public transport network. (p. 33)

Carmona also argues that these place qualities are useful because they can be measured and specified in policies, aiding the likelihood that they will be delivered. The next chapter moves down in spatial scales from the neighbourhood to the building, exploring how the indoor environment affects health and wellbeing. Housing affordability and tenure security are also discussed as key requirements for urban health.

References

Agyeman, J. (2013). *Introducing just sustainabilities: Policy, planning and practice, Just sustainabilities*. Zed Books.

Agyeman, J., & Erickson, J. S. (2012). Culture, recognition, and the negotiation of difference: Some thoughts on cultural competency in planning education. *Journal of Planning Education and Research, 32*, 358–366. https://doi.org/10.1177/0739456X12441213

Alidoust, S., & Bosman, C. (2015). Planning for an ageing population: Links between social health, neighbourhood environment and the elderly. *Australian Planner, 52*, 177–186.

Arup. (2017). *Oxford City Council, Barton NHS Healthy New Town and Underhill Circus redevelopment, health impact assessment*. Ove Arup & Partners Ltd.

Audrey, S., & Batista-Ferrer, H. (2015). Healthy urban environments for children and young people: A systematic review of intervention studies. *Health & Place, 36*, 97–117.

Barnett, D. W., Barnett, A., Nathan, A., Van Cauwenberg, J., Cerin, E., & on behalf of the Council on Environment and Physical Activity (CEPA)—Older Adults working group. (2017). Built environmental correlates of older adults' total physical activity and walking: A systematic review and meta-analysis. *International Journal of Behavioral Nutrition and Physical Activity, 14*, 103.

Bassett, E. (2018). Designing for health: fostering social capital formation through public space. In T. Beatley, C. Jones, & R. M. Rainey (Eds.), *Healthy environments, healing spaces: Practices and directions in health, planning, and design*. University of Virginia Press.

Beaulac, J., Kristjansson, E., & Cummins, S. (2009). A systematic review of food deserts, 1966–2007. *Preventing Chronic Disease, 6*, A105.

Behbehani, F., Dombrowski, E., & Black, M. (2019). Systematic review of early child care centers in low- and middle-income countries and health, growth, and development among children aged 0–3 years (P11-052-19). *Current Developments in Nutrition, 3*.

Black, C., Moon, G., & Baird, J. (2014). Dietary inequalities: What is the evidence for the effect of the neighbourhood food environment? *Health & Place, 27*, 229–242.

Black, P., & Sonbli, T. E. (2019). *The urban design process*. Lund Humphries.

Brainard, J., Cooke, R., Lane, K., & Salter, C. (2019). Age, sex and other correlates with active travel walking and cycling in England: Analysis of responses to the Active Lives Survey 2016/17. *Preventive Medicine, 123*, 225–231.

Brussoni, M., Gibbons, R., Gray, C., Ishikawa, T., Sandseter, E., Bienenstock, A., Chabot, G., Fuselli, P., Herrington, S., Janssen, I., Pickett, W., Power, M., Stanger, N., Sampson, M., & Tremblay, M. (2015). What is the relationship between risky outdoor play and health in children? A systematic review. *International Journal of Environmental Research and Public Health, 12*, 6423–6454.

Burbeck, A. (2020). *Chicana/o murals, placemaking, and the threat of street art in Denver's La Alma-Lincoln Park* (M.A.). University of Colorado at Boulder, United States, Colorado.

Burman, M., Brown, J., Tisdall, K., & Batchelor, S. (2000). *View from the girls: Exploring violence and violent behaviour*. ESRC End of Award Report. Economic and Social Research Council, Swindon.

Burton, E. J., Mitchell, L., & Stride, C. B. (2011). Good places for ageing in place: Development of objective built environment measures for investigating links with older people's wellbeing. *BMC Public Health, 11*, 839.

Cancer Research UK. (2019). Obese people outnumber smokers two to one. Retrieved April 19, 2021, from https://www.cancerresearchuk.org/about-us/cancer-news/press-release/2019-07-03-obese-people-outnumber-smokers-two-to-one

Carmona, M. (2019). Place value: Place quality and its impact on health, social, economic and environmental outcomes. *Journal of Urban Design, 24*, 1–48.

Carmona, M. (2021). *Public places urban spaces: The dimensions of urban design* (3rd ed.). Routledge.

Carver, A., Timperio, A., & Crawford, D. (2008). Playing it safe: The influence of neighbourhood safety on children's physical activity—A review. *Health & Place, 14*, 217–227.

Caspi, C. E., Sorensen, G., Subramanian, S. V., & Kawachi, I. (2012). The local food environment and diet: A systematic review. *Health & Place, 18*, 1172–1187.

Chanan, G., Morton, K., & Harris, K. (2019). *Barton healthy new town concluding evaluation: Three key activities*. Health Empowerment Leverage Project.

Chastin, S. F. M., Abaraogu, U., Bourgois, J. G., Dall, P. M., Darnborough, J., Duncan, E., Dumortier, J., Pavón, D. J., McParland, J., Roberts, N. J., & Hamer, M. (2021). Effects of regular physical activity on the immune system, vaccination and risk of community-acquired infectious disease in the general population: Systematic review and meta-analysis. *Sports Medicine, 51*, 1673–1686.

Chatterjee, H. J., Camic, P. M., Lockyer, B., & Thomson, L. J. M. (2018). Non-clinical community interventions: A systematised review of social prescribing schemes. *Arts Health, 10*, 97–123.

Clements, R. (2004). An investigation of the status of outdoor play. *Contemporary Issues in Early Childhood, 5*, 68–80.

Corburn, J. (2013). *Healthy city planning: From neighbourhood to national health equity* (Planning, History and Environment Series). Routledge.

Costa, S., Benjamin-Neelon, S. E., Winpenny, E., Phillips, V., & Adams, J. (2019). Relationship between early childhood non-parental childcare and diet, physical activity, sedentary behaviour, and sleep: A systematic review of longitudinal studies. *International Journal of Environmental Research and Public Health, 16*, 4652.

Department for Transport. (2020). Travel to school. Retrieved February 25, 2021, from https://www.ethnicity-facts-figures.service.gov.uk/culture-and-community/transport/travel-to-school/latest

Dixon, B. N., Ugwoaba, U. A., Brockmann, A. N., & Ross, K. M. (2020). Associations between the built environment and dietary intake, physical activity, and obesity: A scoping review of reviews. *Obesity Reviews, 22*, 13171.

Eime, R. M., Young, J. A., Harvey, J. T., Charity, M. J., & Payne, W. R. (2013). A systematic review of the psychological and social benefits of participation in sport for adults: Informing development of a conceptual model of health through sport. *International Journal of Behavioral Nutrition and Physical Activity, 10*, 135.

Esteban-Cornejo, I., Carlson, J. A., Conway, T. L., Cain, K. L., Saelens, B. E., Frank, L. D., Glanz, K., Roman, C. G., & Sallis, J. F. (2016). Parental and adolescent perceptions of neighborhood safety related to adolescents' physical activity in their neighborhood. *Research Quarterly for Exercise and Sport, 0*, 1–9.

Fiscella, N. A., Case, L. K., Jung, J., & Yun, J. (2021). Influence of neighborhood environment on physical activity participation among children with autism spectrum disorder. *Autism Research, 14*, 560–570.

Fontaine, K. R. (2000). Physical activity improves mental health. *The Physician and Sportsmedicine, 28*, 83–84.

Gallent, N., Morphet, J., Chiu, R. L. H., Filion, P., Fischer, K. F., Gurran, N., Li, P., Li, P., Schwartzf, A., & Stead, D. (2020). International experience of public infrastructure delivery in support of housing growth. *Cities, 107*, 102920.

Gaster, S. (1991). Urban children's access to their neighborhood: Changes over three generations. *Environment and Behavior, 23*, 70–85.

Gehl Institute. (2018). *Inclusive healthy places: A guide to inclusion & health in public space: Learning globally to transform locally*.

Gehl, J. (1986). "Soft edges" in residential streets. *Scandinavian Housing & Planning Research, 3*, 89–102.

Gill, T. (2021). *Urban playground: How child-friendly planning and design can save cities*. RIBA Publishing.

Groshong, L., Wilhelm Stanis, S. A., Kaczynski, A. T., & Hipp, J. A. (2020). Attitudes about perceived park safety among residents in low-income and high minority Kansas City, Missouri, Neighborhoods. *Environment and Behavior, 52*, 639–665.

Hahn, R. A., & Truman, B. I. (2015). Education improves public health and promotes health equity. *International Journal of Health Services, 45*, 657–678.

Heise, T. L., Romppel, M., Molnar, S., Buchberger, B., Berg, A. v. d., Gartlehner, G., & Lhachimi, S. K. (2017). Community gardening, community farming and other local community-based gardening interventions to prevent overweight and obesity in high-income and middle-income countries: Protocol for a systematic review. *BMJ Open, 7*, e016237.

Hippensteel, C. L., Sadler, R. C., Milam, A. J., Nelson, V., & Debra Furr-Holden, C. (2019). Using zoning as a public health tool to reduce oversaturation of alcohol outlets: An examination of the effects of the new "300 Foot Rule" on packaged goods stores in a Mid-Atlantic City. *Prevention Science, 20*, 833–843.

Joseph, M. (2019). *Mixed-income communities as a strategic lever to impact health equity: Lessons from the field and implications for strategy and investment*. CSSP.

Keeble, M., Burgoine, T., White, M., Summerbell, C., Cummins, S., & Adams, J. (2019). How does local government use the planning system to regulate hot food takeaway outlets? A census of current practice in England using document review. *Health & Place, 57*, 171–178.

Kern, L. (2020). *Feminist city: Claiming space in a man-made world*. Verso.

Kraft, A. N., Thatcher, E. J., & Zenk, S. N. (2020). Neighborhood food environment and health outcomes in U.S. Low-socioeconomic status, racial/ethnic minority, and rural populations: A systematic review. *Journal of Health Care for the Poor and Underserved, 31*, 1078–1114.

Lam, T. M., Vaartjes, I., Grobbee, D. E., Karssenberg, D., & Lakerveld, J. (2021). Associations between the built environment and obesity: An umbrella review. *International Journal of Health Geographics, 20*, 7.

Lambert, A., Vlaar, J., Herrington, S., & Brussoni, M. (2019). What is the relationship between the neighbourhood built environment and time spent in outdoor play? A systematic review. *International Journal of Environmental Research and Public Health, 16*, 3840.

Lee, I.-M., Shiroma, E. J., Lobelo, F., Puska, P., Blair, S. N., & Katzmarzyk, P. T. (2012). Effect of physical inactivity on major non-communicable diseases worldwide: An analysis of burden of disease and life expectancy. *The Lancet, 380*, 219–229.

Madanipour, A. (Ed.). (2010). *Whose public space? International case studies in urban design and development*. Routledge.

Mindell, J. S., & Karlsen, S. (2012). Community severance and health: What do we actually know? *Journal of Urban Health, 89*, 232–246.

Mitchell, L., Burton, E., Raman, S., Blackman, T., Jenks, M., & Williams, K. (2003). Making the outside world dementia-friendly: Design issues and considerations. *Environment and Planning B: Planning and Design, 30*, 605–632.

Mithun. (2010). *Executive summary*, South Lincoln Redevelopment Master Plan Final Report. Mithun, Seattle, WA.

Napier, D., Depledge, M., Knipper, M., Lovell, R., Ponarin, E., Sanabria, E., & Thomas, F. (2017). *Culture matters: Using a cultural contexts of health approach to enhance policy-making*. World Health Organization.

Natural England. (2009). *Childhood and nature: A survey on changing relationships with nature across generations*. Natural England, Warboys, England.

Newlin, K., Dyess, S. M., Allard, E., Chase, S., & Melkus, G. D. (2012). A methodological review of faith-based health promotion literature: Advancing the science to expand delivery of diabetes education to Black Americans. *Journal of Religion and Health, 51*, 1075–1097.

NHS England. (2019). *Putting health into place: Develop and provide healthcare services*. NHS England, London.

NHS London Healthy Urban Development Unit. (2019). *HUDU rapid health impact assessment tool*.

Noonan, R. J. (2021). Family income matters! Tracking of habitual car use for school journeys and associations with overweight/obesity in UK youth. *Journal of Transport and Health, 20*, 100979.

Ohly, H., Gentry, S., Wigglesworth, R., Bethel, A., Lovell, R., & Garside, R. (2016). A systematic review of the health and well-being impacts of school gardening: Synthesis of quantitative and qualitative evidence. *BMC Public Health, 16*, 286.

Oja, P., Titze, S., Kokko, S., Kujala, U. M., Heinonen, A., Kelly, P., Koski, P., & Foster, C. (2015). Health benefits of different sport disciplines for adults: Systematic review of observational and intervention studies with meta-analysis. *British Journal of Sports Medicine, 49*, 434–440.

Oliveira, A., Monteiro, Â., Jácome, C., Afreixo, V., & Marques, A. (2017). Effects of group sports on health-related physical fitness of overweight youth: A systematic review and meta-analysis. *Scandinavian Journal of Medicine & Science in Sports, 27*, 604–611.

Peinhardt, K., & Storring, N. (2019). *A playbook for inclusive placemaking: Community process*. Project for Public Spaces. Retrieved May 19, 2021, from https://www.pps.org/article/a-playbook-for-inclusive-placemaking-community-process

Petrokofsky, C., & Davis, A. (2016). *Working together to promote active travel: A briefing for local authorities*. Public Health England.

Pineo, H., Glonti, K., Rutter, H., Zimmermann, N., Wilkinson, P., & Davies, M. (2018a). Urban health indicator tools of the physical environment: A systematic review. *Journal of Urban Health, 95*, 613–646.

Planning Out. (2019). LGBT+ Placemaking Toolkit. Retrieved December 10, 2021, from https://becg.com/lgbt-placemaking-toolkit-2/

Powell-Wiley, T. M., Ayers, C. R., Lemos, J. A. d., Lakoski, S. G., Vega, G. L., Grundy, S., Das, S. R., Banks-Richard, K., & Albert, M. A. (2013). Relationship between perceptions about neighborhood environment and prevalent obesity: Data from the Dallas heart study. *Obesity, 21*, E14–E21.

Pretty, J., & Barton, J. (2020). Nature-based interventions and mind–body interventions: Saving public health costs whilst increasing life satisfaction and happiness. *International Journal of Environmental Research and Public Health, 17*, 7769.

Rapkins, C. (2017). *Barton healthy new town*. Oxford City Council, Final Report.

Ross, S. E., Flynn, J. I., & Pate, R. R. (2016). What is really causing the obesity epidemic? A review of reviews in children and adults. *Journal of Sports Sciences, 34*, 1148–1153.

Rudge, G. M., Suglani, N., Saunders, P., & Middleton, J. (2013). OP24 are fast food outlets concentrated in more deprived areas? A geo-statistical analysis of an urban area in Central England. *Journal of Epidemiology and Community Health, 67*(A14), 1–A14.

Saunders, P., Saunders, A., & Middleton, J. (2015). Living in a 'fat swamp': Exposure to multiple sources of accessible, cheap, energy-dense fast foods in a deprived community. *The British Journal of Nutrition, 113*, 1828–1834.

Savoie-Roskos, M. R., Wengreen, H., & Durward, C. (2017). Increasing fruit and vegetable intake among children and youth through gardening-based interventions: A systematic review. *Journal of the Academy of Nutrition and Dietetics, 117*, 240–250.

Thompson, C., Cummins, S., Brown, T., & Kyle, R. (2013). Understanding interactions with the food environment: An exploration of supermarket food shopping routines in deprived neighbourhoods. *Health & Place, 19*, 116–123.

Tiesdell, S., & Oc, T. (1998). Beyond 'Fortress' and 'Panoptic' Cities—Towards a Safer Urban Public Realm. *Environment and Planning B: Planning and Design, 25*, 639–655.

Townshend, T., & Lake, A. (2017). Obesogenic environments: Current evidence of the built and food environments. *Perspectives in Public Health, 137*, 38–44.

UNESCO. (2001). *UNESCO universal declaration on cultural diversity*.

Utzet, M., Valero, E., Mosquera, I., & Martin, U. (2020). Employment precariousness and mental health, understanding a complex reality: A systematic review. *International Journal of Occupational Medicine and Environmental Health, 33*, 569–598.

van der Noordt, M., IJzelenberg, H., Droomers, M., & Proper, K. I. (2014). Health effects of employment: A systematic review of prospective studies. *Occupational and Environmental Medicine, 71*, 730–736.

Vincent, J. M. (2014). Joint use of public schools: A framework for promoting healthy communities. *Journal of Planning Education and Research, 34*, 153–168.

WHO. (2007). *Global age-friendly cities: A guide*. WHO, Geneva, Switzerland.

WHO. (2016a). *Global report on urban health: Equitable, healthier cities for sustainable development*. WHO, Geneva, Switzerland.

WHO. (2016b). *Waste and human health: Evidence and needs*. WHO Meeting Report, 5–6 November 2015, Bonn, Germany. WHO Regional Office for Europe, Copenhagen, Denmark.

WHO. (2018a). *Global action plan on physical activity 2018–2030: More active people for a healthier world*. WHO, Geneva, Switzerland.

WHO. (2018b). *The global network for age-friendly cities and communities: Looking back over the last decade, looking forward to the next*. WHO, Geneva, Switzerland.

Wilkie, S., Townshend, T., Thompson, E., & Ling, J. (2018). Restructuring the built environment to change adult health behaviors: A scoping review integrated with behavior change frameworks. *Cities Health, 2*, 198–211.

Won, J., Lee, C., Forjuoh, S. N., & Ory, M. G. (2016). Neighborhood safety factors associated with older adults' health-related outcomes: A systematic literature review. *Social Science & Medicine, 165*, 177–186.

World Bank. (2020). *Handbook for gender-inclusive urban planning and design*. World Bank Group.

Yang, K., & Victor, C. (2011). Age and loneliness in 25 European nations. *Ageing and Society, 31*, 1368–1388.

Chapter 7
Local Health: Building Scale

7.1 Introduction

Literature about healthy buildings often starts with the rationale that most people spend 80–90% of their time indoors. Buildings as a key health exposure, or set of exposures, are made more significant by their large impact on the wider environment, affecting ecosystem and planetary health. Energy use in buildings, for instance, makes up 17.5% of global greenhouse emissions (see Chap. 4), and the figure is much higher in some countries (Ritchie & Roser, n.d.). Some people spend higher portions of their time in buildings, such as the elderly and women; whilst others spend much of their time outdoors due to their profession, such as agricultural workers, or because they are unhoused. These differences result in divergent health-related requirements from buildings that vary across projects and locations. As discussed in the previous chapter in relation to neighbourhoods, there are no perfect design solutions or blueprints for healthy buildings because local context and needs have to be considered in every case. Ideally the design process would begin by asking who will inhabit the building and what are their specific health and wellbeing-related needs. Following this assessment, design solutions can be developed and iterated through participatory processes and, for large projects, tested through health impact assessment. Although the design output will vary, there are shared goals for all buildings that are often described in terms of indoor environmental quality (IEQ) which traditionally includes thermal and acoustic comfort, air quality (related to moisture and ventilation) and adequate lighting. A host of other requirements are also made of today's healthy buildings. From a healthy urbanism perspective, the health impact of buildings should be considered through an ecological lens not only for the occupants but also for people who are spatially and temporally distant from the building and wider ecosystems.

Designing buildings that support health, comfort and wellbeing is not a new concept and different histories and approaches can be found for different types of

H. Pineo, *Healthy Urbanism*, Planning, Environment, Cities,
https://doi.org/10.1007/978-981-16-9647-3_7

buildings. Dovjak and Kukec's (2019) book on healthy and sustainable buildings provides a historical account of the Greek philosopher Socrates' sun-tempered house, which created a pleasant environment through passive design measures ensuring the home stayed warm in the winter and cool in the summer. The sanitarians and social reformers of the nineteenth century drew attention to the harmful exposures found in urban housing in the UK, Canada and the USA, driven by lack of sanitation infrastructure, overcrowding and polluting fuel sources. Architects, engineers and planners responded with design approaches at multiple scales to increase fresh air, clean water and sunlight, environmental components that have long been understood as important for health. Healthcare buildings have a particularly interesting history, given their obvious function as places where people go to be healed from a physical or mental health illness. Jeanne Kisacky's (2017) book explores their history in the USA noting how hospital design shifted between goals for the building to be either a tool for health and healing or a tool to facilitate efficient medical interventions. Her work highlights the interrelations between medicine, culture and architecture, meaning that hospital design influenced medical practices and vice versa. Attention to the health effects of office buildings can be traced back to the 1970s when sick building syndrome (SBS) and other problems arose from poor indoor air quality, moisture and other factors (Sundell, 2017). The symptoms of SBS include 'headache, eye, nose, or throat irritation, dry cough, dry or itchy skin, dizziness and nausea, difficulty in concentrating, fatigue and sensitivity to odours' and it is thought to be linked to time spent in a building, but no specific illness or cause can be identified (Dovjak & Kukec, 2019, p. 61). In contrast to SBS, 'building-related illness' (BRI) is the term used when an illness can be directly attributed to airborne building contaminants (ibid.).

As a result of the various factors and environments described above, research on the health and wellbeing impacts of buildings has branched into numerous subfields and specialisms. As shown in Table 7.1, these areas of research are concerned with different building types, population groups, methodologies and outcomes of concern. The 'exposure' in this research may concern the whole building or could relate to one or more of its characteristics, such as the topics listed in Table 7.3 (next section). Multiple systematic reviews have summarised the evidence linking buildings with specific health outcomes (Gibson et al., 2011; Allen et al., 2015; Ige et al., 2018), although some topics are less represented in public health and

Table 7.1 Diversity of research areas within the wider field of healthy buildings

Building types	Population groups	Outcomes	Methodologies
Education Healthcare Offices Prisons Residential (including diverse tenures, typologies, etc.) Retail	Children Disabled Elderly Low income Minority ethnic Patients	Absence/sick days Comfort Health (specific conditions/ symptoms and general measures) Physical Activity Productivity Quality of life Wellbeing	Experimental Modelling Observational (including social research)

epidemiological studies, such as the impact of tenure security. Indoor air quality (see Chap. 5) and thermal comfort are two building risk factors of growing concern that are closely tied to environmental degradation, inequalities and design. Areas of increasing interest relate to psychosocial health, such as through the use of biophilic design in buildings and space for social interaction (Beatley et al., 2018). Supporting physical activity in buildings is an emerging research area that could be quite impactful in some building types (see the 'irresistible staircase' in the Bullitt Center case study in Chap. 4).

The importance of housing for health is well established in public health literature and international law, yet it is not necessarily ensured by existing models for housing development. The right to adequate housing is recognised in the 1948 Universal Declaration of Human Rights as part of the right to an adequate standard of living (Office of the United Nations High Commissioner for Human Rights and UN Habitat, 2009). The right includes *freedoms* and *entitlements* alongside multiple criteria that must be met. The *freedoms* include protection against forced eviction and the right to choose one's residence, while the *entitlements* include security of tenure and equal and non-discriminatory access to adequate housing, for example. The minimum criteria are described in the following themes: security of tenure; availability of services, materials, facilities and infrastructure; affordability; habitability; accessibility; location; and cultural adequacy (ibid.). Rather than solely focusing on hazards within homes, healthy urbanism takes a broader perspective to incorporate what Carole Després (1991) called the 'meaningful' aspects of housing, later adopted by James Dunn (2002) for research on housing and health inequalities.

Mary Shaw's (2004) conceptual model of housing and health builds on Dunn's discussion of material and meaningful aspects of housing, dividing the exposures into 'hard' and 'soft' mechanisms and direct and indirect pathways to health effects (Table 7.2). The soft factors are less frequently studied in epidemiology, yet they are connected to the hard factors. If a household is in debt with low financial security, they may lack the ability to maintain the physical house, possibly leading to excess damp or other hazards. Similarly, a home with many physical problems could cause

Table 7.2 Conceptual model of housing and health

Mechanisms of health effects	Direct pathways	Indirect pathways	
		Household level	Neighbourhood level
Hard/physical/ material	Physical effects (e.g. temperature, moisture, indoor air quality)	Socio-economic factors (e.g. income and wealth)	Availability of services and facilities
		Proximity to services and facilities	Natural and built environment features
Soft/social/ meaningful	Financial and mental health effects of poor housing (e.g. insecurity and debt)	Cultures and behaviours	Cultures and behaviours
	Feeling of 'home', social status and ontological security		Community, social capital and social fragmentation

Source: Adapted from Shaw (2004, p. 398)

health effects that are costly and reduce occupants' ability to study or work, further reducing their economic security. Measures were included in Shaw's model where there was a known association between certain risks and health outcomes, such as low socio-economic status. She described the existing evidence regarding housing tenure and health, noting that several studies had established links, but that these may have resulted from 'residualisation' whereby the observed effect of poorer health in social housing tenures was a result of the occupants' pre-existing health conditions, not a result of the housing itself. Shaw also presented emerging evidence that housing tenure could be an indicator of health hazards in the home or that tenure could affect occupants' feelings of control and autonomy. Her conceptual model thus integrated these diverse possibilities and provided a holistic understanding of the interconnected material and meaningful factors that affect health in housing. These factors continue to be relevant today. The soft side of housing and health remains relatively understudied and is becoming increasingly problematic due to growing housing affordability crises (Baker et al., 2020). This challenge is further described in the sections on affordability and tenure security.

7.2 Achieving Local Health Framework Goals Through Buildings

A wide range of resources now exist to support design and property professionals with financing, designing and measuring healthy buildings. These resources have many similarities, but they also provide indications of the subtle differences between conceptualisations of the healthy building concept across sectors and nations. Joseph Allen and John Macomber's (2020) book *Healthy Buildings: How Indoor Spaces Drive Performance and Productivity* encapsulates current scientific evidence for designing healthy buildings alongside clear exposition of the business case, particularly for commercial and public sector buildings in the USA. They use the '9 Foundations for a Healthy Building' that Allen views as the 'handful of things we need to do to make a building healthier' in a bid to simplify what Allen and Macomber viewed as an overly complex set of messages in this field (ibid., p. 86). The World Green Building Council, an umbrella body for national green building councils, has also produced numerous reports on healthy buildings including a *Health & Wellbeing Framework* (World GBC, 2020) setting out six principles for a healthy and sustainable built environment: (1) protect health, (2) prioritise comfort, (3) harmony with nature, (4) facilitate healthy behaviour, (5) create social value and (6) take climate action. There are also healthy building frameworks for specific building uses, such as offices, homes and schools. The WHO (2018a) *Housing and Health Guidelines* provides an evidence-based summary of the key factors affecting health in housing, which is now complemented by an international review of policies, regulations and legislation (WHO, 2021). The British Council for Offices (2018) critically reviewed existing standards and produced a health and wellbeing framework for offices informed by Derek Clements-Croome et al.'s (2019) Flourish model.

Looking at these frameworks together, there are many common themes, yet it is difficult to distil the important factors down to an easily digestible list. Table 7.3

Table 7.3 Comparison of themes covered in healthy building frameworks

9 Foundations of a healthy building (Allen & Macomber, 2020)	Health & Wellbeing Framework (World GBC, 2020)	Housing and Health Guidelines (WHO, 2018a)	Health and Wellbeing Themes (BCO, 2018)
	Access to Nature		Outside
	Active Design		(Outside)
	Adaptation and Resilience	*Integrated in thermal comfort*	Inside
Air Quality	Air Quality	Air Quality, Radon, Tobacco Smoke	Breathe
	Biodiversity, Nature-based Solutions		
	Biophilic Design		(Inside)
	Climate Change Mitigation	*Integrated in thermal comfort*	
	Community Health		
	Construction Worker Health		Touch
Dust & Pests		(Air Quality)	Clean
	Ergonomic		(Inside)
		Household Crowding	
	Human Rights	*Objectives and Rationale*	
	Inclusive Design	Housing Accessibility	(Outside)
	Infectious Disease	(Household Crowding)	(Clean)
Lighting & Views	Lighting, Visual		See, (Inside)
	Material Health	Asbestos, Lead	(Clean), (Touch)
	Mental Health, Social Connectivity, Social Value		Feel (i.e. overall life satisfaction)
Moisture		(Air Quality)	(Breathe)
	Nutrition		Nourish
Noise	Acoustics	Noise	Hear
	Resource efficiency		
Safety & Security		Injury Hazards	(Outside)
			Sense (i.e. ability to control environment)
Thermal Health	Thermal Comfort	Low Indoor Temperatures and Insulation, High Indoor Temperatures	(Sense)
Ventilation	(Air Quality)	(Air Quality)	(Breathe)
Water Quality	Water Quality, Water Efficiency Hydration	Water, (Lead)	(Clean), (Touch)

Note: Themes in brackets are listed more than once for comparison purposes. Topics in italics are covered in the framework, but not as a core theme compares the aforementioned frameworks and highlights that air quality, noise, thermal comfort and ventilation are consistently viewed as important. Differences across the frameworks could be attributed to specific building types (e.g. household overcrowding), explicit links to sustainability (e.g. resource efficiency) or the inclusion of topics which other frameworks encompass in wider themes (e.g. dust is often covered under air quality). As Pineo (2020) established, frameworks for healthy urban design and planning rarely address issues of equity, inclusion and sustainability. The WorldGBC is moving in that direction through the inclusion of environmental and social issues that are not normally covered (e.g. biodiversity and community health).

Given the high number of healthy building guidance documents, this chapter focuses on the topics where further understanding and integration of concepts is required. The chapter also avoids repeating topics that have been described at other scales of health impact. Biodiversity and resource efficiency, for example, are relevant to healthy buildings but they were covered through the planetary health chapter. The design and planning goals listed in THRIVES at the decision-making scale most relevant for buildings include acoustic and thermal comfort, affordability, tenure security, lighting and space.

7.2.1 Acoustic Comfort

Noise pollution is a serious environmental health threat that is persistent in cities where traffic, industrial processes and other sources harm people and wildlife. Exposure to noise pollution can cause permanent hearing loss and impairments and it can increase anxiety, annoyance, sleep disturbance (see Box 7.5) and psychological distress (Babisch, 2006; Lekaviciute et al., 2013). Traffic noise is also a risk factor for cardiovascular diseases, which are believed to occur as a result of elevated stress hormone levels and oxidative stress (Hahad et al., 2019). As with air pollution and other environmental burdens, studies have found that low-income and minority ethnic groups are exposed to higher noise pollution levels (EEA, 2020). Furthermore, children, the elderly, pregnant women and people with pre-existing conditions, among others, may be more susceptible to negative impacts of noise.

The health effects of noise are typically covered by legislation (e.g. the EU 2002 Environmental Noise Directive) that aims to drive down noise sources and exposures. Although there is not a clear cut-off for safe and unsafe acoustic environments, the WHO (2018b) has produced noise guidelines for Europe which detail recommended exposure cut-offs for different noise sources, such as 53 dB L_{den} (day-evening-night equivalent sound level) for road traffic. Noise maps are produced in European countries (and elsewhere) to guide urban planning processes. City planning and building layout and orientation can be used to reduce exposure to noise, for example, by separating housing (or bedrooms) from major roads and industrial activities. Acoustic barriers and insulation can be applied, including the use of vegetation and trees, which creates co-benefits for air pollution reduction and other objectives. People's perceptions of noise and other environmental conditions are complex. For instance, a view to outdoor vegetation has been shown to reduce noise annoyance for people living next to busy roads (Van Renterghem & Botteldooren, 2016).

Sound in itself is not bad and it can be harnessed to improve quality of life and wellbeing through acoustic design that appreciates the subjective perception of sounds and urban quiet spaces. The concept of 'soundscapes' refers to the 'acoustic environment as perceived or experienced and/or understood by a person or people, in context' (ISO, 2014, p. 1). A systematic review found that positive perceptions of soundscapes have been associated with positive health effects, yet further research is needed (Aletta et al., 2018). The potentially transformative perspective of soundscapes is advocated by Antonella Radicchi et al. (2021) who call for public participation in design projects, noise action plans and policies that will pursue 'acoustic environmental quality for health and the environment' (p. 2). Rather than adopting an approach that seeks to reduce all noise from cities, the soundscape approach values sounds as resources. To integrate the soundscape approach, Radicchi et al. (2021) propose

> combinatory actions—reducing noise, overcoming risky thresholds, protecting areas of high acoustic quality, creatively and collectively designing and managing the soundscape. This must be implemented through an interdisciplinary approach embracing participatory urban design and planning, psychology, ecology and the environment, slow mobility, mobile and digital technology, and sound art. (p. 8)

There are several methods for gathering public perceptions of acoustic environments that can inform design and planning, alongside future research. For example, soundwalking and the use of mobile apps, such as Radicchi's Hush City app (Radicchi et al., 2016), may help planners understand how population groups are affected by sounds in specific times and places.

7.2.2 *Thermal Comfort*

Excess heat and cold in buildings have always been harmful to health, but the climate crisis has amplified the intensity and frequency of extreme weather events creating serious health threats in cities, particularly for people who lack the means to adapt thermal conditions in their homes and who are vulnerable due to age or pre-existing health conditions. The concept of thermal comfort reflects the subjective and individual nature of perceiving temperature, which varies by age, gender and other factors. As explained by Noël Djongyang et al. (2010) 'the judgment of comfort is a cognitive process involving many inputs influenced by physical, physiological, psychological, and other factors' (p. 2627). Reference to the concept of 'comfort' does not lessen the importance of this topic, nor does it imply lower severity of the health issues involved, because comfort and health are inter-related and can have diverse effects on physical and mental health impacts over time. In housing, excess heat or cold can cause respiratory and cardiovascular diseases, including emergency hospitalisations and death (WHO, 2018a). Thermal comfort is also likely to have social impacts that relate to health. Thomson and Thomas (2015) found that warmth improvements in homes resulted in increased usable space; greater opportunities for studying and social interaction, including hospitality; and reduced time off work or school. In non-residential buildings, excess heat or cold can reduce productivity and ability to concentrate, disrupting the primary activities in schools (de Dear et al., 2015), offices (Allen & Macomber, 2020) and other settings. Action to increase thermal comfort should be taken at multiple urban scales and building lifecycle stages.

The risk of overheating within urban buildings and public spaces is exacerbated by the urban heat island (UHI) effect, whereby impervious surfaces in cities absorb heat resulting in higher temperatures than surrounding rural areas. Globally, temperature differentials between cities and rural hinterlands can vary between 3.5 °C and 12 °C and the use of mechanical air conditioning to alleviate extreme heat creates a vicious cycle that further heats urban areas (Wong et al., 2013). Mortality increases are observed even at small temperature increases and extreme heat waves cause more fatalities in the USA than hurricanes, floods and tornadoes combined (ibid.). The European heatwave of August 2003 was responsible for 35,118 deaths (ibid.). In major American cities, the summer daytime temperature is higher for people of colour and the poor compared to White and wealthier populations (Hsu et al., 2021). Intra-city heat variations in the USA are linked to historical planning and housing policies, whereby higher temperatures are experienced in neighbourhoods that had historically been 'redlined' by the Home Owners' Loan Corporation (HOLC) in the 1930s (Hoffman et al., 2020). Neighbourhoods were ranked by the HOLC in maps delineating areas between 'best' and 'hazardous' (across four categories) real estate investments that were largely based on the racial profile of the neighbourhood. This practice was later banned but is widely blamed for disparities in environmental conditions within US cities. According to Jeremy Hoffman et al., the potential causes of the overheating now occurring in redlined neighbourhoods include the significant road and building construction projects that were concentrated there during the mid-1900s, alongside the relative lack of green infrastructure. The materials used in those highways and building projects (e.g. asphalt, concrete, cinder blocks and bricks) store heat and prevent these areas from cooling overnight. The authors recommend planning policies that require green infrastructure in new development, particularly taking account of the 'historic and present-day drivers that generate these asphalt-rich and tree canopy-poor land uses on intra-urban heat, and local communities' (ibid., p. 10).

Breaking the vicious cycle of urban heat and reducing excess morbidity and mortality requires consideration of how heat is distributed across a city and where interventions at multiple scales are best deployed to support health. Planners would benefit from UHI vulnerability assessments that map risk according to surface level temperatures, building characteristics, access to greenspace, crime, noise (as people may avoid opening windows to cool buildings) and other locally informed variables (Mavrogianni et al., 2022). This data can be used to combat UHI health risks by

Box 7.1 Policy to reduce overheating in London
Strategic planning in London is the shared responsibility of the Mayor of London, 32 London boroughs and the Corporation of the City of London. The Mayor is required to produce a spatial planning strategy (the 'London Plan') and to keep it under review. Chapter 5 of the London Plan sets out a range of policies for climate change mitigation and adaptation (Mayor of London, 2016). Policy 5.9 *Overheating and cooling* states that major development proposals should reduce potential overheating and reliance on air conditioning systems and demonstrate this in accordance with the following cooling hierarchy: minimise internal heat generation through energy-efficient design; reduce the amount of heat entering a building in summer through orientation, shading, albedo, fenestration, insulation and green roofs and walls; and manage the heat within the building through exposed internal thermal mass and high ceilings, passive ventilation, mechanical ventilation and active cooling systems (ensuring they are the lowest carbon options). Albedo refers to designing buildings that will reflect solar radiation, such as 'cool roofs' which use reflective paint, tiles or other materials.

informing policies that increase green and blue infrastructure and influence building layout, orientation and cooling strategies; and by determining high-risk neighbourhoods to target for retrofit solutions or other measures. For instance, in London, UK, the city adopted a hierarchy for reducing overheating in new development that seeks to maximise zero-carbon solutions (Box 7.1). While New York created a network of community assets that can be used during extreme heat events to provide free access to air-conditioned spaces in schools, community centres and other non-profit organisations called 'Cooling Centres', among other initiatives that target areas with high overheating risk. The report *Beating the Heat: A Sustainable Cooling Handbook for Cities* (UNEP, 2021) details numerous urban cooling strategies applicable in countries around the world.

Improving thermal comfort in existing housing can be achieved through adaptations to homes paid by government grants, housing providers or individual households. Fuel poverty (also called energy poverty) has traditionally referred to households with low incomes that could not be kept warm at a reasonable cost; but this concept should be broadly defined to encompass excess heat and cold (Sanchez-Guevara et al., 2019). The mechanisms to remedy fuel poverty (and increase thermal comfort) will vary depending on the housing stock, energy supply and climate. For example, in the UK and New Zealand, grants have been provided for heating system upgrades and insulation, alongside setting standards such as the UK's Decent Homes Standard for social housing and New Zealand's Residential Tenancies (Healthy Homes Standards) Regulations 2019 for rental accommodation (WHO, 2021). Upgrades to home energy efficiency and heating systems are important health equity measures that can improve health through multiple pathways (Thomson & Thomas, 2015). Decreasing energy consumption in buildings is also

Box 7.2 A systems thinking view of household energy efficiency and health

Clive Shrubsole et al. (2014) examined studies of the impacts of increasing energy efficiency in homes related to increased airtightness, replacing uncontrolled ventilation with purpose provided ventilation (PPV) and installing insulation. They found over 100 unintended consequences from these measures, many of which related to health. This paper demonstrates the complexity of the topic with interventions producing numerous potential negative and positive effects. For example, increasing airtightness may result in quieter indoor environments, which may support wellbeing and child development. However, airtightness may also lead to increased relative humidity (which leads to risk of dust mites, mould and asthma severity for example) and exposure to indoor air pollutants. Similarly, increased insulation can reduce excess cold in the winter but may lead to higher risk of overheating in warm months, leading to excess deaths. These consequences can be reduced or avoided through integrated design and policy approaches that consider indoor environmental quality alongside energy efficiency strategies.

an important carbon reduction strategy. However, there may be unintended health consequences of increased energy efficiency in homes when integrated design is not applied (see Box 7.2).

Increasing thermal comfort in new buildings and public spaces requires design strategies that take into account physical and personal variables alongside other design goals, such as climate change mitigation. Existing standards and engineering approaches for thermal comfort have been informed by four physical variables (air temperature, air velocity, relative humidity and mean radiant temperature) and two personal variables (clothing insulation and activity level, i.e. metabolic rate), yet these factors are not always adequately considered by design teams (Djongyang et al., 2010). Such standards have been criticised for driving a narrow range of acceptable indoor temperatures that are not suited to climate change mitigation objectives and do not meet the needs of all building occupants. Moving forward, there is a need to rethink building standards and planning policies to ensure a more equitable distribution of building and urban features that are protective against extreme heat and cold, whilst also avoiding mechanical cooling systems that contribute to climate change where possible. Mumovic and Santamouris (2019) provide guidance about integrated design for thermal comfort and sustainability in buildings.

7.2.3 *Affordability*

Housing affordability and tenure security are two inter-related goals for healthy housing that vary across nations and require due consideration to local context. At a basic level, affordable housing is important for health in terms of providing shelter and ensuring that occupants have enough remaining income for food, fuel and other necessities. More broadly, affordable housing is important for health and spatial equity. Planning policies often aim for mixed income/tenure housing to avoid spatially segregating people by socio-economic status. Homes in desirable urban neighbourhoods are more expensive partly because they have amenities that people value such as transport nodes, good schools and parks. Buying or renting a home at market rates in these neighbourhoods is often unaffordable for public sector or low paid workers, such as teachers and nurses, retired people or those relying on social welfare. Without planned provision of affordable housing, these population groups are relegated to poorer quality 'naturally' affordable housing, which is often located further away from places of work and school. As previous sections of this book have demonstrated, environmental health burdens tend to be concentrated in areas of deprivation, meaning that separating people based on income will result in health inequities. A healthy city is one that provides high-quality affordable housing that is spatially distributed.

Unaffordable housing, tenure insecurity and other forms of housing disadvantage have been shown to harm physical and mental health (Singh et al., 2019). Researchers in Australia have contributed substantial evidence on affordability and mental health. Using a longitudinal sample of 20,906 Australians, Emma Baker et al. (2020) showed that people with low mental health scores were more vulnerable to the negative health effects of unaffordable housing (measured by Housing Affordability Stress) which refers to a household 'in the lowest 40% of the equivalised disposable income distribution and paying >30% of gross household income in rent or mortgage costs' (p. 3). Another study by this research group found that when private tenants' homes became unaffordable, they experienced a small decline in mental health that was not found among homeowners whose housing became unaffordable (Bentley et al., 2016). In the USA, Janette Downing's (2016) systematic review documents the increased anxiety and depression associated with the foreclosure crisis following the housing bubble burst beginning in 2007, which led to one in 45 homes being subject to foreclosure filing. Her research highlighted that neighbourhoods with low-income and minority ethnic populations were disproportionately affected by sub-prime lending and foreclosures, resulting in health inequities. The review also showed that experiencing unaffordable housing was associated with poor self-reported health.

In policy terms, the concept of affordable housing differs internationally with regard to the percentage of household income spent on housing, the difference between market and affordable rates and the mechanisms for providing lower-cost housing. Policy definitions for affordable housing are often linked to tenures. For instance, the National Planning Policy Framework in England defines affordable housing as including the following types: affordable housing for rent, starter homes, discounted market sales housings and other affordable routes to homeownership (MHCLG, 2021). Affordable housing should be at least 20% lower than local market rates, whether they are sold or rented, and the English system relies on public and private sector organisations to build and manage such properties. The Swedish system focuses on rental accommodation for low-income residents, which comprise

Box 7.3 A new model for affordable housing in Gothenburg, Sweden
City leaders in Gothenburg, Sweden, found that historical affordable housing models were insufficient to deal with issues of social segregation driven by increasing immigration (Pineo, 2015). The city experimented with a new approach whereby developers were invited to build on city land in the old port neighbourhood of Frihamnen if they committed to provide a range of affordable rental units. The rent levels were based on research of local incomes and household costs with four bands ranging from 'very affordable' at 56% of the average market rate to 'luxury' at 121%. The city planners explicitly aimed to ensure that the redevelopment of Frihamnen was an inclusive community that would attract the city's growing population of immigrants and Swedish-born residents. Gothenburg's model aimed to avoid some of the negative health and social impacts that can be created through urban change.

roughly half of the units in new developments (see Box 7.3). The country has a system of non-profit public housing, described as *allmännytta* 'for the benefit of everyone', which is available to anybody regardless of need (Hedman, 2008). Landlords and tenant unions negotiate rents using a 'utility value system', whereby rents are based on quality, location and management of the property, and this has influenced market rent rates (ibid.); however, this has not been sufficient to address housing affordability challenges in some cities, as described in Box 7.3. In the USA, Dolores Acevedo-Garcia et al. (2004) examined the health effects of housing mobility policies. Rental assistance (e.g. through housing vouchers) is the main form of government housing assistance for low-income residents. Housing vouchers aim to deconcentrate poverty and racially desegregate poverty through housing mobility, where residents can leave deprived neighbourhoods such as through the Moving to Opportunity (MTO) programme. Their review found that housing mobility policies resulted in better self-reported health and reduced asthma attacks in children, which could have been attributed to better household or neighbourhood environmental quality. A later evaluation of MTO also found that participants had reduced diabetes, obesity, physical limitations and psychological distress (Sanbonmatsu et al., 2012).

7.2.4 *Tenure Security*

Tenure security relates to health in terms of both the material and meaningful aspects of housing described earlier in this chapter, and it is closely linked to housing affordability. Homeowners are more likely to have the financial means and the rights to maintain and modify their homes to meet their material housing and health-related needs, such as extending a home's size as a family grows or installing shutters to increase thermal comfort. In some contexts, renters may be less likely to achieve the meaningful aspects of homes, such as social status and ontological security, that relate to mental health and wellbeing (Shaw, 2004). The concept of tenure is not binary between ownership and rental; it includes a variety of categories that vary internationally and create different potential risks for health. The term 'tenure' describes the relationship between the household and the housing unit (Lee & Van Zandt, 2019) and it indicates the 'legal status of a person's right to occupy their home' (Angel & Gregory, 2021, p. 3). Bentley et al. (2016) state that tenure is not intrinsically good or bad for health, but instead it is 'the combination of social, legal, economic and cultural dynamics surrounding individual tenures in particular places [that] exposes individual households to stress, and consequent declines in mental health' (p. 218). Lee and Van Zandt have argued that renters are more vulnerable to disasters as their homes tend to suffer greater damage and they have fewer resources and rights in recovery stages. The term tenure security is sometimes described as housing security and it relates to the 'ability to retain a home, and owners have a much stronger right to stay in the home than do renters', although such rights vary internationally (Lee & Van Zandt, 2019, p. 158).

There are clear relationships between tenure insecurity and health and wellbeing, even after accounting for other financial hardship factors. As described under affordability, some population groups are disproportionately impacted by tenure insecurity, including children, people with pre-existing mental health conditions, and low-income and minority ethnic groups. Household insecurity has been linked to behavioural problems in children (Krieger & Higgins, 2002). Threat of eviction is a main category of housing insecurity according to the European Typology on Homelessness and Housing Exclusion (ETHOS) and it includes difficulty making rent or mortgage payments, alongside other insecurity issues of living under the threat of violence or in insecure accommodation (Vásquez-Vera et al., 2017). The threat of forced eviction results in 'negative mental (e.g. depression, anxiety, psychological distress, and suicides) and physical (poor self-reported health, high blood pressure, and child maltreatment) health outcomes' (ibid., p. 205). Levels of homelessness are rising in high-income countries and being unhoused is associated with poor health and disability (Aubry et al., 2020). Providing permanent supportive housing has been shown to increase housing stability, food security and subjective quality of life that can end homelessness (ibid.). Comparative analyses have helped to understand the effects of housing and other public policies on tenure security and health. For instance, a study found higher levels of wellbeing among Austrian mortgaged owners compared to those in the UK, which the authors stated could be explained in part by 'the decline in UK pension provision and the increasing reliance on housing as an asset to finance future welfare consumption' (Angel & Gregory, 2021, p. 16).

Urban regeneration and gentrification are often seen as causes of tenure insecurity and housing affordability problems that relate to health, as explored in Box 7.4. While government-led housing renewal programmes may be offered as a solution for housing affordability, they must be managed carefully to avoid unintended consequences, such as the displacement of residents living in the private rented sector. The solutions for housing insecurity will vary internationally and should be focused on the groups at highest risk. Vásquez-Vera et al. (2017) describe strategies and interventions for improving housing systems, increasing social rental

Box 7.4 Regeneration, gentrification and health

Upgrades to urban infrastructure, whether planned or naturally occurring, can affect health in many ways. When the process is led by urban or national policy, it is often called urban revitalisation, renewal or regeneration. These changes may lead to gentrification where low-income residents who previously inhabited the upgraded neighbourhood are displaced by wealthier incomers. Research on the health impacts of planned neighbourhood renewal provide mixed results (Thomson et al., 2006). Thomson and Thomas (2015) use systematic review data to develop a theory about housing improvements and health. They explain why the theoretical health improvements that should follow area-wide housing and infrastructure improvements are not necessarily detected by research studies. For instance, changes could be small or may require longer timescales of measurement. There could also be negative health effects from neighbourhood renewal or gentrification processes.

As introduced in Chap. 3, the negative health impacts of gentrification can result from loss of income and support networks and residents' perception that they no longer 'belong' in a place. Residents may be forced to move to less expensive and lower quality neighbourhoods, potentially increasing their exposure to health risks. A systematic review about the health impacts of gentrification and urban development highlights the complex political, social and economic factors that influence this relationship, meaning that these development processes are not necessarily 'good' or 'bad' for health (Schnake-Mahl et al., 2020). The authors found that community representation in the process, pre-existing spatial and racial inequities, levels of affordable housing and other factors could mediate the potential health outcomes. However, they also highlighted deficiencies in the existing evidence base that should be overcome to understand these factors more fully.

Planners and other stakeholders involved in leading urban change are often caught between competing interests and voices. It can simultaneously be the case that the health of a specific community is threatened by poor-quality housing and poor infrastructure, yet the solutions to remedy these risks would cause broader neighbourhood changes that would also threaten health, such as loss of power in local decision-making or increased rents. Participatory processes are essential to understand and respond to residents' concerns with urban change. The financial value created through new development and improvements to a neighbourhood should benefit existing residents to prevent unintended health disbenefits. Planners should also ensure that residents are not displaced due to changes in affordability or tenure security caused by the upgrades to housing and infrastructure. The specific mechanisms for achieving this vary, but they could include rent caps, replacement of social housing units and provision of additional affordable or shared-ownership units.

accommodation and providing universal social protection. With regard to disaster resilience in the USA, Lee and Van Zandt (2019) propose increasing the stock and quality of affordable housing and rent control policies to stop rent increases after disasters, which are likely to increase due to climate change.

7.2.5 *Lighting*

Artificial light and daylight can impact health and wellbeing by regulating sleep, improving mood and even increasing recovery times following medical procedures. Productivity, mood, sleep quality and concentration in schools, offices and for night-shift workers can be increased through the use of blue-enriched white light (Sletten et al., 2017; Keis et al., 2014; Viola et al., 2008). Lighting is also important for safety within homes and public spaces. Existing building standards have focused on visual performance, comfort, aesthetics and safety, alongside energy efficient (Cedeño-Laurent et al., 2018a), leaving a gap regarding the quality of light for health purposes. Artificial lighting risks negative health impacts from energy use, light pollution (harming people and ecosystems) and sleep disruption.

The evidence base demonstrating specific health and wellbeing outcomes from lighting is relatively weak, with a great breadth of topics covered, but a lack of consistent measures allowing synthesis of effects and an over-reliance on self-reported measures. Table 7.4 lists the results of a systematic review by Oluwapelumi Osibona et al. (2021) of the health impacts from natural and artificial light exposures. They highlight that 860 million people do not have access to electricity, instead they rely on kerosene and oil light sources which create risks for respiratory diseases and burns. Goal 7 of the UN SDGs aims to address this problem: 'Ensure access to affordable, reliable, sustainable and modern energy for all' (UN General Assembly, 2015).

Both artificial and natural lighting play an important role in sleep by synchronising human circadian rhythms (Cedeño-Laurent et al., 2018a) and exposure to both has changed substantially during processes of industrialisation and urbanisation (see Box 7.5). There are also short-term effects of lighting that can inhibit melatonin

Table 7.4 Health and wellbeing outcomes associated with diverse lighting exposures

Natural light	Artificial light
Sunlight exposure associated with reduced risk of tuberculosis. Morning illumination associated with improved sleep and better depression scores. Lack of natural light associated with increased reporting of depression, injuries from falls in the home and risk of leprosy.	Kerosene- and oil-fuelled lights associated with acute respiratory infections and burns, including for children. Exposure to cooler, blue enriched 17,000 K light (compared to warmer 4000 K light) associated with lower reported daytime anxiety levels and worsened subjective sleep quality. Light at night exposure (≥3 lux) associated with increased risk of dyslipidemia, body mass index and abdominal obesity, and at levels ≥5 lux higher night-time blood pressure and depression.

Source: Osibona et al. (2021)

Box 7.5 Sleep and the built environment

The amount and quality of sleep that people have each night is increasingly understood as important for health, wellbeing and performance. Cappuccio et al. (2018) set out some alarming statistics about the health impacts of changing sleep patterns in society in their book on *Sleep, Health and Society: From Aetiology to Public Health.* Whilst the epidemiology of sleep remains a relatively new field, there is evidence that reduced sleep is a strong predictor of cardiovascular disease and sleep deprivation is associated with chronic conditions. Worryingly, they reported that sleep disorders are increasing. In the USA, the number of people with this problem is predicted to double over the next 20 years.

The close links between sleep and the built environment are described by Cedeño-Laurent et al. (2018a), with the three main factors being lighting, thermal conditions and noise. Lighting affects neuro-endocrine and neuro-behavioural responses and it helps to synchronise human circadian rhythms, which have not evolved to the constant light exposure of modern society. The process of thermoregulation is closely tied to sleep regulation. Being in a room that is too warm or too cold can disrupt sleep, with studies showing that a temperature range of 20–29 °C is considered thermoneutral, depending on factors such as clothing and bedding. Noise is a significant cause of sleep disruption that is expected to increase with urbanisation and the growth of air travel. Cedeño-Laurent et al. point out that nocturnal noise is more strongly associated with cardiovascular disease than daytime noise, meaning that bedroom noise conditions are particularly important. They also highlight that all of these factors need to be considered holistically, particularly as people may have to make trade-offs between noise, thermal comfort, safety and other factors.

As with other environmental health burdens, low-income and minority ethnic populations are more likely to live in places where environmental factors that contribute to poor sleep are higher. Cedeño-Laurent et al. state, 'Environmental inequalities in combination with other psychosocial stressors in poor communities are associated with insufficient sleep or precarious sleep hygiene, resulting in increased risk for presenting cardiovascular and other chronic diseases' (p. 8). In addition to high occupancy density, noise exposures and poor housing quality, these populations also have higher exposures to indoor air pollutants which have been associated with reduced sleep duration and quality. They outline a number of built environment solutions that should be 'universally accessible' to avoid sleep and health inequities, including: lighting sources that mimic natural daily variation in the sun's intensity and spectrum (including smartphones and tablets); applying a new lighting metric for melanopic lux in building standards; increased attention to thermal conditions for sleep in building design, including the potential for smart technologies to balance thermal conditions with personal sleep efficiency metrics; and using building orientation/layout and acoustical dampening technologies to reduce noise exposure (ibid.).

production and therefore impact sleep. The principle that can be adopted by design professionals is to ensure a consistent '24-hour light–dark cycle each day, with daylight or high-intensity blue-enriched electric light in the day and sleep in darkness at night' (ibid., p. 3).

Patients' access to daylight in healthcare facilities may improve wellbeing, pain management and time in hospital. A systematic review focused on cancer management found that natural lighting and noise reduction were associated with patients' wellbeing, while access to nature (for patients, visitors and staff) was associated with better outcomes (Gharaveis & Kazem-Zadeh, 2018). An American study of spinal surgery patients found that greater exposure to sunlight after surgery was associated with reduced stress, pain and pain relief medication, compared to patients with lower exposures (Walch et al., 2005). While a study in Korea found that increased daylight could reduce patients' average length of stay in hospital (Choi et al., 2012). There is a growing body of research on healthcare facility design and patient health and wellbeing demonstrating the importance of layout, access to nature, lighting, noise, ventilation and many other factors (Halawa et al., 2020; Lledó, 2019; Ulrich et al., 2008).

Although the evidence base is still growing, a number of design and policy recommendations can be formed from existing knowledge. Cedeño-Laurent et al. (2018a) argue that lighting quality is often disregarded in building standards, with a focus on visual performance, comfort, aesthetics and safety, alongside energy efficient lighting. They propose the development of a new lighting metric to inform design for improved sleep (see Box 7.5). Osibona et al. (2021) suggest that 'stronger emphasis should be placed on the physiological impacts of lighting in homes' (p. 16). Integrated design methods are needed to balance the benefits of light with negative impacts from glare or overheating, for example. Lighting must be considered at the scale of blocks and urban policy rather than simply individual buildings, as neighbouring units may block light indoors, in private gardens or in public spaces, and lighting standards have had significant impacts on urban form around the world (Carmona, 2021). There is conflicting guidance available to planners and design professionals about the amount of natural light that should be achieved within buildings. Further evidence about the health impacts of lighting may support more informed design guidance, but there is also a need to apply a health equity perspective. For instance, developers and planners should consider whether the homes with the poorest quality lighting in a new development are the allocated units for affordable housing that will disproportionately impact people on low incomes and ethnic minorities.

7.2.6 Space

The importance of adequate space for health relates to overcrowding, but it is also important to enable privacy, study, play, cooking and storage. Measures of overcrowding (e.g. persons per room) assess whether there is adequate dwelling space for occupants' needs related to shelter, space and privacy (Shannon et al., 2018).

Household overcrowding is linked to the spread of infectious diseases, such as tuberculosis (Braubach et al., 2011) and COVID-19 (Aldridge et al., 2021), and psychological distress (Krieger & Higgins, 2002). Reducing overcrowding by increasing usable space in homes (e.g. through household extensions or warmth interventions) has been shown to reduce stress, illness and mess, whilst improving use of the kitchen leading to better diet, studying and leisure opportunities and family functioning (Thomson & Thomas, 2015).

Children may be particularly disadvantaged by lack of space in homes. James Dunn's (2020) review of child health and housing highlights the potential for a psychologically protective effect of adequate space which allows children space for self-regulation from social or environmental stimuli (e.g. noise and light), building on Kaplan's Attention Restoration Theory (see Chap. 5). Matt Barnes et al. (2011) investigated the health impacts of living in poor-quality housing for children using data from a representative longitudinal study in Britain. Homes in England are among the smallest in Europe, owing partly to the lack of legal space requirements that are present in other European countries (Madeddu et al., 2015). Barnes et al.'s study found that longer durations of living in poor housing increased children's vulnerability to poor outcomes. Living in overcrowded housing increased children's risks of feeling unhappy about their health, not having a quiet space to study and being suspended/excluded from school. Persistently living in overcrowded housing resulted in reduced physical activity in comparison with children not living in poor-quality housing. These negative effects may trigger a chain of events that affect children's health throughout their lives.

Adequate space can be required or encouraged through healthy housing standards, building regulations or other government policies. For example, there is a national space standard that can be adopted by local planning authorities in England where there is evidence for the need to do so (DCLG, 2015). In the USA, the American Public Health Association has been working to improve the standard of housing for health since 1938. Their National Healthy Housing Standard (NCHH and APHA, 2014) sets out minimum space requirements, such as a minimum floor area for habitable rooms of 70 ft^2 (6.5 m^2). A type of housing that avoids these standards is known as micro-housing (also called micro-apartments and micro-flats). These units are typically a single room (with a kitchenette, toilet and shower/bath) that is roughly 150 ft^2 (14 m^2) and units may be clustered around a communal space, such as a shared kitchen. Micro-housing is market rate housing that targets young professionals or students living in cities where larger accommodation would be prohibitively expensive. Some potential challenges with micro-housing were raised by Seattle-based interview participants in Pineo and Moore's (2021) exploratory study of integrating health and wellbeing into new development. Participants discussed how small efficiency dwelling units (SEDUs), as they are known in Seattle, were not being designed to high standards (better examples were provided in Los Angeles) and were seen as driving potential wellbeing issues for residents. The units are not covered by environmental and design review processes in Seattle and have been criticised for increasing density in neighbourhoods without adequate provision of infrastructure (Haines, 2020). The health and wellbeing impacts of micro-housing need to be further researched from the multi-scalar perspective of healthy urbanism.

7.3 Monitoring Indicators for Local Health at the Building Scale

Measuring the health and sustainability impact of buildings is comparatively easier than evaluating these factors at the neighbourhood and urban scale. Nevertheless, post-occupancy evaluations are still a rare occurrence even though they can identify easily fixable design issues. The indicators in Table 7.5 were gathered from Pineo et al.'s (2018a) systematic review and they are not only relevant to the building scale. Philomena Bluyssen (2010) provides detailed information on measuring the health and comfort impacts of buildings, including indicators for thermal comfort and lighting, acoustical and air quality.

7.4 Using Sustainable Design for Health in Nightingale Housing

Improving the sustainability and health impacts of residential buildings involves a step change in current development models. Despite the existence of sustainability standards, such as BREEAM and LEED for three decades, a stubbornly small portion of new residential development exceeds minimum building regulation requirements for energy use in line with the reductions required to avoid global environmental breakdown. The Nightingale Housing model from Melbourne,

Table 7.5 Selected monitoring indicators for local health impacts at building scale

Goal	Monitoring indicators
Acoustic comfort	% of population exposed to noise pollution measured at Lden >55 dB.
	% of residents exposed to noise levels higher than 35 dB during the night
	% of residents exposed to noise levels higher than 45 dB during the day
	% of residents who report noise annoyance.
Thermal comfort	Urban heat index
	% of residents in fuel poverty
	% of households using wood (or coal) fires for home heating
Affordability	% of households paying over 30% of income on rent or mortgage
	Ratio of the citywide median household income to median home values in the area
	% respondents who are satisfied with the state of repair of their home
Tenure security	% of households with access to secure tenure (owned or rented)
	% of housing which is operated as social housing
	Rate of homelessness per 10,000 people
Lighting	% of people who feel that streets and public spaces are well-lit at night
	% of people exposed to inadequate lighting
Space	Occupancy per area (m^2) or per habitable room
	% of households with one or more rooms too few

Australia, is challenging the status quo of residential development with the goal of creating a scalable approach for sustainable housing. The project began with The Commons which was built on land purchased in 2007 by six architects with a vision to shake up the concept of affordable and sustainable housing (Doyon & Moore, 2019). Although their full goals were disrupted by the global financial crisis, they completed the project in 2013 and then moved on to create Nightingale 1 across the street from The Commons (Fig. 7.1) as the first demonstration of their new housing model under the newly created Nightingale Housing Pty. Ltd., a non-profit organisation (Moore & Doyon, 2018). Sustainability and affordability are their driving goals and these are explicitly linked to health, social interaction and community through the non-profit's guiding principles: affordability, transparency, sustainability, deliberative design and community contribution. The Nightingale Housing model provides internationally relevant lessons with regard to its guiding principles and ways of working, as explored by Andréanne Doyon and Trivess Moore (ibid.). This case study is based on Doyon and Moore's articles and the author's site visit to Nightingale 1 and tour with a project architect and resident in April 2018. The sustainability accolades for both The Commons and Nightingale 1 are highlighted in Table 7.6 and both projects have received awards.

There are two key lessons from this case study of relevance to healthy urbanism. First, the Nightingale Housing model aims to be replicable and scalable to transform what the organisation's founder, Jeremy McLeod, described as the unaffordable and low-quality status quo of the Australian housing industry. This goal is

Fig. 7.1 Front facades of The Commons (left) and Nightingale 1 (right) apartment buildings with ground floor commercial uses. (Source: Author, 2018)

Table 7.6 Sustainability achievements at The Commons and Nightingale 1

The Commons	Nightingale 1
7.5 Star NatHERS rating, equal to predicted thermal energy load of 68 MJ/m²/year (40% than regulatory requirements) On-site renewables (5 kW solar PV, solar hot water, hydronic heating boiler) No mechanical air conditioning or toxic indoor materials Undercover bicycle parking with 70 spaces and car share scheme Passive design uses double-glazed windows, timber doors, exposed thermal mass (concrete), rainwater collection Locally manufactured and recycled materials Communal spaces in rooftop garden and laundry	8.2 Star NatHERS rating, equal to predicted thermal energy load of 48 MJ/m²/year. On-site renewables (e.g. 18 kW solar PV) and the building is 100% fossil-fuel free Improved rooftop garden design from The Commons 42 bicycle parking spaces and 3 car spaces Rainwater collection for irrigation and the common-area toilets Covenant applied to building which caps sales price of apartments to the average price rise of the area

Note: NatHERS is the Nationwide House Energy Rating System
Source: Moore and Doyon (2018), Doyon and Moore (2019)

achieved by providing transparency regarding design, specifications and costs, alongside the use of 'tried and tested' building approaches that do not require specialist knowledge in construction. Architects can become licensed to develop Nightingale projects (for a fee) and they receive help with financing and access to the waiting list of prospective apartment owners, who are, in turn, involved in project decision-making. In 2018, Nightingale Housing adopted the German Baugruppen model of community-led housing (see Chap. 8) which will be used on a large project of seven buildings in Melbourne. Doyon and Moore (2019) reported that there were 20 licensed architects and 12 completed or current projects, mainly in Victoria but also Western Australia. Embedded in this partnership approach is what Doyon and Moore (2019) described as 'learning-by-doing and learning from others so the newer Nightingale Housing developments push boundaries further than its predecessors' (p. 22). This reflexive approach was also identified as important for increasing healthy urbanism practices by Pineo and Moore (2021), especially given the need for experimentation in this area.

A second key lesson from the Nightingale Housing model relates to its combined affordability and sustainability attributes, with multiple benefits for health. Pineo and Moore (2021) reported that built environment stakeholders found the economic constraints and risks of healthy urban development to constrain developers' willingness to adopt healthy design measures at different scales. Planners and design teams note that quality design is balanced with targets for affordable housing contributions and compromises are usually made. The Nightingale Housing model reduces the cost of purchasing and running its apartments by diverging from industry trends, including the following: not offering onsite car parking, second bathrooms, individual laundry, mechanical air conditioning; capping profit for investors at 15% and avoiding marketing and real estate agent costs. This means that new units are cheaper to purchase, and through adoption of the Baugruppen model, they aim to recuperate profits for apartment owners. Moore and Doyon (2018) note that apartment owners are targeted as those who are more aligned to sustainable living and at Nightingale 1 the architects

> wanted the building to be occupied by owners rather than renters in the hope that this would foster a more stable community and avoid property speculation. This was addressed through a selection process (e.g., interviews/surveys with perspectives [prospective] occupants) to ensure an understanding of what occupants wanted and their financial capacity to afford the property, and through creation of a covenant on the building restricting apartments from being on-sold for more than the average price rise of the local area for 20 years, therefore locking in affordability gains for future owners. (p. 10)

The problem of property speculation reducing sense of community in apartment buildings has also been reported in cities in the USA, the UK and Canada. However, there is a risk that this exclusive and selective model contributes to the existing segregation of socio-economic groups within society based on those who can afford to purchase property, although the model aims to produce affordable property. The Nightingale model is still in its infancy and it should be evaluated over time to understand the implications for health equity.

There are numerous sustainability measures in Nightingale 1 and The Commons that contribute to its benefits across multiple scales of health impact. Rainwater harvesting is used for irrigation on private balconies (Fig. 7.2) and the rooftop garden and common-area toilets. Grape vines grow down the north-facing facade and balconies, providing shading during the summer and allowing the sun to warm the apartments in winter. Other passive design measures, primarily thermal mass from concrete, help to provide thermal comfort and residents report that apartments

Fig. 7.2 A private balcony in the Nightingale 1 building shows irrigation and grape vines. (Source: Author, 2018)

Fig. 7.3 Roof-top communal laundry facilities and social space for residents on Nightingale 1. (Source: Author, 2018)

remain cool during heatwaves. Communal spaces, such as the shared laundry facilities and shared outdoor kitchen (Figs. 7.3 and 7.4), support social interaction and environmental goals. The shared washing machines are integrated into the greywater recycling system and save space in individual units. Residents can dry their laundry in a communal hanging space, reducing excess moisture in homes. The buildings are adjacent to a train station and they contain ample bicycle storage, reducing the need for private car ownership.

A concluding point about the value of Nightingale Housing relates to its position as a pioneer of a new type of housing, demonstrating to the wider industry and the public what can be achieved. Doyon and Moore (2019) describe this as an 'unprotected niche', or a form of experimentation in a living lab, adopting conceptualisations from the sustainability transitions literature (see Chap. 8). By 'unprotected', they mean that Nightingale Housing did not benefit from the use of public land, subsidies or other support that other pilot and demonstration projects typically receive. That the model has succeeded without such support is a testament to its value, scalability and replicability in other parts of Australia, New Zealand and possibly further afield in the future. Further evaluations of the Nightingale Model from a healthy urbanism and sustainability transition lens would help to establish whether this approach has disrupted the Australian housing market (and elsewhere) for the better and to what extent the developments support health through sustainability, equity and inclusion.

Fig. 7.4 Communal social space on the roof of Nightingale 1 and the facade of The Commons visible across the street. (Source: Author, 2018)

7.5 Conclusion

A noticeable shift has occurred in the development industry over the past decade with increasing attention to health and wellbeing driven by many factors, not least growing awareness of the urgency of dealing with global challenges such as climate change and rising chronic diseases. Compared to a decade ago, there are now vastly different expectations about how buildings meet occupants' health and wellbeing needs; however, there is a risk that many people in society will be left behind in the drive for healthier buildings. It is therefore essential that debates about healthy buildings are adequately balanced with consideration of inclusion, equity and sustainability. There has also been recognition of a new market opportunity to sell and lease properties as 'healthy' with multiple reports attesting to the added value of such features in office and residential markets (Bolden, 2020; Clifford, 2018). The COVID-19 pandemic provided additional momentum for that shift by focusing the attention of building operators and occupants on ventilation, social distancing and adaptable spaces. The next chapter will delve into the issue of implementing healthy urban development through urban policy, design and management.

References

Acevedo-Garcia, D., Osypuk, T. L., Werbel, R. E., Meara, E. R., Cutler, D. M., & Berkman, L. F. (2004). Does housing mobility policy improve health? *Housing Policy Debate, 15*, 49–98.

Aldridge, R. W., Pineo, H., Fragaszy, E., Eyre, M., Kovar, J., Nguyen, V., Beale, S., Byrne, T., Aryee, A., Smith, C., Devakumar, D., Taylor, J., Katikireddi, S. V., Fong, W. L. E., Geismar, C., Patel, P., Shrotri, M., Braithwaite, I., Navaratnam, A. M. D …, Hayward, A., & on behalf of Virus Watch Collaborative. (2021). Household overcrowding and risk of SARS-CoV-2: Analysis of the Virus Watch prospective community cohort study in England and Wales. *medRxiv*. 2021.05.10.21256912.

Aletta, F., Oberman, T., & Kang, J. (2018). Associations between positive health-related effects and soundscapes perceptual constructs: A systematic review. *International Journal of Environmental Research and Public Health, 15*, 2392.

Allen, J. G., MacNaughton, P., Laurent, J. G. C., Flanigan, S. S., Eitland, E. S., & Spengler, J. D. (2015). Green buildings and health. *Current Environmental Health Reports, 2*, 250–258.

Allen, J. G., & Macomber, J. D. (2020). *Healthy buildings: How indoor spaces can drive performance and productivity*. Harvard University Press.

Angel, S., & Gregory, J. (2021). Does housing tenure matter? Owner-occupation and wellbeing in Britain and Austria. *Housing Studies, 0*, 1–21.

Aubry, T., Bloch, G., Brcic, V., Saad, A., Magwood, O., Abdalla, T., Alkhateeb, Q., Xie, E., Mathew, C., Hannigan, T., Costello, C., Thavorn, K., Stergiopoulos, V., Tugwell, P., & Pottie, K. (2020). Effectiveness of permanent supportive housing and income assistance interventions for homeless individuals in high-income countries: A systematic review. *The Lancet Public Health, 5*, e342–e360.

Babisch, W. (2006). Transportation noise and cardiovascular risk: Updated review and synthesis of epidemiological studies indicate that the evidence has increased. *Noise & Health, 8*, 1.

Baker, E., Pham, N. T. A., Daniel, L., & Bentley, R. (2020). New evidence on mental health and housing affordability in cities: A quantile regression approach. *Cities, 96*, 102455.

Barnes, M., Butt, S., & Tomaszewski, W. (2011). The duration of bad housing and children's wellbeing in Britain. *Housing Studies, 26*, 155–176.

Beatley, T., Jones, C., & Rainey, R. M. (Eds.). (2018). *Healthy environments, healing spaces: Practices and directions in health, planning, and design*. University of Virginia Press.

Bentley, R. J., Pevalin, D., Baker, E., Mason, K., Reeves, A., & Beer, A. (2016). Housing affordability, tenure and mental health in Australia and the United Kingdom: A comparative panel analysis. *Housing Studies, 31*, 208–222.

Bluyssen, P. M. (2010). Towards new methods and ways to create healthy and comfortable buildings. *Building and Environment, 45*, 808–818.

Bolden, K. (2020). Sustainable, healthy buildings and real estate. *Ernst Young*. Retrieved January 02, 2011, from https://www.ey.com/en_us/real-estate-hospitality-construction/sustainable-healthy-buildings-meeting-real-estate-expectations

Braubach, M., Jacobs, D., & Ormandy, D. (2011). *Environmental burden of disease associated with inadequate housing: A method guide to the quantification of health effects of selected housing risks in the WHO European region*. World Health Organization, Regional Office for Europe, Copenhagen.

British Council for Offices. (2018). *Wellness matters: Health and wellbeing in offices and what to do about it*.

Cappuccio, F., Miller, M. A., Lockley, S. L., & Rajaratnam, S. M. W. (2018). Sleep, health and society. In F. Cappuccio, M. A. Miller, S. W. Lockley, & S. M. W. Rajaratnam (Eds.), *Sleep, health, and society: From aetiology to public health* (pp. 1–9). Oxford Scholarship Online. Oxford University Press.

Carmona, M. (2021). *Public places urban spaces: The dimensions of urban design* (3rd ed.). Routledge.

Cedeño-Laurent, J. G., Allen, J., & Spengler, J. D. (2018a). The built environment and sleep. In F. Cappuccio, M. A. Miller, S. W. Lockley, & S. M. W. Rajaratnam (Eds.), *Sleep, health, and society: From aetiology to public health, Oxford Scholarship Online*. Oxford University Press.

Choi, J.-H., Beltran, L. O., & Kim, H.-S. (2012). Impacts of indoor daylight environments on patient average length of stay (ALOS) in a healthcare facility. *Building and Environment, 50*, 65–75.

Clements-Croome, D., Turner, B., & Pallaris, K. (2019). Flourishing workplaces: A multisensory approach to design and POE. *Intelligent Buildings International, 0*, 1–14.

Clifford, M. (2018). Investors see returns in healthy buildings. *Jones Lang LaSalle*. Retrieved January 02, 2021, from https://www.jll.co.uk/en/trends-and-insights/investor/investors-see-returns-in-healthy-buildings

de Dear, R., Kim, J., Candido, C., & Deuble, M. (2015). Adaptive thermal comfort in Australian school classrooms. *Building Research and Information, 43*, 383–398.

Department for Communities and Local Government. (2015). *Technical housing standards: Nationally described space standard.* Department for Communities and Local Government, London.

Després, C. (1991). The meaning of home: Literature review and directions for future research and theoretical development. *Journal of Architectural and Planning Research, 8*, 96–115.

Djongyang, N., Tchinda, R., & Njomo, D. (2010). Thermal comfort: A review paper. *Renewable and Sustainable Energy Reviews, 14*, 2626–2640.

Dovjak, M., & Kukec, A. (2019). *Creating healthy and sustainable buildings: An assessment of health risk factors*. Springer International Publishing.

Downing, J. (2016). The health effects of the foreclosure crisis and unaffordable housing: A systematic review and explanation of evidence. *Social Science & Medicine, 162*, 88–96.

Doyon, A., & Moore, T. (2019). The acceleration of an unprotected niche: The case of Nightingale Housing, Australia. *Cities, 92*, 18–26.

Dunn, J. (2002). Housing and inequalities in health: A study of socioeconomic dimensions of housing and self reported health from a survey of Vancouver residents. *Journal of Epidemiology and Community Health, 56*, 671–681.

Dunn, J. R. (2020). Housing and healthy child development: Known and potential impacts of interventions. *Annual Review of Public Health, 41*, 381–396.

European Environment Agency. (2020). Environmental noise in Europe. Publications Office, Luxembourg.

Gharaveis, A., & Kazem-Zadeh, M. (2018). The role of environmental design in cancer prevention, diagnosis, treatment, and survivorship: A systematic literature review. *HERD Health Environments Research & Design Journal., 11*, 18–32.

Gibson, M., Petticrew, M., Bambra, C., Sowden, A. J., Wright, K. E., & Whitehead, M. (2011). Housing and health inequalities: A synthesis of systematic reviews of interventions aimed at different pathways linking housing and health. *Health Place*, Health Geographies of Voluntarism, *17*, 175–184.

Hahad, O., Kröller-Schön, S., Daiber, A., & Münzel, T. (2019). The cardiovascular effects of noise. *Deutsches Ärzteblatt International, 116*, 245–250.

Haines, T. (2020). Micro-housing in Seattle update: Combating "Seattle-ization". *Seattle University School of Law Review Supra 43*, 6.

Halawa, F., Madathil, S. C., Gittler, A., & Khasawneh, M. T. (2020). Advancing evidence-based healthcare facility design: A systematic literature review. *Health Care Management Science, 23*, 453–480.

Hedman, E. (2008). *A history of the Swedish system of non-profit municipal housing*. The Swedish Board of Housing, Building and Planning.

Hoffman, J. S., Shandas, V., & Pendleton, N. (2020). The effects of historical housing policies on resident exposure to intra-urban heat: A study of 108 US urban areas. *Climate, 8*, 12.

Hsu, A., Sheriff, G., Chakraborty, T., & Manya, D. (2021). Disproportionate exposure to urban heat island intensity across major US cities. *Nature Communications, 12*, 2721.

Ige, J., Pilkington, P., Orme, J., Williams, B., Prestwood, E., Black, D., Carmichael, L., & Scally, G. (2018). The relationship between buildings and health: A systematic review. *Journal of Public Health, 41*(2), e121–e132.

International Organization for Standardization. (2014). *ISO 12913-1:2014 Acoustics—soundscape—Part 1: Definition and conceptual framework*. ISO, Geneva, Switzerland.

Keis, O., Helbig, H., Streb, J., & Hille, K. (2014). Influence of blue-enriched classroom lighting on students' cognitive performance. *Trends Neurosci. Educ., 3*, 86–92.

Kisacky, J. S. (2017). *Rise of the modern hospital: An architectural history of health and healing, 1870–1940*. University of Pittsburgh Press.

Krieger, J., & Higgins, D. L. (2002). Housing and health: Time again for public health action. *American Journal of Public Health, 92*, 758–768.

Lee, J. Y., & Van Zandt, S. (2019). Housing tenure and social vulnerability to disasters: A Review of the evidence. *Journal of Planning Literature, 34*, 156–170.

Lekaviciute, J., Kephalopoulos, S., Clark, C., & Stansfeld, S. (2013). *Institute for health and consumer protection, 2013*. Final Report: ENNAH European Network on Noise and Health. Publications Office, Luxembourg.

Lledó, R. (2019). Human centric lighting, a new reality in healthcare environments. In T. P. Cotrim, F. Serranheira, P. Sousa, S. Hignett, S. Albolino, & R. Tartaglia (Eds.), *Health and social care systems of the future: Demographic changes, digital age and human factors, advances in intelligent systems and computing* (pp. 23–26). Springer International Publishing.

Madeddu, M., Gallent, N., & Mace, A. (2015). Space in new homes: Delivering functionality and liveability through regulation or design innovation? *The Town Planning Review, 86*, 73–95.

Mavrogianni, A., Taylor, J., Symonds, P., Oikonomou, E., Pineo, H., Zimmermann, N., & Davies, M. (2022). Cool cities by design: Shaping a healthy and equitable London in a warming climate. In G. McGregor & C. Ren (Eds.), *Urban climate science for planning healthy cities*. Springer.

Mayor of London. (2016). The London plan, Chapter 3. *London's People*.

Ministry of Housing, Communities and Local Government. (2021). National planning policy framework.

Moore, T., & Doyon, A. (2018). The uncommon nightingale: Sustainable housing innovation in Australia. *Sustainability, 10*, 3469.

Mumovic, D., & Santamouris, M. (2019). *A handbook of sustainable building design and engineering: An integrated approach to energy, health and operational performance* (2nd ed.). Routledge.

National Center for Healthy Housing, American Public Health Association. (2014). National Healthy Housing Standard.

Office of the United Nations High Commissioner for Human Rights, UN Habitat. (2009). *The right to adequate housing*. Fact Sheet No. 21/Rev.1. Office of the United Nations High Commissioner for Human Rights, Geneva, Switzerland.

Osibona, O., Solomon, B. D., & Fecht, D. (2021). Lighting in the home and health: A systematic review. *International Journal of Environmental Research and Public Health, 18*, 609.

Pineo, H. (2015). Housing for inclusive communities—an experimental model in Gothenburg. *Town & Country Planning, 84*, 566–571.

Pineo, H. (2020). Towards healthy urbanism: Inclusive, equitable and sustainable (THRIVES)—an urban design and planning framework from theory to praxis. *Cities Health, 0*, 1–19. https://doi.org/10.1080/23748834.2020.1769527

Pineo, H., Glonti, K., Rutter, H., Zimmermann, N., Wilkinson, P., & Davies, M. (2018a). Urban health indicator tools of the physical environment: A systematic review. *Journal of Urban Health, 95*, 613–646.

Pineo, H., & Moore, G. (2021). Built environment stakeholders' experiences of implementing healthy urban development: An exploratory study. *Cities Health, 0*, 1–15. https://doi.org/10.1080/23748834.2021.1876376

Radicchi, A., Henckel, D., & Memmel, M. (2016). Citizens as smart, active sensors for a quiet and just city. The case of the "open source soundscapes" approach to identify, assess and plan "everyday quiet areas" in cities. *Noise Mapping, 5*, 1–20.

Radicchi, A., Yelmi, P. C., Chung, A., Jordan, P., Stewart, S., Tsaligopoulos, A., McCunn, L., & Grant, M. (2021). Sound and the healthy city. *Cities Health, 5*, 1–13.

Ritchie, H., & Roser, M. (n.d.). Emissions by sector [WWW Document]. *Our World Data*. Retrieved April 16, 2021, from https://ourworldindata.org/emissions-by-sector

Sanbonmatsu, L., Marvakov, J., Potter, N. A., Yang, F., Adam, E. K., Congdon, W. J., Duncan, G. J., Gennetian, L. A., Katz, L. F., Kling, J. R., Kessler, R. C., Lindau, S. T., Ludwig, J., & McDade, T. W. (2012). The long-term effects of moving to opportunity on adult health and economic self-sufficiency. *City, 14*, 109–136.

Sanchez-Guevara, C., Núñez Peiró, M., Taylor, J., Mavrogianni, A., & Neila González, J. (2019). Assessing population vulnerability towards summer energy poverty: Case studies of Madrid and London. *Energy and Buildings, 190*, 132–143.

Schnake-Mahl, A. S., Jahn, J. L., Subramanian, S. V., Waters, M. C., & Arcaya, M. (2020). Gentrification, neighborhood change, and population health: A systematic review. *Journal of Urban Health, 97*, 1–25.

Shannon, H., Allen, C., Clarke, M., Dávila, D., Fletcher-Wood, L., Gupta, S., Keck, K., Lang, S., & Allen Kahangire, D. (2018). Web Annex A: Report of the systematic review on the effect of household crowding on health, In *WHO housing and health guidelines*. World Health Organization, Geneva, Switzerland.

Shaw, M. (2004). Housing and public health. *Annual Review of Public Health, 25*, 397–418.

Shrubsole, C., Macmillan, A., Davies, M., & May, N. (2014). 100 Unintended consequences of policies to improve the energy efficiency of the UK housing stock. *Indoor and Built Environment, 23*, 340–352.

Singh, A., Daniel, L., Baker, E., & Bentley, R. (2019). Housing disadvantage and poor mental health: A systematic review. *American Journal of Preventive Medicine, 57*, 262–272.

Sletten, T. L., Ftouni, S., Nicholas, C. L., Magee, M., Grunstein, R. R., Ferguson, S., Kennaway, D. J., O'Brien, D., Lockley, S. W., & Rajaratnam, S. M. W. (2017). Randomised controlled trial of the efficacy of a blue-enriched light intervention to improve alertness and performance in night shift workers. *Occupational and Environmental Medicine, 74*, 792–801.

Sundell, J. (2017). Reflections on the history of indoor air science, focusing on the last 50 years. *Indoor Air, 27*, 708–724.

Thomson, H., Atkinson, R., Petticrew, M., & Kearns, A. (2006). Do urban regeneration programmes improve public health and reduce health inequalities? A synthesis of the evidence from UK policy and practice (1980–2004). *Journal of Epidemiology and Community Health, 60*, 108–115.

Thomson, H., & Thomas, S. (2015). Developing empirically supported theories of change for housing investment and health. *Social Science & Medicine, 124*, 205–214.

Ulrich, R. S., Zimring, C., Zhu, X., DuBose, J., Seo, H.-B., Choi, Y.-S., Quan, X., & Joseph, A. (2008). A review of the research literature on evidence-based healthcare design. *HERD Health Environ. Res. Des. J., 1*, 61–125.

United Nations Environment Programme. (2021). *Beating the heat: A sustainable cooling handbook for cities*. UNEP, Nairobi, Kenya.

United Nations General Assembly. (2015). Resolution adopted by the general assembly on 25 September 2015: Transforming our world: The 2030 Agenda for Sustainable Development.

Van Renterghem, T., & Botteldooren, D. (2016). View on outdoor vegetation reduces noise annoyance for dwellers near busy roads. *Landscape and Urban Planning, 148*, 203–215.

Vásquez-Vera, H., Palència, L., Magna, I., Mena, C., Neira, J., & Borrell, C. (2017). The threat of home eviction and its effects on health through the equity lens: A systematic review. *Social Science & Medicine, 175*, 199–208.

Viola, A. U., James, L. M., Schlangen, L. J., & Dijk, D.-J. (2008). Blue-enriched white light in the workplace improves self-reported alertness, performance and sleep quality. *Scandinavian Journal of Work, Environment & Health, 34*, 297–306.

Walch, J. M., Rabin, B. S., Day, R., Williams, J. N., Choi, K., & Kang, J. D. (2005). The effect of sunlight on postoperative analgesic medication use: A prospective study of patients undergoing spinal surgery. *Psychosomatic Medicine, 67*, 156–163.

WHO. (2018a). *WHO housing and health guidelines*. WHO, Geneva, Switzerland.

WHO. (2018b). *Environmental noise guidelines for the European Region*. WHO Regional Office for Europe, Copenhagen, Denmark.

WHO. (2021). *Policies, regulations and legislation promoting healthy housing: A review*. WHO, Geneva, Switzerland.

Wong, K. V., Paddon, A., & Jimenez, A. (2013). Review of world urban heat islands: Many linked to increased mortality. *Journal of Energy Resources Technology, 135*, 11.

World GBC. (2020). *Health & wellbeing framework: Six principles for a healthy, sustainable built environment*. Executive Report.

Chapter 8
Practising Healthy Urbanism

8.1 Introduction

The challenge of creating healthy cities is not solely a technical problem that can be solved through design innovations, standards and up-skilling of the relevant professionals. Instead, this challenge is largely social, rooted in issues of governance, politics and power. An important starting point towards achieving healthy urbanism is creating greater understanding among built environment professionals and decision-makers of the social and environmental determinants of health, particularly the structural barriers that are shaped by decisions about how health-promoting resources are distributed in society. Based on lessons from the history of urban planning and public health, Corburn (2017) argues that 'an over-reliance on technological solutions and physical designs, without accompanying institutional change, fails to protect the most vulnerable population groups' (p. 38). As societies recover from the economic and social impacts of COVID-19, broader acknowledgement of the interconnections between environmental degradation, social inequalities and health will need to be converted into action in the form of policies, infrastructure provision and good design. This will require planners and others to reconsider whose health is supported or harmed in processes of urban change using diverse forms of knowledge. Corburn calls for planners to

> question critically the adequacy of existing norms and institutions that help determine how practitioners use or abuse power, respond to or even resist market forces, work to empower some groups and dis-empower others, promote multi-party decision-making, or simply rationalize decisions already made. (ibid., p. 36)

This chapter contributes to this call for change by describing specific policy processes, financing mechanisms and models of development that could support health, whilst pointing to potential challenges. The focus of this chapter is about converting the theory and evidence presented in the previous chapters into practical actions for healthy urbanism.

H. Pineo, *Healthy Urbanism*, Planning, Environment, Cities,
https://doi.org/10.1007/978-981-16-9647-3_8

The implementation focus of this chapter addresses the type and scale of action encompassed by the THRIVES framework's design and planning goals and scales of decision-making—meaning that urban policies and design at multiple scales are relevant. Since there can be large variation in the systems governing such urban change, it is useful to take a moment to define what kinds of policy and development are discussed here. The focus is on the strategies, policies and plans arising from city planning departments, produced in collaboration with other public sector services (including public health) and the community. Planning is an ill-defined profession and planning systems vary internationally. A description by Parker and Street (2021) states that planning systems are

> *decision-making frameworks* that draw on knowledge inputs in order to inform choices about how land is used and how a variety of positive and negative impacts are managed. Creating formal plans is but one expression of this. (p. 4)

Urban plans will typically include sections on transport, housing, green infrastructure, education, employment, waste and so on. Although each of these subjects has specialised knowledge, professional bodies and sometimes their own city government departments, they should be considered holistically in integrated planning processes and are thus relevant in this chapter's focus on implementation. Changes to the urban built environment in the form of development are highly complex with many stakeholders responding to the effects of each other's actions over time. Rydin (2010) states that urban development is 'the process of physically producing the built environment, by bringing together multiple actors from construction companies to development financiers to local planners and others' (p. 15). The scale of such development may be small or large, including incremental changes to existing buildings and major urban renewal programmes (ibid.).

This chapter draws on the author's experience and research with practitioners to describe how healthy urbanism principles can be applied in policy-making and development with a focus on overcoming recognised challenges. A goal of this chapter is to outline the ways in which competing interests in urban development can be managed in pursuit of health. Such considerations are not technical or solvable by decision-support tools, instead they require open dialogue with interested parties (e.g. politicians, community members and developers). For these conversations to be successful, all parties need to have some understanding about the context in which the proposed urban change may or may not occur, such as financial constraints and historic burdens faced by communities. The following sections will unpack these contextual issues in more detail.

8.1.1 *Systems Thinking and Urban Governance for Health*

The transition from the current state of art in planning and urban design practice to one which routinely prioritises health and wellbeing involves making changes in complex systems. Chapter 3 introduced the concept of systems thinking as it relates to urban health, including a description of seven characteristics of complex systems: dynamic, high number of elements, interconnected, non-linear structure, feedback, counter-intuitive and emergent behaviour. The theoretical foundations for conceptualising urban health as a complex system and managing change within such systems has been linked to 'complexity thinking' or 'systems thinking' (Glouberman et al., 2006; Sterman, 2006; Rydin et al., 2012) and 'adaptive management' or 'adaptive systems science' (Corburn, 2013). The field of sustainability transitions or socio-technical transitions draws upon systems science and is aligned to existing approaches for healthy urban design and planning. Sustainability transitions theory 'focuses on advancing the understanding of highly institutionalised processes that constrain sustainable innovations in their attempts of leapfrogging the prevailing unsustainable alternatives—thereby constraining path-breaking and wide-scale changes' (Savaget et al., 2019, p. 879). Distilling out the common actions for policy and decision-makers described in this literature reveals the following themes:

1. Urban change should be informed by a wide *range of perspectives*, including residents' knowledge and diverse disciplines.
2. Policy and design solutions should be informed by *local context* and are otherwise subject to failure given the significant variations in urban systems and geographies.
3. Numerous localised policy and design approaches should be trialled with an *experimental* approach that includes *close monitoring* and adaptation where necessary to adjust unexpected problems.
4. Monitoring and evaluating the health and wellbeing impacts of policy and design measures must include a range of knowledge types, including *objective measures and lived experience*, alongside opportunities for dialogue.
5. Information from monitoring should inform *reflexive practices* of policy and design that can feed 'double-loop learning' (i.e. using feedback to change one's understanding of cause-and-effect relations).

These actions are not yet standard practice for urban governance and significant change will need to occur in professional education and practice to make these ways of working a reality. A focus on the political nature of healthy urbanism challenges is essential and again, drawing on Corburn (2017), this will require that 'decision-making processes and institutions that shape places are altered to focus on equity', and equitable distribution of resources (p. 32). This chapter provides examples of how this may be achieved in policy and infrastructure delivery, alongside sections on incorporating community knowledge and monitoring and evaluation.

The process of physically producing the built environment to support health is examined in this chapter with a focus on finances and models of development. Research on socio-technical transitions in sustainable housing policies (e.g. Moore et al., 2014) and developments (e.g. Moore & Doyon, 2018; Doyon & Moore, 2019) provides useful insights for the comparatively under-researched area of healthy urban development. For instance, the importance of individual champions, pilot/demonstration projects and the need for reflexive practice were described by participants in Pineo and Moore's (2021) exploratory study of the experiences of built environment professionals working on healthy urban development and those findings are widely acknowledged in socio-technical transitions (STT) accounts of sustainable housing. To understand how experimentation (the third theme above) can practically be applied in healthy urban planning and design, this chapter draws upon a review by Sengers et al. (2019) of socio-technical experiments which defines these as 'an inclusive, practice-based and challenge-led initiative designed to promote system innovation through social learning under conditions of uncertainty and ambiguity' (p. 161). While there are many examples of such experiments that aim to improve sustainability, there is a lack of literature evaluating such experiments for healthy urban planning and design from an STT perspective.

8.2 Policy-Making for Healthy Urbanism

The opportunities and methods for integrating health and wellbeing into the planning policy process are described by Pineo et al. (2019) as ideally occurring in a cycle. Figure 8.1 shows an adapted version of their simplified process of urban planning with an outer circle depicting constraints and influences on planning, such as higher-tier policies and environmental factors. The inner circle demonstrates how community and other stakeholder participation should influence all stages of planning, from the gathering of diverse knowledge about health and place to the monitoring and review stage after projects are implemented. Health impact assessment is shown as informing policy and decision-making. This is a normative description of policy-making as an adaptive process that involves learning and adjusting policies and decisions over time, building on frameworks such as Corburn's (2013) Adaptive Urban Health Justice approach. Corburn describes five components: (1) democratic participation, (2) integrated decision-making, (3) multi-dimensional monitoring, (4) social learning and (5) adjustment and innovation.

Research about the process of healthy urban planning has demonstrated that this idealised learning and decision-making cycle does not match reality on the ground (Pineo et al., 2020c; Pineo & Moore, 2021). In practice, the extent of democratic participation in planning varies considerably across nations and at the level of specific urban policies and programmes. Some population groups are routinely underrepresented in planning on the basis of income, race/ethnicity, age, gender, sexuality and other characteristics, reducing the representation of diverse views in two components of the process outlined in Fig. 8.1 ('Knowledge and evidence gathering' and 'Community and stakeholder participation') and resulting in places that do not meet everybody's health needs. Another aspect of this diagram that frequently fails in practice is the monitoring and review stage. Monitoring and review is not well-resourced in local government and there is a long-standing challenge with using the information produced in monitoring reports to inform policy and decision-making (see Innes, 1998; Innes & Booher, 2010). The inability of governments to evaluate and learn from policy impact creates a significant weakness for the effectiveness of

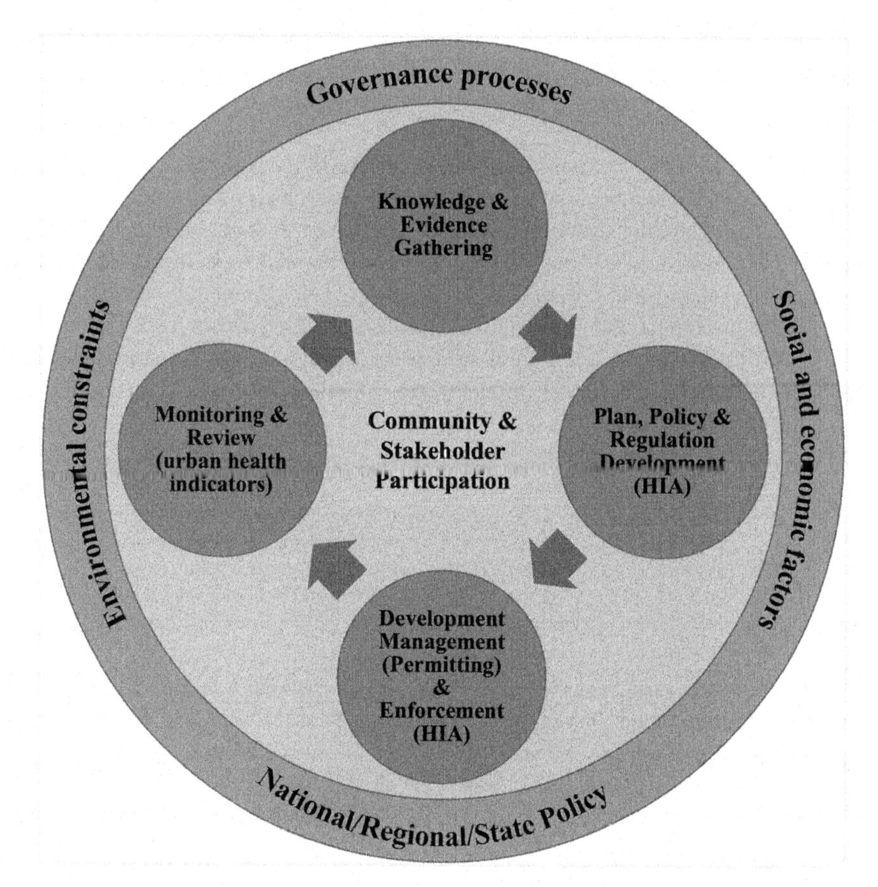

Fig. 8.1 Simplified version of healthy urban planning process. (Source: Adapted from Pineo et al., 2019. HIA: health impact assessment)

health and sustainability governance due to the complexity described earlier in this chapter.

In addition to a lack of monitoring and representative community participation, there are numerous other gaps between prescriptive approaches to healthy urban planning and planners' experiences. Research with practitioners in the USA, the UK and Australia has identified multiple challenges affecting delivery of healthy places (Carmichael et al., 2012, 2019; Fazli et al., 2017; Lowe et al., 2018; Ige-Elegbede et al., 2020; Pineo et al., 2020c). The urban health evidence base is difficult for planners to access, interpret and translate into policy. There is a lack of localised evidence about the economic benefits of healthy planning for different stakeholders. Policy- and decision-makers do not have a shared understanding of health and how it can be influenced by the built environment. Silo working and competing interests across government departments (within and across tiers) inhibit collaborative and integrated policy-making, which goes against the health in all policies (HiAP) and whole-of-society approaches advocated by the WHO. Delivery mechanisms (infrastructure funding and planning departments, for instance) are under-resourced. Underlying all of these challenges are the fundamental difficulties inherent in planning for urban change—those of power dynamics, under-represented voices and conflicting interests.

In response to these challenges, lessons from the WHO Healthy Cities initiative and other projects have been drawn together in numerous guidance documents for healthy urban planning, which continue to grow in number and geographical spread.[1] The value of new guidance is often through context-specific strategies and case studies that suit particular political, legislative, environmental and social settings. The UN-Habitat and WHO (2020) publication, *Integrating Health in Urban and Territorial Planning: a sourcebook,* provides broad lessons about process that can be widely applied, alongside links to case studies and resources. It calls for the following strategic healthy urban planning actions:

- Political commitment and leadership across civil society and the built environment and public health professions.
- Stakeholder commitment to develop a shared vision for healthier and more equitable placemaking and policy decisions with territorial and spatial implications.
- Establishment of new organizational structures, relationships and ways of working—including the way we train built environment and public health practitioners.
- Organizational investment to establish health in UTP [urban and territorial planning] as a norm. (p. 5)

Numerous case studies in this book have shown how these strategic actions are implemented in practice. For example, changes to Seattle's planning regulations in the Bullitt Center example (Chap. 4) required commitment and leadership across public and private sector stakeholders. The Healthy China 2030 plan (Box 2.2) strongly represents political commitment that has helped local government teams outside of health to implement their part of the national goals. The NHS England Healthy New Towns Programme (Box 2.3) and the relocation of public health teams from the NHS into local government (resulting from the Health and Social Care Act 2012) created new organisational structures, relationships and ways of working.

8.2.1 Health Impact Assessment

Prior to initiating a new project, plan, programme or policy, the relevant decision-making body should review its health impact. The process of health impact assessment (HIA) has been used since at least the 1980s. Some countries or regions require HIAs to be conducted for proposed development as part of the planning consent process. HIA has also been required by the World Bank and International Finance Corporation prior to investment (Cave, 2015). Other assessments may include

---

[1] An online database of healthy urban planning and design guidance documents is available at http://www.healthyurbanism.net/guidance-for-healthy-urbanism/. The database can be filtered by country, urban decision-making scales and other factors.

health considerations, such as Strategic Environmental Assessments (SEA) and Environmental Impact Assessments (EIA), although they are unlikely to be as comprehensive with regard to health and wellbeing impacts as an HIA. The quantitative component of HIA may focus on health protection from noise or pollution, for example, and this is likely to be informed by existing indicator data published by public health organisations. The qualitative component of HIA incorporates knowledge from affected communities. Participatory workshops or surveys can be used to gather detail about health impacts that are not covered in existing datasets, such as quality of local greenspace or perceived safety. The Barton Park case study in Chap. 6 provides an example of HIA for a major development. The THRIVES framework could support the HIA process by informing which health impacts should be measured and expanding the boundaries of typical assessments. In addition to HIA, there are other decision-support tools that calculate the potential health and wellbeing impacts of diverse urban development options such as the CENSE tool (Combined Environmental Stressors' Exposure) created by Vlachokostas et al. (2014).

The value of an HIA depends on how well it is conducted and whether it informs improvements to policies, plans and projects, rather than being a tick-box exercise. Corburn (2017) contends that HIA can be valuable in preventing urban problems from being displaced and in bridging disciplinary divides (see the section on monitoring and evaluation later in this chapter). An international analysis of HIA practitioners' experiences by Harris et al. (2014) concluded that the assessment process facilitated forms of social learning and practical reasoning 'linking facts and values, within policy formulation' (p. 52). Experience in the UK has shown that even though HIA is required by many local authorities as part of planning applications for large developments, there are gaps in terms of implementing recommendations to improve health and monitoring impact. This implementation gap is part of the reason why planners may rely upon standards. For instance, a new standard called Livewell was adopted by Essex Planning Officers Association in England to enable early engagement between public health, planning and master-planning teams and to have greater clarity about which design measures will be implemented to improve health.

8.2.2 *Healthy Built Environment Standards*

Policy-makers may use standards to ensure that new development is supportive of health and sustainability objectives. New healthy built environment standards such as WELL and Fitwel (see Table 8.1) have joined an existing set of sustainable (or green) building standards to inform design and benchmarking achievements across project types and countries. Standards (or assessment tools) are normative frameworks for evaluating the quality of a product or system. Healthy built environment standards contribute to professional knowledge and conceptualisations of how to design healthy places (Callway et al., 2020). The use of green building standards

Table 8.1 Selected healthy built environment standards, year of publication and country and organisation of origin

Standard	Year launched	Organisation	Country of origin
Assessment Standard for Healthy Building (T/ASC02-2016)	2017	Chinese Academy of Building Science	China
Fitwel	2017	Center for Active Design	USA
Regenerative Ecological, Social, and Economic Targets (RESET)	2013	GIGA	China
WELL Building Standard	2014	International Well Building Institute	USA

Box 8.1 Healthy building standards in China

In response to the Chinese government's health agenda and influence from the WELL standard, the Chinese Academy of Building Science published the Assessment Standard for Healthy Building (T/ASC02-2016) in January 2017. The standard defines healthy buildings as those 'that can provide healthier environment, facilities and services, promote users' physical and mental health, and achieve the improvement of health performance, on the basis of fundamental functions'. Themes in the standard include ventilation, indoor air quality, building material pollutants, lighting, exercise and health, elderly oriented design and other factors (Wang et al., 2020). Prior to this voluntary standard, residential buildings were required to comply with CECS 179-2009 Technical Regulations for the Construction of Healthy Housing which covers environmental and social conditions (Wang et al., 2017).

There has been considerable adoption of the Assessment Standard for Healthy Building, yet some scholars do not believe it encompasses the right priorities for healthy urban development in China. Wang et al. (2020) reviewed implementation of the standard on 50 certified projects across 14 provinces and cities, with a total construction area of nearly 5,000,000 m^2, and found that lack of suitable construction products was a barrier to integrating heath into projects. They also noted that an integrated standard system to improve design, construction, maintenance, evaluation and performance testing, covering the whole life cycle of healthy buildings, is needed to support the implementation of projects.

can be required through policy, mandated on government-funded buildings and encouraged through incentives (Retzlaff, 2009), and similar mechanisms are being used to advance health-related standards. China's Assessment Standard for Healthy Building (Box 8.1) is an example of a voluntary standard created by a research institute, while the Livewell standard (see earlier) was created by local government bodies to aid with policy implementation.

Sustainable building standards are likely to be supportive of health because they address design and planning goals across the planetary, ecosystem and local health impacts of THRIVES. Such standards exist in at least 42 countries spanning low-, middle- and high-income development status (Callway et al., 2020) and they often follow the content and format of the first standards: Building Research Establishment Environmental Assessment Method (BREEAM) in 1990 and Leadership in Energy and Environmental Design (LEED) in 2000 (Reed et al., 2009). Despite the proliferation of standards, sustainable buildings are still a relatively small, but growing, part of the building stock in most countries, with greater coverage in the office sector. In Canada, only 4% of new residential buildings are certified green buildings (Canada Green Building Council, 2020). An analysis of certified green office buildings in ten markets (including selected cities in Australia, Canada, Europe and the USA) found that 18.6% of space was certified as green in 2018, compared to 6.4% in 2007 (Holtermans & Kok, 2018).

Indoor environmental quality is a key component of most sustainable building standards, meaning that there are strong theoretical reasons that they should support health (Cedeño-Laurent et al., 2018b). There have been few empirical studies to test the health benefits of certified green buildings. Compared to conventional housing, studies have identified self-reported health benefits from living in certified sustainable housing. After moving to LEED certified housing, occupants reported 47% fewer sick building syndrome symptoms and fewer cases of mould, pests, inadequate ventilation and stuffiness (Colton et al., 2015). Renovation to Enterprise Green Communities resulted in better self-reported health of residents (Breysse et al., 2011).

Studies of certified green office buildings have found improved cognitive performance, sleep and physical activity. MacNaughton et al. (2017) studied cognitive performance in ten high-performing office buildings in America, six of which were retrofitted to the LEED standard. They found that participants in certified green office buildings had 'better environmental perceptions, 30% fewer sick building symptoms, 26.4% higher cognitive function scores and 6.4% higher Sleep Scores' after controlling for potentially confounding factors such as annual earnings and level of education (ibid.; p. 183). Although these effects were unexplained, there were environmental differences between the buildings that may have resulted in these benefits, including reduced humidity and improved lighting conditions.

The growth of healthy built environment standards indicates increased interest in the real estate sector in creating healthy spaces (McArthur & Powell, 2020). It is important to consider the extent to which such standards address the sustainability, equity and inclusion components of healthy urbanism and to what extent they may diminish opportunities for local knowledge to influence the form of new development. An analysis of built environment practitioners' experiences of using healthy built environment standards found that these rating tools can influence individual, organisational and market-scale conceptualisations of healthy development (Callway et al., 2020). These standards also influenced design decisions, such as the use of specific building materials and internal fittings. However, Callway et al. argue that health-focused standards risk perpetuating a narrow perspective of health that focuses more on the health of building occupants than wider communities, ecosystems and planetary health. WELL and Fitwel offer clear processes to gain joint certification with sustainability standards including LEED, BREEAM or Living Building Challenge, yet pursuing multiple certifications has financial implications that may reduce uptake. There are likely to be trade-offs if a healthy built environment standard is prioritised over a sustainability standard and these tensions require further research. A multi-scalar perspective of health impacts is needed to ensure that new buildings are supportive of health in terms of sustainability, equity and inclusion.

8.3 Incorporating Community Knowledge of Health and Place

Inclusive design and planning processes are required to create healthy places that respond to contextually and culturally specific requirements. People's relationship with health and place is complex, and as noted by Kevin Lynch (1960), their observations are filtered through memories, values, beliefs and attitudes. There are likely to be conflicting views about how a specific place, such as a park or housing estate, supports or harms health that depend on an individual's own experiences. Lack of diversity in design and planning teams coupled with educational curricula that have not represented the full needs of society are likely to result in development proposals that do not meet the health needs of everybody. Research has demonstrated that even objectively measured 'walkable' built environments or healthy food vendors may not be used by local residents due to memories of violent crime or feelings of discrimination that are unlikely to be anticipated by professionals and will only be surfaced in participatory design and planning activities (Dennis et al., 2009). Sherry Arnstein's (1969) Ladder of Citizen Participation sets out eight levels of non-participation/participation from 'manipulation' at the lowest level to 'citizen control' at the highest level. Her typology highlights the power dynamics in political processes, whereby tokenistic engagement activities may simply be used to tick a box, such as the Citizen Advisory Committees (CACs) of past American urban renewal projects. There is now much greater recognition in built environment professions that technical and community knowledge are both required to create places that work for people, but the reality of community participation remains difficult.

The structural barriers to health experienced by under-represented communities may not be known to city planners and design teams promoting urban change, thus necessitating collaboration with affected residents. Chapter 7 described the concept of cultural competency (Agyeman & Erickson, 2012) which should be developed by built environment practitioners to understand diverse perspectives within a community. One pertinent example is the process that Los Angeles County Metropolitan Transportation Authority (LA Metro) went through with 'Measure M—the Los Angeles County Traffic Improvement Plan' to fund transport in this part of

California, USA. Low-income and racial minority residents in Los Angeles have historically had reduced access to transportation infrastructure, exacerbating other environmental and social burdens that are harmful for health and wellbeing. The regional transport authority worked with community representatives and residents in low-income and racial minority neighbourhoods to inform the provision and design of new transport infrastructure to be funded by 'Measure M'. Voters approved the measure in 2016, which represents a half-cent sales tax that generates roughly US$860 million per year to fund a list of approved transit and highway projects, plus bikeways and greenways. LA Metro worked closely with the Los Angeles County Bicycle Coalition (LACBC) to build support for Measure M by understanding and planning to remedy mobility infrastructure gaps across underserved communities. In this process, the authority created the LA Metro Equity Platform Framework to outline the vision and implementation plan for equitable and healthy transport infrastructure (see Box 8.2) that is guided by residents.

There are many methods to gather residents' views about health and place which could be seen as a starting point for community participation that could expand to more collaborative co-design approaches. It is important to be clear on the scope and purpose of any knowledge gathering activities, including careful consideration of how frequently specific residents are asked to participate in such activities. Methods to gather community views should seek a wide range of perspectives and may be conducted at different times and locations (including online) to facilitate diverse participation. The following methods have been used to explore residents'

Box 8.2 Los Angeles' Metro equity platform

Access to transport infrastructure determines whether people can reach the destinations and services that support daily life, including jobs, education and healthcare facilities. In recognition that there are vast disparities in transport access across Los Angeles, the Los Angeles County Metropolitan Transportation Authority (LA Metro) committed to adopting an equity lens across its service delivery and planning. LA Metro's Executive Management Committee (2018) acknowledges that these disparities are determined by race and class in Los Angeles, and other personal characteristics such as disability and age. To increase access for everybody, LA Metro developed and adopted the Metro Equity Platform Framework. The platform contains four points: (1) define and measure, (2) listen and learn, (3) focus and deliver and (4) train and grow (Los Angeles County Metropolitan Transportation Authority, 2018). The Platform is based on knowledge gained through forums between transportation professionals and the community. LA Metro will use the platform to inform decisions and strategic planning, such as the Long Range Transportation Plan. All of these activities will be supported by equity training within the department.

views about place and health.[2] These approaches could be run by any organisation and adapted for different contexts.

- *Asset mapping*: Residents may identify features of the urban environment or social networks as local assets for health and wellbeing, such as a church, park or playground. These assets can be mapped spatially to identify areas that need to be protected/supported in processes of urban change or gaps that should be filled.
- *Digital engagement*: Apps and websites for digitally gathering residents' views are increasingly used by local governments and developers. Signs with QR codes or URLs can direct people to a website where they can comment on the environment that they are currently experiencing. Residents can upload photos and comments or respond to a series of questions about proposed designs. Online events may also occur alongside in-person workshops, which can support people with disabilities or caring responsibilities to participate without leaving their home.
- *Group street audits and walkabouts*: Planners or community organisers can lead a group of residents in a walk around their neighbourhood to discuss problems, assets and proposed changes. Checklists and audit forms can be used to prompt participants to discuss certain features and record feedback from the group.
- *Participatory mapping*: Spatially mapping residents' views related to health and place can transform their knowledge into a piece of evidence that is regularly used in planning processes. For instance, residents could map out the areas in which they feel safe and unsafe from road traffic or crime.
- *Photo surveys*: Gathering data in the form of images can be a useful way to learn from people who may have difficulty verbalising their opinions, including young people or immigrants. Participants can be given a camera or they can use their smartphone to photograph their neighbourhood over a period of time, returning the images to the organisers. People may also take photos of a group walk. The images can be grouped by themes (e.g. places where people feel relaxed or anxious) or placed on maps. Some digital engagement tools can facilitate the process of gathering and mapping community-sourced images and text.
- *Workshops and events*: There are many formats for in-person workshops and events which can be useful as a format to share information with the community and receive responses back in real time. Voting pads and apps can be used to gather data from the audience about proposals and issues.

Before engaging with residents about urban development, professionals should consider how power and privilege will shape the conversations and debates that arise. This is particularly important to interrogate and address issues of inequality and exclusion that exist or could be created/further embedded through urban change. Ascala Sisk et al. (2020) provide urban planners, public health professionals and

[2] More information and case studies about gathering community knowledge about health and place can be found online at http://www.healthyurbanism.net/community-knowledge/.

others involved in promoting healthy development with a series of steps for examining power and privilege in their work. Step one is about creating or seeking out 'Brave Spaces' (see Arao & Clemens, 2013) for exploring the role of power. Professionals that run workshops and events to gather the views of people who are typically under-represented in planning and design could support their active participation by creating a 'Brave Space' which would involve acknowledging the limits of expertise and holding organisers and participants accountable to practices that support diversity and inclusion, among other actions. Step two is about understanding the role of *power*, which they define as 'the ability to direct laws, policies, and investment that shape people's lives'. In considering a particular project or policy, practitioners are encouraged to question the problem(s) that the work is trying to solve. Prompts include asking who benefits and suffers from the problem remaining unsolved and whether those who are most affected are represented in the decision-making process. The final step is to analyse and challenge *privilege*, which they define as 'the accumulation of benefits of special rights'. Practitioners are invited to examine the areas of their lives where they hold privilege. If conflict arises during conversations about urban change, practitioners can think about a series of prompts to analyse and challenge privilege, such as 'Who or what is blamed for the conflict in the narrative describing the challenge?'

8.4 Funding Healthy Place-Making

Economic justifications for healthy and sustainable urban development are frequently used to persuade decision-makers to improve upon 'business as usual' approaches. Estimates about lost productivity and healthcare costs are produced to demonstrate the societal impacts of unhealthy development. For instance, the medical costs from poor housing in the European Union are estimated to cost 9 billion euros annually (Nicol et al., 2017). The WHO (2018) states that physical inactivity results in annual global costs of INT$14 billion for lost productivity and INT$54 billion for healthcare. These significant sums highlight the distributed costs and savings across public and private sectors related to health and place. Those who pay for the costs of infrastructure and development are not usually the same organisations benefitting from potential future savings, but ultimately, all of society pays for the costs of unhealthy urban environments (Pineo & Rydin, 2018). Demonstrating the 'business case' is an important way to shift decision-making and investment towards healthy and sustainable forms of development. However, it is not always possible to identify a return on investment over short timescales, which can create challenges when seeking to persuade different stakeholders of the benefits of healthy design. This section critically evaluates the financial models for healthy development at different scales.

8.4.1 The Business Case for Healthy Property Development

Growing interest in health and wellbeing from office and residential property developers largely rests on the potential for increased property values and cost savings for occupants. The Urban Land Institute examined 13 American developments of multiple scales and uses and they found increased rental values, property sales, interest from investors/lenders/occupants and sales rates (Kramer et al., 2014). Developers reported that costs to achieve healthier development were minimal or were offset by the benefits. Real estate services companies, such as Jones Lang LaSalle, and consultancy Ernst & Young have described the 'transformation' towards healthy and sustainable buildings as being propelled by rental/sales premiums and the COVID-19 pandemic, with new healthy building standards playing a key role (Bolden, 2020; Clifford, 2018). A report by the Center for Active Design (CfAD), United Nations Environment Programme Finance Initiative (UNEP FI) and BentallGreenOak (2021) found that nearly 90% of surveyed real estate investors planned to update their company's health and wellness strategy in the following year. Tenants' demands were driving this change and respondents reported a 4.4–7% rental premium for healthy buildings (ibid.). Such demand is partly fuelled by evidence of productivity gains in healthy buildings. A study in the USA found that optimising building ventilation and energy use could result in improved work performance by 0.5% and reduced absenteeism by five hours per year, which was quantified to save US employers of small- to medium-large offices between US$28 and US$55 billion, depending on which strategy was used (Ben-David et al., 2017). A pillar of the World Green Building Council's (2014, 2016) campaign for healthy and green office places was their reasoning that 90% of business operating costs are related to staff salaries and benefits, thus improving staff productivity and health makes economic sense.

In the context of deregulation and slow progress on sustainable construction, some organisations have reasoned that health brought a more convincing justification for sustainable design than energy cost savings or the risks of climate change. Around the time of WorldGBC's Better Places for People campaign launch, there was a perception among industry leaders that the economic case for sustainable construction was not enough to persuade property owners and developers to adopt green design measures. For instance, due to lobbying from volume housebuilders, the UK government scrapped its voluntary Code for Sustainable Homes standard in 2015, despite its own consultation showing that the majority of homebuilders observed costs to be dropping and processes to be smoother due to increased standardisation (DCLG, 2011). The essence of the WorldGBC campaign was that sustainable building standards could be leveraged to improve employee health, thus reducing costs for employers and creating a new form of value for real estate. The organisation's more recent *Health & Wellbeing Framework* (see Chap. 7) takes a broader view and seeks to create long-term social value for communities. Despite some shifts in industry towards creating social value, it remains the case that proponents of healthy and sustainable development need clear evidence of the business case because economic viability drives decision-making in new development.

8.4.2 Development Viability and Planning Gain

Demonstrating the business case for healthy development requires evidence that any associated costs will not adversely affect 'viability' in development economic terms. Development viability refers to the profit margin remaining after a developer pays the cost of development, including land, construction, regulatory requirements (e.g. land remediation) and planning gain (Rydin, 2013). Figure 8.2 shows where the costs associated with healthy development may fit into a typical viability calculation, potentially increasing the market value of development as higher-quality places are worth more. Developers will rely on evidence from surveyors to demonstrate that upfront costs can be recuperated at the point of sale. Infrastructure and community benefits created through planning gain are funded by the increase in land value generated by planning permission. Outcomes of planning gain, such as affordable housing and green infrastructure, are competing for limited resources and the former is frequently cited as dominating negotiations thereby reducing money spent on other features that would benefit health and sustainability (Moore et al., 2014; Carmichael et al., 2019; Pineo et al., 2020c). In the UK, reduced public spending following the Great Recession (and associated austerity measures) also reduced power within local government to deliver or negotiate health-promoting infrastructure as other objectives have taken priority (Carmichael et al., 2019; Mell, 2019).

Planning gain is only one model to fund urban infrastructure, as described in Chap. 6. Nick Gallent et al. (2020) reviewed the process for public infrastructure

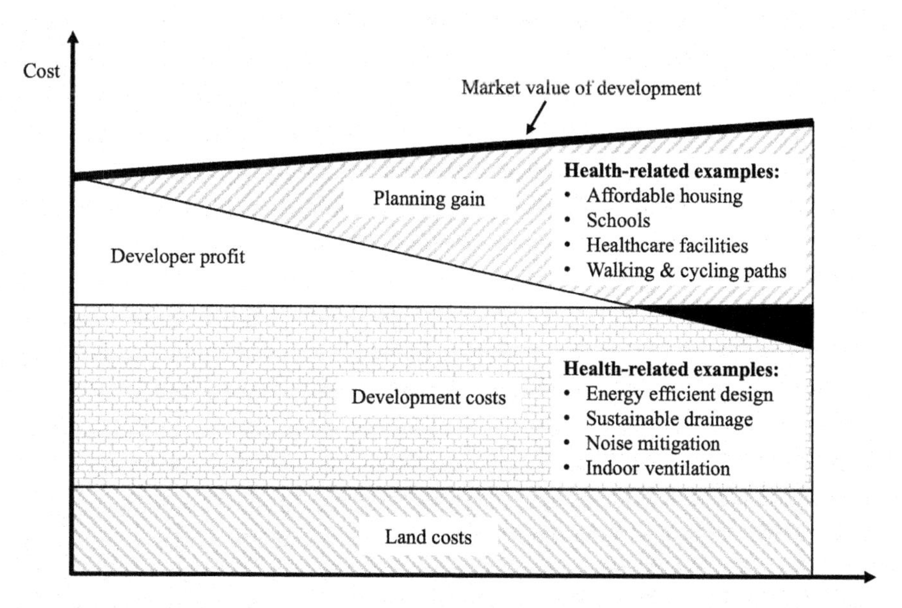

Fig. 8.2 Planning gain and development viability with example health-related costs. (Source: Adapted from Rydin, 2013)

delivery in support of housing growth in a comparative study of the UK, Australia, Germany, the Netherlands, Canada, the US and Hong Kong/Mainland China. Their analysis demonstrates that planning, funding and delivering infrastructure is a complex process that is time consuming, crosses multiple government boundaries, involves many stakeholders and requires negotiation and compromise. In the neoliberal city, public infrastructure delivery relies on private finances and 'user pays' models. The responsibility for coordinating infrastructure and the impact of any failures or delays often falls on local government. Some of the models reviewed by Gallent et al. include direct government provision (e.g. funded by local taxes), developer contributions and impact fees (i.e. planning gain) and joint ventures for land development (between public and private bodies).

Through the lens of health, there are two points raised by Gallent et al. that are framed here as opportunities and risks for funding health-promoting infrastructure. First is that 'strong government involvement in land assembly and development (acquiring land at rural or industrial prices, providing infrastructure and services, and recouping costs through final sales) is likely to promote more contained urban form, as well as efficient and equitable servicing' (ibid., p. 3). This approach would support health because compact urban form enables active travel and reduces pollution. The equity of servicing means that low-income residents would not be disadvantaged in accessing services, such as high-quality schools and public transport nodes. The second point is that infrastructure delivery through a regulatory planning framework, such as charging development impact fees, can create unwanted outcomes. Gallent et al. give the example of 'development charging regimes which impose levies on the basis of housing density (i.e. per dwelling unit) rather than land area (i.e. per hectare/acre), [which] may unintentionally promote urban sprawl and encourage developers to prioritize high end, detached homes able to attract a high profit margin' (p. 3). This type of development would create a heavy environmental burden; it would not support physical activity and social interaction, nor would it benefit low-income residents. No single infrastructure delivery model is championed by Gallent and colleagues, but instead they acknowledge the challenges and contextual nature of each system. However, it is clear that infrastructure funding is an important instrument for determining the extent to which health-promoting resources are distributed equitably in society.

8.4.3 Economic Decision-Support Tools

Decision-support tools and project evaluations with quantified economic benefits of healthy urban infrastructure provide powerful evidence for planning processes. The WHO's Health Economic Assessment Tool (HEAT) for Walking and Cycling is an example decision-support tool that models the economic benefits of active travel and can be used to compare policy and infrastructure options. Existing methods for transport cost benefit analysis have widely recognised deficiencies that underestimate costs such as environmental pollution and transport-related injuries (Næss et al., 2012). For instance, road construction tends to increase overall traffic volumes, known as 'induced traffic', which is often downplayed in cost-benefit analyses (ibid.). The WHO HEAT tool determines the economic value of reduced mortality from active transport infrastructure due to physical activity, air pollution and collisions, alongside carbon emission reductions (Kahlmeier et al., 2011; Götschi et al., 2020). Transport planners and other users input local data and the online tool produces outputs, allowing comparison between the current situation and the impact of proposed transport infrastructure. The HEAT methodology was developed by health experts using epidemiological evidence about the health impact of walking and cycling. There are limitations to the tool such as its exclusion of morbidity-related health impacts and young/elderly populations (among others); however, it still provides valuable evidence for policy-makers in a simple format (Götschi et al., 2020). Data about the successes of past projects are a useful form of evidence for decision-makers, such as the health and economic benefits of mobility infrastructure interventions in Bogotà, Colombia, and London, England (see Box 8.3).

Box 8.3 Two cases of mobility infrastructure to support public health

The *TransMilenio bus rapid transit (BRT) system in Bogotà, Colombia,* began in 2000 and operates articulated buses with a capacity of 160 passengers each that travel on dedicated routes. BRTs are a cost-effective and efficient form of urban transport that reduce traffic congestion and pollution. Bogotà also invested in a network of cycle routes (Ciclorutas), pedestrian-friendly design measures, car-free days and car-use restrictions. Evaluations found higher levels of physical activity among people living near BRT stations and cycle routes (Becerra et al., 2013). An analysis of phase 1 and 2 of the BRT system estimated a value of US$167.5 million for 'reduced injuries, deaths and losses due to road crashes' and US$114.5 million for 'positive impact on health due to reduced emissions of air pollutants' (Hidalgo et al., 2013, p. 134). Monitoring showed that sulphur dioxide declined by 43%, nitrogen oxide by 18% and particulate matter by 12% (Turner et al., 2012). The wider co-benefits resulting from the TransMilenio system include: increased accessibility of affordable and reliable transport; shorter waiting times for passengers within stations; improved quality of life of urban residents through improved cleanliness, safety and efficiency of Bogotá's transport networks; creation of additional employment opportunities for transit workers; and increased ability for residents to make informed choices about sustainable transport (UNFCCC, 2010).

Following the demonstrated benefits of *London's Congestion Charging Zone*, similar restrictions have been used in Singapore, Oslo and Stockholm to reduce traffic congestion. The CCZ was introduced in 2003 and covered an

(continued)

Box 8.3 (continued)

area of 21 km^2. In 2020, the Ultra Low Emission Zone (ULEZ) expanded the scheme and emissions criteria. Evaluation of the CCZ found that the greatest health improvement was found in the wards within the charging zone (with a population of roughly 200,000 people) where a small increase in life expectancy could be attributed to the charge (Tonne et al., 2008). The income from the charging zone far exceeds its costs and the £1.7 billion net revenue from 2003 to 2017 was reinvested in transport, including roughly £1.3 billion on bus network improvements, £196 million on roads and bridges, £80 million on road safety, £90 million on local transport/borough plans and £64 million on sustainable transport and the environment (Transport for London, 2017). The co-benefits of this scheme have been increased funding for sustainable transport initiatives, improved air quality, reduced greenhouse gas emissions and reduced air pollution related to illnesses in the charging zone.

8.4.4 Risks of Relying on a Business Case for Health

Proponents of healthy place-making look for added value that can be leveraged to pay for infrastructure and good design; however, there are risks with this approach. The rationale for finding a 'business case' for healthy development is analogous to economic arguments for sustainable construction which are usually supported by the following premises: resource efficiency produces savings, reduced energy demand lowers operating costs, new firms and supply chains are created, and reputational gains from Environmental, Social and Governance (ESG) actions translate into increased profit. Stuart Green (2018) argues that this logic is dangerous because it relies on existing ways of working (ensuring profitability) to make the business case stack up rather than a fundamental shift in business models. An example of how this risk relates to health is the frequently cited increase in property values that healthy planning and design creates (Pineo, 2016; Chang, 2018; Carmona, 2019). An analysis commissioned by CEOs for Cities in America compared urban house values in neighbourhoods with varying levels of walkability. The houses in areas with the highest walkability scores had property values between US$4000 and US$34,000 higher than houses in average score areas (Cortright, 2009). Homeowners will benefit from the increased value of their properties while governments will gain in property tax receipts. Housebuilders could also benefit from higher sales values if new properties are in walkable neighbourhoods or if large-scale developments are designed to support walking through street design, connectivity and other measures. From a healthy urbanism perspective, the business case for healthy design needs to address the structural barriers to accessing healthy environments experienced by low-income and other marginalised communities. Increased property values in healthy homes and neighbourhoods will exclude the most disadvantaged in society and further exacerbate existing inequities. Relying on added value in a market-led model of healthy development is unlikely to solve city leaders' growing problems from the effects of the climate crisis, childhood obesity, mental illness and other chronic diseases.

In closing, there are several challenges with using economic arguments to persuade developers to build healthier places. First, developers and investors are only likely to consider the direct benefits accrued from healthy design measures (e.g. increased property sales) rather than wider economic benefits to society (Henneberry et al., 2011; Moore et al., 2014), and such premiums are not necessarily consistent across all property types and tenures. Second, it is not clear that the design measures that create economic value are necessarily the most beneficial for health. For example, biophilia may improve wellbeing and occupants' perceptions of attractive office spaces. Yet improved ventilation, for example, may be undervalued by occupants even though it is likely to have greater health and wellbeing benefits. Relying on consumer demand is not necessarily sufficient to produce health-promoting environments. Third, if developers can charge higher values for healthier properties, this will result in disbenefits for low-income residents, exacerbating existing inequalities (as described earlier). Finally, even in cases where developers have a broader view of investment returns, such as public sector-led development, they still may have siloed budgets that cannot account for the savings achieved in other departments (e.g. health and social care) and they may still require low-cost designs in terms of construction and maintenance, which would likely be at odds with health objectives. The next section describes models for development that are more compatible with the inclusion, equity and sustainability principles of healthy urbanism.

8.5 Models for Healthy Development

There are no silver bullet solutions to the challenges of planning and financing healthy development, but there are several models that have potential for greater use with appropriate adaptation to local needs and context. This section describes various opportunities for healthy urban development, including supporting outcome-driven developers, community-led projects and pilot programmes. These opportunities add to the more specific case studies in Chaps. 2, 3, 4, 5, 6 and 7.

8.5.1 Outcome-Driven Developers

This book has argued that status quo models for development are falling very short of urban health and sustainability needs. But there are examples of good practice that can be scaled up. Participants in Pineo and Moore's (2021) study described developers who are doing things differently because they are 'mission driven', 'values based' and 'enlightened' (p. 6). Based on participants' experiences, Pineo and Moore created a general typology to draw distinctions between 'outcome-driven'

Table 8.2 Characteristics of outcome- and output-driven developers from Pineo and Moore (2021)

Outcome-driven developers	Output-driven developers
Longer-term goals and interests	Shorter-term goals and interests
Broader and interconnected impacts	Narrow and focused impacts
Focus on finances	Focus on finances
Interests in societal impacts (e.g. supporting local economies, sustainability and health)	Interests in economic impacts (e.g. property sales value, ease of leasing property)
Go beyond regulations	Meet regulations
Pilot, experiment	Follow known practice

developers who have adopted ways of working that support going beyond minimum regulations and experimenting with innovative healthy design approaches, versus 'output-driven' developers who remain in a traditional risk averse mode that focuses on short-term financial gains (Table 8.2). It should be stressed that both types of developers were focused on finances and economic viability, but the outcome-driven organisations would actively pursue long-term positive impacts. Being an outcome- or output-driven developer was not simply a result of specific markets, property sectors or organisation types (public or private), but instead indicates an ethos or mindset within the organisation that may be more amenable to healthy and sustainable design. Developments that produce positive societal and environmental impacts can be seen as important drivers of incremental changes to expectations and norms about urban environments that will influence policy-makers, consumers and the wider development sector. Outcome-driven developers could be supported to achieve positive impact through financial incentives, simplification of the planning consent process or supportive regulations that allow innovations (as discussed in the Bullet Center case study in Seattle).

Grow Community on Bainbridge Island in Washington, USA, is an example of residential development by developers who could be called outcome-driven (Pineo & Moore, 2021). The investors were seen as catalysing the project's ambition for high sustainability performance and affordable homes for a broad range of people. The development is certified to the One Planet Living standard and many of the design features are supportive of social interaction among occupants (Fig. 8.3). For instance, rows of detached properties face each other with shared footpaths and gardens between homes. Apartment blocks are facing communal gardens and the community centre. Residents say that this configuration makes social interaction with neighbours easy and a daily occurrence, which particularly suits the high number of older residents. A diversity of housing types (detached and semi-detached houses and apartments) and tenures (rental and owned) were provided and a participatory design process was used.

The Nightingale Housing model in Australia (described in Chap. 7) also fits the model of outcome-driven developers with its emphases on health, sustainability and affordability. The team behind this model also wanted to develop a sense of community and took efforts to ensure that occupants would align to the overall ethos of shared spaces and sustainable living. They were inspired by the Baugruppen (group

Fig. 8.3 Grow Community in Bainbridge, Washington, USA. (Source: Author, 2019)

build) model from Germany, but a key difference is that the Australian model involves traditional finance mechanisms rather than future occupants financing the project.

8.5.2 Community-Led Urban Change

Community-led initiatives and group-build developments have the potential to support health and sustainability objectives. The Transitions Town movement is an example grassroots movement where community groups lead local sustainability projects, such as renewable energy companies and garden sharing. Such initiatives could potentially create awareness of environmental issues beyond the individuals involved and this diffusion of knowledge could lead to systemic changes (Sengers et al., 2019). Group self-build housing (or Baugruppen in German) is a community-led initiative that forms the primary method of development in some of the best-known places for sustainable urbanism, such as the Vauban suburb of Freiburg, Germany. Groups of friends or strangers come together and build their homes cooperatively, with varying levels of construction or project management contribution, sometimes with the help of professionals. According to Iqbal Hamiduddin and Nick Gallent (2016), the Baugruppen model offers a solution for three key policy challenges: (1) *affordability*, homes are more affordable due to the cost savings of developing multiple homes at once, (2) *social interaction*, residents cooperate throughout the development process and build social capital, and (3) *inclusive design*, homes meet the needs of occupants because they influence the whole design process. In Freiburg, as with other German cities, Baugruppen projects are supported by planners in several ways. Public control of land enables the city to freeze land prices and control planning permission. The city makes serviced plots available and groups of interested people bid for a plot with their submission containing details on design and the group's finances. Hamiduddin and Gallent found that this early gateway stage resulted in the exclusion of some potential residents who could not obtain financing or who were perceived to be 'bad neighbours'.

The broader concept of co-housing, where people intentionally live together but may not build or finance the accommodation themselves, is similarly viewed as good for quality of life, health and wellbeing. A review by Julie Carrere et al. (2020) notes that numerous studies of co-housing projects found evidence of 'increased social support, sense of community and physical, emotional and economic security, as well as reduced social-isolation' (p. 23). These projects have grown in popularity internationally, with examples of intergenerational and elderly co-housing developments in the USA, Europe, Australia, New Zealand and Japan. For example, the Marmalade Lane Cohousing in Cambridge, UK, was built in 2017 and includes 42 sustainable homes with shared community facilities and gardens. Carrere and colleagues cautioned that co-housing could create unintended harms by increasing inequalities within society because disadvantaged populations have had unequal access to these projects. This is similar to the criticisms of group-build projects, which warn that they may be exclusive, often suited to young and affluent people (Hamiduddin, 2015), and that they may lead to gentrification (Droste, 2015). These community-driven models could be adjusted to allow for wider participation and this requires further investigation and experimentation.

Beyond housing development, other forms of community-led urban change could support health and wellbeing. Bottom-up, top-down and hybrid forms of temporary urbanism (also called tactical, guerrilla, insurgent and pop up urbanism) can be used to create pocket parks, mini libraries, mobility infrastructure and more (Andres et al., 2021; Carmona, 2021), as shown in Fig. 8.4. The COVID-19 pandemic spurred numerous temporary urbanism projects around the world, including temporary hand-washing facilities, outdoor eating and pedestrian-only zones. Lauren Andres et al. (2021) note that different types of temporary urbanism are all connected in their ability to support adaptability, activation of multiple forms of value and transitioning within larger processes of urban transformation. This can be useful in supporting health and sustainability in terms of rapid urban interventions to crises (e.g. opening air-conditioned spaces to the public during heat waves) and 'meanwhile uses' during longer processes of urban change. They provide examples of temporary gardens and a pond that was open for swimming during the redevelopment of King's Cross St Pancras in London, UK, and the Hope for Communities aerial water project that created a safe water source for people in Kibera, Nairobi, Kenya.

Fig. 8.4 Examples of temporary urbanism. Left: A temporary public space in Wynyard Quarter, Auckland, New Zealand (Source: Author, 2018); Right: Temporary traffic barriers in Los Angeles, California, USA, as part of the Slow Streets LA project during the COVID-19 pandemic

8.5.3 Pilot and Demonstration Projects

Claiming that a new development will be good for health and adopting innovative healthy design measures creates an element of risk that can be managed through pilot or demonstration projects (Pineo & Moore, 2021). Green (2018) states that 'the construction sector's competitive model is primarily governed by risk management' which he described as inhibiting uptake of sustainable design and construction methods. Governments use pilot and demonstration projects to offer protection against financial, reputational and other risks associated with innovative construction. An experimental approach can also be driven by developers and their design team. A Seattle-based architect described the challenges encountered when trying to use an innovative waste-water treatment system because it was not allowed through local regulations. They said, 'We called it an "experimental system" which then provided cover for everybody to monitor extremely carefully, but to allow [it] to proceed and then prove itself' (Pineo & Moore, 2021, p. 8). Many of the case studies in this book were experimental and aimed to test and prove new ways of working towards healthy and sustainable development, including the Bullitt Center, Vila Viva, Barton Park and Nightingale Housing. Mobility infrastructure is often changed through temporary trialling and monitoring, such as a new roundabout or traffic restriction, to gauge local support and effectiveness of the intervention.

The success and scalability of sustainable development projects has been investigated in the socio-technical transitions literature through the lens of 'niche' developments, experiments and urban living labs. Moore and Higgins (2016) list examples of demonstration projects, including BedZED in London, UK, and zHome in Issaquah, Washington, USA, both of which were enabled by the local authority making land available for development below the market rate. They examined the value of a demonstration project called the Nicholson in Melbourne, Australia, in terms of performing as a model for sustainable urban development that can have wider influence in the industry. Some of the innovative elements of the project, specifically its mixed-tenure housing and use of modular construction methods, were successful in demonstrating the potential for these approaches. Stakeholders gained 'real world feedback and market testing without them having to take the financial and reputational risk themselves' (ibid., p. 14). Other components of the Nicholson demonstration project were not seen as influencing wider practice, including its environmental sustainability, mixed use and onsite governance. Moore and Higgins concluded that government-backed demonstration projects had value and could be improved with greater dissemination of learning to industry, which requires formal monitoring and evaluation. This call for ongoing evaluation and reflexive learning is widely seen as important for scaling up the successes of demonstration projects (Sengers et al., 2019).

8.6 Monitoring and Evaluation

Before, during and after a project or policy initiative, monitoring and evaluation processes should be used to guide decision-making, test impact against a set of pre-agreed goals and capture lessons for future initiatives. Indicators are a key part of this process, such as those listed in Chaps. 4, 5, 6 and 7. Monitoring progress in urban health is important because it is a complex system, there are weaknesses in the evidence base, design intentions are not always achieved in built environment projects and organisations should be held accountable for failures (Corburn, 2013; Rydin et al., 2012). Ecosocial epidemiology highlights the need for monitoring to identify intra-city inequities, legitimate multiple sources of knowledge and attribute responsibility for positive and negative health outcomes (Corburn, 2013). This section describes key considerations for selecting and using indicators in policy-making and on development projects.

8.6.1 Urban Health Indicators in Policy Processes

Indicators form part of the evidence base about urban health challenges, including who is affected and how impacts change over time. Planners can access online dashboards or public health reports for data about air pollution concentrations, noise levels, tree canopy coverage and other factors. Pineo et al. (2018b) worked with policy-makers in London, UK, and Dubai, UAE, to develop urban health indicators and an international index. They found that indicators can help stakeholders across the public and private sectors to gain awareness of urban health issues and to identify and explore shared responsibilities and competing interests. The use of indicators in long-standing collaborations across public health and planning services in San Francisco, USA, and Victoria and New South Wales, Australia, helped to build inter-sectoral relationships with far-reaching positive impacts on healthy planning processes (Pineo et al., 2020c). Collections of urban health indicators (mainly the San Francisco Indicators Project and Community Indicators Victoria) were developed to inform policy-making, but their influence was primarily indirect and occurred through creating opportunities to build and maintain inter-sectoral relationships, reframe knowledge, change professional norms, support advocates to challenge business-as-usual approaches and represent community interests and equity.

The process of developing the indicator tools in San Francisco, Victoria and New South Wales was collaborative and allowed diverse actors to engage in dialogue about urban health over time. Pineo analysed the value and use of indicators in these settings through a systems thinking lens to understand important interconnections and feedback relations (Pineo, 2019; Pineo et al., 2020c). The indicator tools were useful for multiple stages of healthy planning (as described earlier in this chapter). For instance, the indicators showed health and environmental inequities within cities, which informed the knowledge and evidence gathering stage. Diverse actors within the cities (e.g. charities and community groups) used specific indicators in reports and other communications to influence policy and decision-makers. Planners and public health professionals in local government used the indicators as evidence to negotiate with developers on specific proposals. The value of these indicator tools is largely attributable to their role in bringing people together to learn about, discuss and deliberate local health and place issues. Box 8.4 summarises the key transferable lessons about these urban health indicator projects (and others) that could be applied elsewhere to inform healthy planning policy.

Box 8.4 Supporting healthy planning with urban health indicator tools
Using metrics for healthy urban policy and development requires several considerations. Corburn (2013) suggests 'multi-dimensional monitoring' using indicators linked to Krieger's pathways of embodiment (see Chap. 3) from a range of data sources. Compared to traditional health indicators, Corburn and Cohen (2012) argue that urban health equity indicators need to be more dynamic, collaborative and transparent. Pineo et al. (2020a) synthesised studies of ten urban health indicator tools and their use by local government policy-makers in transport, urban planning, housing and regeneration. The analysis suggested that 'useful' indicator tools have the following characteristics:

- Indicator data are related to policy and they measure both policy inputs and outputs.
- Data are available (ideally mapped) at small geographic scales (i.e. neighbourhoods).
- Social and built environment factors are measured, including subjective and objective metrics.
- Longitudinal data are provided (thus ongoing data collection costs should be considered when selecting indicators).
- Indicator tools are produced in collaborative projects with community and professional expertise.

8.6.2 *Monitoring Development Projects*

Indicators can be used to monitor a particular urban development, but there is a need to consider timescales of impact and data collection. For short-term exposures such as noise and air pollution created during construction, regular monitoring and transparent reporting should occur. If problems arise, mitigation measures can be taken to avoid harm. Box 8.5 describes a good example of public reporting of indicator data in a development in Auckland, New Zealand. Measuring long-term impacts, such as residents' physical activity or wellbeing, can be informed by existing public health indicators. However, advanced planning is required to gather accurate measures of impact. The people living in the area before and after development are not always the same, so any uplift in health may not relate to local populations. Ideally, baseline measures of residents who will be affected by the development can be taken to show a 'before and after' story. Yet, any differences in the environment or health indicators may not have resulted from the development itself. Changes could be caused by other factors, such as national or local air pollution control policies. It is also important to recognise the potential for a large time lag before improvements to health can be detected.

One of the key challenges for sustainable and healthy buildings is the difference between simulated and actual design outcomes, known as the performance gap. Monitoring data helps to identify such gaps, but wider measurement and evaluation tools can be used to avoid the gap occurring. Sylvia Coleman et al. (2018) describe

Box 8.5 Monitoring urban development impacts in Auckland, New Zealand

Monitoring indicators are an important part of innovation for urban health and sustainability, yet they are rarely applied effectively in practice. Wynyard Quarter in Auckland, New Zealand, is a harbourside regeneration project with a strong approach to ongoing monitoring and review. The project's Sustainable Development Framework outlines how the development will move beyond net zero energy, water and waste impacts towards positively contributing energy and water and being 'restorative'. Among other requirements, homes and commercial buildings are required to meet New Zealand's sustainable building standards of 7 Homestar rating and 5 Green Star rating, respectively. The planning documents emphasise the importance of these strong sustainability objectives for human health and wellbeing. The Wynyard Quarter Smart website provides a continually updated report on progress against these sustainability goals (https://www.wynyard-quarter.co.nz/wqsmart/home). Users can explore data for categories including: Environmental quality; Economic vitality; Resource efficiency; Transport, movement and connectivity; and Vision. Each category contains additional topics with descriptions about the development strategy and progress.

the performance gap broadly, looking at a series of modelled, measured and perceived environmental and human comfort goals. Their conceptual framework for built environment performance assessment includes three types of gaps: (1) Predicted Gap (e.g. the difference between modelled and measured energy and water consumption), (2) Expectations Gap (e.g. gaps between expectations for occupant comfort and post-occupancy evaluations) and (3) Outcomes Gap (e.g. the difference between environmental monitoring data and occupant survey results). Coleman et al. call for a new emphasis on the human performance and wellbeing aspects of environmentally sustainable buildings, aiming for net positive contributions towards health and environmental goals, which they call 'regenerative sustainability', building on the work of Robinson and Cole (2015). Coleman et al.'s approach requires continuous monitoring and analysis throughout design and occupancy stages, including under-used techniques like pre- and post-occupancy evaluations (Pre-OE and POE), scenario analysis and backcasting.

8.7 Conclusion

Healthy built environment policy development and implementation are affected by diverse political priorities, perceived risks, competing objectives, knowledge gaps and other factors. This chapter has described promising solutions and ways of working that have been successful and could be scaled up with appropriate adaptations in new contexts. Transformative change from business-as-usual methods will require increased awareness of the wider determinants of health among the general public, which can then lead to shifts in political agendas. Vested interests, such as pharmaceutical companies, will continue to frame health in biomedical terms which diverts attention from the structural and environmental barriers to health that are created through existing modes of urban planning and development. Integrating health into built environment education curriculum is part of the solution. Governments, community groups, journalists, charities and the private sector also have a role to play in shifting current understandings of health and wellbeing. Working together, professionals and the public can identify design and policy solutions that meet multiple health, social, economic and environmental objectives simultaneously, reducing cost and maximising value. The next chapter will look to the future of healthy urbanism by considering disaster recovery and prevention strategies, incremental and transformative change to stop environmental breakdown, framing complex challenges and the promise of new technological solutions.

References

Agyeman, J., & Erickson, J. S. (2012). Culture, recognition, and the negotiation of difference: Some thoughts on cultural competency in planning education. *Journal of Planning Education and Research, 32*, 358–366. https://doi.org/10.1177/0739456X12441213

Andres, L., Bryson, J. R., & Moawad, P. (2021). Temporary urbanisms as policy alternatives to enhance health and well-being in the post-pandemic city. *Current Environmental Health Reports, 8*(2), 167–176.

Arao, B., & Clemens, K. (2013). From Safe Spaces to Brave Spaces, a new way to frame dialogue and diversity and social justice. In L. M. Landreman (Ed.), *The art of effective facilitation; Reflections from social justice educators* (pp. 135–150). Stylus Publishing.

Arnstein, S. R. (1969). A ladder of citizen participation. *Journal of the American Planning Association, 85*, 24–34.

Becerra, J. M., Reis, R. S., Frank, L. D., Ramirez-Marrero, F. A., Welle, B., Arriaga Cordero, E., Paz, F. M., Crespo, C., Dujon, V., Jacoby, E., Dill, J., Weigand, L., & Padin, C. M. (2013). Transport and health: A look at three Latin American cities. *Cadernos de Saúde Pública, 29*, 654–666.

Ben-David, T., Rackes, A., & Waring, M. S. (2017). Alternative ventilation strategies in U.S. offices: Saving energy while enhancing work performance, reducing absenteeism, and considering outdoor pollutant exposure tradeoffs. *Building and Environment, 116*, 140–157.

Bolden, K. (2020). Sustainable, healthy buildings and real estate. *Ernst Young*. Retrieved January 02, 2011, from https://www.ey.com/en_us/real-estate-hospitality-construction/sustainable-healthy-buildings-meeting-real-estate-expectations

Breysse, J., Jacobs, D. E., Weber, W., Dixon, S., Kawecki, C., Aceti, S., & Lopez, J. (2011). Health outcomes and green renovation of affordable housing. *Public Health Reports, 126*, 64–75.

Callway, R., Pineo, H., & Moore, G. (2020). Understanding the role of standards in the negotiation of a healthy built environment. *Sustainability, 12*, 9884.

Canada Green Building Council. (2020). Canada's green building engine: Market impact & opportunities in a critical decade.

Carmichael, L., Barton, H., Gray, S., Lease, H., & Pilkington, P. (2012). Integration of health into urban spatial planning through impact assessment: Identifying governance and policy barriers and facilitators. *Environmental Impact Assessment Review, 32*, 187–194.

Carmichael, L., Townshend, T. G., Fischer, T. B., Lock, K., Petrokofsky, C., Sheppard, A., Sweeting, D., & Ogilvie, F. (2019). Urban planning as an enabler of urban health: Challenges and good practice in England following the 2012 planning and public health reforms. *Land Use Policy, 84*, 154–162.

Carmona, M. (2019). Place value: Place quality and its impact on health, social, economic and environmental outcomes. *Journal of Urban Design, 24*, 1–48.

Carmona, M. (2021). *Public places urban spaces: The dimensions of urban design* (3rd ed.). Routledge.

Carrere, J., Reyes, A., Oliveras, L., Fernández, A., Peralta, A., Novoa, A. M., Pérez, K., & Borrell, C. (2020). The effects of cohousing model on people's health and wellbeing: A scoping review. *Public Health Reviews, 41*, 22.

Cave, B. (2015). Assessing the potential health effects of policies, plans, programmes and projects. In H. Barton, S. Thompson, M. Grant, & S. Burgess (Eds.), *The Routledge handbook of planning for health and well-being: Shaping a sustainable and healthy future* (pp. 371–385). Taylor and Francis.

Cedeño-Laurent, J. G., Williams, A., MacNaughton, P., Cao, X., Eitland, E., & Spengler, J. D. (2018b). Building evidence for health: Green buildings, current science, and future challenges. *Annual Review of Public Health, 39*, 291–308.

Center for Active Design, United Nations Environment Programme Finance Initiative, BentallGreenOak. (2021). *A new investor consensus: The rising demand for healthy buildings.*

Chang, M. (2018). *Securing constructive collaboration and consensus for planning healthy developments: A report from the Developers and Wellbeing project*. Town and Country Planning Association.

Clifford, M. (2018). Investors see returns in healthy buildings. *Jones Lang LaSalle*. Retrieved January 02, 2021, from https://www.jll.co.uk/en/trends-and-insights/investor/investors-see-returns-in-healthy-buildings

Coleman, S., Touchie, M. F., Robinson, J. B., & Peters, T. (2018). Rethinking performance gaps: A Regenerative sustainability approach to built environment performance assessment. *Sustainability, 10*, 4829.

Colton, M. D., Laurent, J. G. C., MacNaughton, P., Kane, J., Bennett-Fripp, M., Spengler, J., & Adamkiewicz, G. (2015). Health benefits of green public housing: Associations with asthma morbidity and building-related symptoms. *American Journal of Public Health, 105*, 2482–2489.

Corburn, J. (2013). *Healthy city planning: From neighbourhood to national health equity* (Planning, History and Environment Series). Routledge.

Corburn, J. (2017). Equitable and healthy city planning: Towards healthy urban governance in the century of the city. In E. De Leeuw & J. Simos (Eds.), *Healthy cities: The theory, policy, and practice of value-based urban planning* (pp. 31–41). Springer.

Corburn, J., & Cohen, A. K. (2012). Why we need urban health equity indicators: Integrating science, policy, and community. *PLoS Medicine, 9*, e1001285.

Cortright, J. (2009). *Walking the walk: How walkability raises home values in US cities*. CEOs for Cities.

Dennis, S. F., Gaulocher, S., Carpiano, R. M., & Brown, D. (2009). Participatory photo mapping (PPM): Exploring an integrated method for health and place research with young people. *Health & Place, 15*, 466–473.

Department for Communities and Local Government. (2011). *Cost of building to the code for sustainable homes: Updated cost review*. Department for Communities and Local Government, London.

Doyon, A., & Moore, T. (2019). The acceleration of an unprotected niche: The case of Nightingale Housing, Australia. *Cities, 92*, 18–26.

Droste, C. (2015). German co-housing: An opportunity for municipalities to foster socially inclusive urban development? *Urban Research and Practice, 8*, 79–92.

Executive Management Committee. (2018). *Metro board report: Metro equity platform framework* (Board Report). Los Angeles County Metropolitan Transportation Authority, Los Angeles, CA.

Fazli, G. S., Creatore, M. I., Matheson, F. I., Guilcher, S., Kaufman-Shriqui, V., Manson, H., Johns, A., & Booth, G. L. (2017). Identifying mechanisms for facilitating knowledge to action strategies targeting the built environment. *BMC Public Health, 17*, 1.

Gallent, N., Morphet, J., Chiu, R. L. H., Filion, P., Fischer, K. F., Gurran, N., Li, P., Li, P., Schwartzt, A., & Stead, D. (2020). International experience of public infrastructure delivery in support of housing growth. *Cities, 107*, 102920.

Glouberman, S., Gemar, M., Campsie, P., Miller, G., Armstrong, J., Newman, C., Siotis, A., & Groff, P. (2006). A framework for improving health in cities: A discussion paper. *Journal of Urban Health, 83*, 325–338.

Götschi, T., Kahlmeier, S., Castro, A., Brand, C., Cavill, N., Kelly, P., Lieb, C., Rojas-Rueda, D., Woodcock, J., & Racioppi, F. (2020). Integrated impact assessment of active travel: Expanding the scope of the health economic assessment tool (HEAT) for walking and cycling. *International Journal of Environmental Research and Public Health, 17*, 7361.

Green, S. (2018). Sustainable construction. In *Sustainable futures in the built environment to 2050* (pp. 172–193). John Wiley & Sons, Ltd.

Hamiduddin, I. (2015). Social sustainability, residential design and demographic balance: Neighbourhood planning strategies in Freiburg, Germany. *Town Planning Review, 86*, 29–52.

Hamiduddin, I., & Gallent, N. (2016). Self-build communities: The rationale and experiences of group-build (Baugruppen) housing development in Germany. *Housing Studies, 31*, 365–383.

Harris, P., Sainsbury, P., & Kemp, L. (2014). The fit between health impact assessment and public policy: Practice meets theory. *Social Science & Medicine, 108*, 46–53.

Henneberry, J., Lange, E., Moore, S., Morgan, E., & Zhao, N. (2011). Physical-financial modelling as an aid to developers' decision-making. In S. Tiesdell & D. Adams (Eds.), *Urban Design in the real estate development process* (pp. 219–235). John Wiley & Sons, Ltd.

Hidalgo, D., Pereira, L., Estupiñán, N., & Jiménez, P. L. (2013). TransMilenio BRT system in Bogota, high performance and positive impact—Main results of an ex-post evaluation. *Research in Transportation Economics*, THREDBO 12: Recent developments in the reform of land passenger transport, *39*, 133–138.

Holtermans, R., & Kok, N. (2018). *International green building adoption index*. Finance, GSBE Theme Sustainable Development.

Ige-Elegbede, J., Pilkington, P., Bird, E. L., Gray, S., Mindell, J. S., Chang, M., Stimpson, A., Gallagher, D., & Petrokofsky, C. (2020). Exploring the views of planners and public health practitioners on integrating health evidence into spatial planning in England: A mixed-methods study. *Journal of Public Health, 43*, 664–672.

Innes, J. E. (1998). Information in Communicative Planning. *Journal of the American Planning Association, 64*, 52–63.

Innes, J. E., & Booher, D. E. (2010). *Planning with complexity: An introduction to collaborative rationality for public policy*. Routledge.

Kahlmeier, S., & World Health Organization, Regional Office for Europe. (2011). *Health economic assessment tools (HEAT) for walking and for cycling: Methodology and user guide: Economic assessment of transport infrastructure and policies*. World Health Organisation, Regional Office for Europe, Copenhagen.

Kramer, A., Lassar, T. J., Federman, M., & Hammerschmidt, S. (2014). *Building for wellness: The business case*. Urban Land Institute.

Los Angeles County Metropolitan Transportation Authority. (2018). Metro equity platform framework.

Lowe, M., Whitzman, C., & Giles-Corti, B. (2018). Health-promoting spatial planning: Approaches for strengthening urban policy integration. *Planning Theory and Practice, 19*, 180–197.

Lynch, K. (1960). *The image of the city*, Publication of the Joint Center for Urban Studies. MIT Press.

MacNaughton, P., Satish, U., Laurent, J. G. C., Flanigan, S., Vallarino, J., Coull, B., Spengler, J. D., & Allen, J. G. (2017). The impact of working in a green certified building on cognitive function and health. *Building and Environment, 114*, 178–186.

McArthur, J. J., & Powell, C. (2020). Health and wellness in commercial buildings: Systematic review of sustainable building rating systems and alignment with contemporary research. *Building and Environment, 171*, 106635.

Mell, I. (2019). The impact of austerity on funding green infrastructure: A DPSIR evaluation of the Liverpool Green & Open Space Review (LG&OSR), UK. *Land Use Policy, 91*, 104284.

Moore, T., & Doyon, A. (2018). The uncommon nightingale: Sustainable housing innovation in Australia. *Sustainability, 10*, 3469.

Moore, T., & Higgins, D. (2016). Influencing urban development through government demonstration projects. *Cities, 56*, 9–15.

Moore, T., Horne, R., & Morrissey, J. (2014). Zero emission housing: Policy development in Australia and comparisons with the EU, UK, USA and California. *Environmental Innovation and Societal Transitions, 11*, 25–45.

Næss, P., Nicolaisen, M. S., & Strand, A. (2012). Traffic forecasts ignoring induced demand: A Shaky fundament for cost-benefit analyses. *European Journal of Transport and Infrastructure Research, 12*, 291–309.

Nicol, S., Roys, M., Ormandy, D., & Ezratty, V. (2017). *The cost of poor housing in the European Union*. BRE Press Briefing Paper.

Parker, G., & Street, E. (2021). Conceptualising the contemporary planning profession. In G. Parker & E. Street (Eds.), *Contemporary planning practice: Skills, specialisms and knowledges* (pp. 1–11). Red Globe Press.

Pineo, H. (2016). The value of healthy places—for developers, occupants and society. *Town and Country Planning*, Securing Outcomes from United Action, *85*, 477–480.

Pineo, H. (2019). *The value and use of urban health indicator tools in the complex urban planning policy and decision-making context* (Doctoral Thesis). University College London, London.

Pineo, H., Glonti, K., Rutter, H., Zimmermann, N., Wilkinson, P., & Davies, M. (2020a). Use of urban health indicator tools by built environment policy- and decision-makers: A systematic review and narrative synthesis. *Journal of Urban Health, 97*, 418–435.

Pineo, H., & Moore, G. (2021). Built environment stakeholders' experiences of implementing healthy urban development: An exploratory study. *Cities Health, 0*, 1–15. https://doi.org/10.1080/23748834.2021.1876376

Pineo, H., & Rydin, Y. (2018). *Cities, health and well-being*. Royal Institution of Chartered Surveyors.

Pineo, H., Zimmermann, N., Cosgrave, E., Aldridge, R. W., Acuto, M., & Rutter, H. (2018b). Promoting a healthy cities agenda through indicators: Development of a global urban environment and health index. *Cities Health, 2*, 27–45.

Pineo, H., Zimmermann, N., & Davies, M. (2019). Urban planning: Leveraging the urban planning system to shape healthy cities. In S. Galea, C. K. Ettman, & D. Vlahov (Eds.), *Urban health* (pp. 198–206). Oxford University Press.

Pineo, H., Zimmermann, N., & Davies, M. (2020c). Integrating health into the complex urban planning policy and decision-making context: A systems thinking analysis. *Palgrave Communications, 6*, 1–14

Reed, R., Bilos, A., Wilkinson, S., & Schulte, K.-W. (2009). International comparison of sustainable rating tools. *Journal of Sustainable Real Estate, 1*, 1–22.

Retzlaff, R. C. (2009). The use of LEED in planning and development regulation: An exploratory analysis. *Journal of Planning Education and Research, 29*, 67–77.

Robinson, J., & Cole, R. J. (2015). Theoretical underpinnings of regenerative sustainability. *Building Research and Information, 43*, 133–143.

Rydin, Y. (2010). *Governing for sustainable urban development* (1st ed.). Earthscan.

Rydin, Y. (2013). *The future of planning: Beyond growth dependence*. Policy Press.

Rydin, Y., Bleahu, A., Davies, M., Dávila, J. D., Friel, S., De Grandis, G., Groce, N., Hallal, P. C., Hamilton, I., Howden-Chapman, P., Ka-Man Lai, C. J., Lim, J. M., Osrin, D., Ridley, I., Scott, I., Taylor, M., Wilkinson, P., & Wilson, J. (2012). Shaping cities for health: Complexity and the planning of urban environments in the 21st century. *The Lancet, 379*, 2079–2108.

Savaget, P., Geissdoerfer, M., Kharrazi, A., & Evans, S. (2019). The theoretical foundations of sociotechnical systems change for sustainability: A systematic literature review. *Journal of Cleaner Production, 206*, 878–892.

Sengers, F., Wieczorek, A. J., & Raven, R. (2019). Experimenting for sustainability transitions: A systematic literature review. *Technological Forecasting and Social Change, 145*, 153–164.

Sisk, A., MacLeish-White, O., Gavin, V., Butler, T., Ogbu, L., Davis, V. O., Chaudhury, N., Hamdi, H., Worden, K., Kabane, N., Poticha, S., & Pathuis, H. (2020). Confronting power and privilege for inclusive, equitable, and healthy communities. *The BMJ*. Retrieved June 23, 2021, from https://blogs.bmj.com/bmj/2020/04/16/confronting-power-and-privilege-for-inclusive-equitable-and-healthy-communities/

Sterman, J. D. (2006). Learning from evidence in a complex world. *American Journal of Public Health, 96*, 505–514.

Tonne, C., Beevers, S., Armstrong, B., Kelly, F., & Wilkinson, P. (2008). Air pollution and mortality benefits of the London Congestion Charge: Spatial and socioeconomic inequalities. *Occupational and Environmental Medicine, 65*, 620–627.

Transport for London. (2017). FOI request detail. Retrieved April 16, 2021, from https://www.tfl.gov.uk/corporate/transparency/freedom-of-information/foi-request-detail

Turner, M., Kooshian, C., & Winkelman, S. (2012). *Colombia's Bus Rapid Transit (BRT) development and expansion*. Report produced for Mitigation Action Implementation Network (MAIN) by CCAP.

UNFCCC. (2010). *CDM project co-benefits in Bogotá*. Rapid and reliable bus transport for urban communities, Colombia.

Vlachokostas, C., Banias, G., Athanasiadis, A., Achillas, C., Akylas, V., & Moussiopoulos, N. (2014). Cense: A tool to assess combined exposure to environmental health stressors in urban areas. *Environment International, 63*, 1–10.
Wang, Q., Deng, Y., Li, G., Meng, C., Xie, L., Liu, M., & Zeng, L. (2020). The current situation and trends of healthy building development in China. *Chinese Science Bulletin, 65*, 246–255.
Wang, Q., Meng, C., & Li, G. (2017). Development demands and prospect of healthy buildings. *Heating, Ventilation and Air-Conditioning, 47*, 32–35.
WHO. (2018). *Global action plan on physical activity 2018–2030: More active people for a healthier world*. WHO, Geneva, Switzerland.
WHO. (2020). *Personal interventions and risk communication on air pollution*. WHO, Geneva, Switzerland.
World GBC. (2014). *Health, wellbeing and productivity in offices—The next chapter for green building*.
World GBC. (2016). *Building the business case: Health, wellbeing and productivity in Green Offices*. World Green Building Council, Toronto.

Chapter 9
Looking to the Future

9.1 Introduction

In presenting the theory and principles of healthy urbanism, this book has aimed to reframe current thinking about how urban environments affect health. Issues of intergenerational and international equity from the sustainable development discourse have been positioned as central to conceptualisations of healthy place-making, alongside the imperative for inclusive processes and outcomes. Through this focus on sustainability, equity and inclusion as defining features of healthy urban change, this book has argued that current practices of urban design and planning are insufficient to protect and improve health. It has also argued that relying on the existing 'business case' to propel healthy development is a flawed strategy and fundamental shifts in models for development and place-making are required.

There are workable models within existing urban governance systems to manage the associated risks and rewards of healthy development, such as community-led development and public sector support for innovation. This chapter will consider whether these incremental change models are sufficient in the face of the urgent and existential threat of environmental breakdown. International responses to COVID-19 and the extreme weather events of recent years offer a glimpse of how governments and societies perceive and react to public health emergencies, which are predicted to increase in the coming years. Models of recovery and prevention need to take a holistic view of health risks in cities which can be informed by the THRIVES approach. The chapter will critically consider these issues before turning to a discussion of strategies to positively frame challenges and solutions. Transformative models for healthy urban development can be achieved with the knowledge and technology that exists today, yet this is unlikely to occur without changes to political agendas and decision-making, which, in turn, may require public demand for change. As outlined in Chap. 3, recent societal trends are quickly shifting political agendas, including growing awareness and support for policies that support climate

H. Pineo, *Healthy Urbanism*, Planning, Environment, Cities,
https://doi.org/10.1007/978-981-16-9647-3_9

change mitigation and adaptation. These changes are creating momentum which should be harnessed by built environment and public health professionals to facilitate cross-sector support for healthy and sustainable models of urban development. Collaboration across sectors and with the public will be key to making the necessary shifts in policy and decision-making, as will positive framing of the problems and solutions.

9.2 Disaster Recovery and Prevention

The unequal burden of COVID-19 experienced by people living in areas of high overcrowding, poor air quality and other environmental burdens have demonstrated that creating healthy cities means equal consideration of inclusion, equity and sustainability. The pandemic's inequitable health impacts were predictable in the sense that these populations are generally likely to experience layers of disadvantage that make them more vulnerable to health threats. Diez Roux et al. (2020) explain, 'Strong residential segregation by social class, ethnicity, and migration status leads to large differences in these physical and social environments within cities, resulting in spatially patterned social and biological vulnerability to COVID-19' (p. 2). Low-income and minority ethnic residents are more likely to live in neighbourhoods with high environmental burdens and they are also more likely to have non-communicable disease (NCDs), partly attributable to the poorer quality of their environment, which put them at greater risk of severe COVID-19 outcomes and other health impacts. Disasters caused by climate change, such as extreme weather events, are also predicted to disproportionately affect vulnerable population groups. Many of the associated health impacts could be prevented by changing the inequitable distribution of built environment features that protect and promote health, such as high-quality housing, greenspaces, nutritious and affordable food, safe and active mobility infrastructure and others. These changes would benefit everybody in society and they would create cities that are more resilient to the impacts of climate change or new diseases, but they will require significant changes in how urban spaces are planned and maintained.

A multi-scalar and systems thinking perspective about the health impacts of urban developments must be part of plans for COVID recovery and future disaster prevention. Throughout the COVID-19 pandemic, there has been much focus on buildings and the extent to which they adequately support health in terms of disease transmission, a hitherto rare consideration in modern building design in high-income countries. This followed a decade of increased attention to health in property development, particularly in terms of wellbeing and productivity in offices. The design and engineering measures to achieve these outcomes include optimised lighting systems that support circadian rhythms and biophilic design, among others. There is nothing wrong with this aspiration to maximise health and wellbeing; however, it should not be forgotten that many people in society are living, studying or working in buildings that directly harm their health, such as through exposure to mould, dampness or polluted air. It should not be the case that 'healthy' buildings are the preserve of the wealthy, nor should those buildings harm health through their impact on wider ecosystem, planetary health or social systems. There is a need to bring all buildings up to a minimum standard to ensure that health inequalities in society are not exacerbated, particularly during health emergencies. There are two examples that demonstrate the complexity of disaster recovery and prevention in the built environment, including the tensions between short- and long-term health priorities, the limitations of voluntary action and the need for systems thinking.

The first potential concern relates to the important drive to improve ventilation in buildings to reduce exposure to indoor and outdoor pollutants and infectious diseases. Morawska et al. (2021) call for improvements to building ventilation to reduce indoor respiratory infections, including but not limited to COVID-19. They caution that 'although building designs should optimize indoor environment quality in terms of health and comfort, they should do so in an energy efficient way in the context of local climate and outdoor air pollution' (ibid., p. 690). This point is very important, given the current portion of global emissions arising from energy use in buildings that would potentially be significantly increased if many buildings were to increase ventilation using energy-intensive technologies. Energy-efficient ventilation and air-cleaning systems are available, such as mechanical ventilation with heat recovery, but others are energy intensive. In the face of these multiple challenges, a report by the UK's National Engineering Policy Centre (NEPC) calls for integrated thinking:

> There is an urgent need for greater regulatory focus on buildings in operation, not just to improve resilience to infection and indoor air quality but also to address the pressing need to deliver significant carbon emission savings from our buildings. These two ambitions need to be driven forward in tandem and efforts across government need to be fully coordinated. (2021, p. 7)

There are barriers to achieving this integrated approach to ventilation raised by the NEPC, including 'modest levels of relevant knowledge, skills, budgets and organisational maturity on crucial issues such as ventilation' among building owners and operators (ibid., p. 2). In the rush to recover from COVID, return to 'life as normal' and reduce a perceived risk of litigation, building managers could install technologies that negatively impact on thermal comfort, accessibility or energy demand and these systems may be expensive to replace. The NEPC report recommends a series of short- and medium-term actions to adapt buildings and infrastructure for increased ventilation that is synergistic with net zero and other goals, including increased training, regulation, enforcement and demonstration projects. Other countries have adopted guidance and funding mechanisms for building ventilation, such as Singapore and Canada, but this is an area with significant knowledge gaps that will need to be addressed with integrated thinking before widespread change will be achieved.

The need for air conditioning to prevent deaths and illness from overheating is another example tension for disaster management and prevention. Extreme heat events are likely to increase due to climate change and they are exacerbated in cities, especially in neighbourhoods with minimal green infrastructure, as described in Chap. 7. Use of mechanical air conditioning is a necessary public health intervention to save lives in many cities where existing buildings cannot maintain thermally safe environments, yet this creates a vicious cycle that further heats urban areas (Wong et al., 2013) and contributes to air pollution and climate change. From a healthy urbanism perspective, mechanical cooling in buildings should consider short- and long-term health impacts across multiple scales. There are passive cooling techniques that can be designed into new buildings or retrofitted during operation, but if these are to be deployed at pace and scale there will be a need for rapid training for contractors and building owners (including homeowners) about appropriate technologies and design measures. Technical experts will need to find ways to effectively communicate with these varying audiences to support the successful adoption of new cooling systems and strategies.

Both the ventilation and air conditioning examples highlight the need for systems thinking and caution against the risks of relying on individual action. If these interventions rely solely on voluntary individual action (by homeowners, businesses and employers), they are likely to primarily benefit affluent populations. The public sector has a role to play through regulation or financial incentives to avoid health inequities occurring in building types where the user does not (or cannot) pay for a higher quality environment, such as in social housing or schools. Governments can also support skills development, technology dissemination and communication strategies that would increase consumer demand for these technologies. The costs to the public purse of proactively addressing these problems will be significantly smaller than the costs to the economy and health systems caused by delaying action.

Equity should drive disaster management and recovery plans to increase long-term resilience to future health risks for everyone. During the early months of the COVID-19 pandemic, public health experts raised concerns about the vulnerabilities of people living in informal settlements, precarious housing and those experiencing homelessness (Corburn et al., 2020; Lewer et al., 2020). These warnings led some governments to implement temporary emergency responses such as accommodation for people experiencing homelessness and eviction bans, but many of these safeguards were removed. Other emergency responses in the built environment have transitioned into permanent infrastructure changes, such as temporary traffic restrictions to enable outdoor eating, play and social distancing. For example, between March and September 2020, London used emergency legislation to implement 72 Low Traffic Neighbourhoods (LTNs) covering around 300,000 Londoners (see Chap. 6). Despite their controversial nature, evaluations of London's LTNs have found that they have been deployed equitably across the city (Aldred et al., 2021), they do not delay emergency vehicles (Goodman et al., 2021) and they reduce traffic injuries (walking, cycling, and driving became roughly 3–4 times safer per trip) (Laverty et al., 2021). Such evaluations provide valuable evidence for city leaders to invest further in solutions that improve health and sustainability.

Potential pandemic management and recovery strategies in the built environment have been trialled and analysed in real-time with commentators doing their best to anticipate the duration and ripple effects of the virus. Drawn from expert insights

(Corburn et al., 2020; Diez Roux et al., 2020; Corburn & Sverdlik, 2017), the following built environment recovery strategies should be considered in terms of informal settlements and more widely in cities for pandemic management, recovery and prevention, but also to achieve wider urban health and sustainability outcomes:

- Recognise the inequitable differences within cities related to the quality of the physical environment and access to employment and services. Target investment in the areas of greatest need to increase resilience for the whole city.
- Avoid making assumptions that high population density is either wholly good or bad. Look for urban design and policy solutions that maximise the positive benefits of density (such as public transport viability) whilst minimising negative impacts (such as overcrowded housing).
- Invest time and resources in participatory processes to guide design and policy responses. This requires professionals to value community knowledge and to adopt new methods for involving diverse members of society.
- Adopt health impact assessment (HIA) and Health in All Policies (HiAP) approaches to develop and improve public policies and development plans.

These strategies can be adopted for future disaster management, such as the extreme weather events occurring from climate change. Importantly, they should be coupled with monitoring and evaluation processes to ensure that investments and interventions have the intended effect.

9.3 Incremental to Transformative Change

The scale and pace of change required to meet the goals of healthy urbanism in the context of environmental breakdown puts in question the adequacy of incremental improvements to policy and development. Melanie Crane et al. (2021) highlight the scale of the problem through the case of London's housing stock where a home must be retrofitted every five minutes for the next 30 years to achieve the city's 2050 net zero commitment. Achieving these housing stock upgrades could be transformational for urban health, particularly if the homes are simultaneously retrofitted to suit London's projected climate change impacts in terms of overheating and flooding with prioritisation of vulnerable residents. This is only one element of the required change in cities to make them healthy and sustainable, which on its own would necessitate significant investment in materials, training and labour. But progress has been slow and a Retrofit Accelerator programme announced in 2020 would only see 1600 homes retrofitted over three years (Mayor of London, 2020b). Mechanisms must be rapidly identified and deployed to move beyond such incremental improvements to fundamental and transformative urban changes for health and sustainability. However, there is an important role for bottom-up and top-down solutions at multiple scales to shift public awareness, industry norms and political agendas.

Market penetration of sustainable buildings over the past 30 years offers an informative parallel for the potential pace and scale of the response to calls for healthy buildings. The first sustainable building standard, BREEAM, was developed in 1990 as a voluntary standard, which could be seen as functioning initially through bottom-up action from innovative developers and design teams. Over a decade later, adoption was also driven by top-down policy requirements to achieve certification as part of planning consent for non-residential developments in many local authorities. There are now 42 similar standards across low-, middle- and high-income countries applied through a mixture of voluntary and mandatory mechanisms (Callway et al., 2020). Despite three decades of supply chain learning, new technologies and subsequent cost reductions, the proportion of new certified sustainable buildings remains quite low, particularly for housing. Only 4% of new Canadian residential buildings are certified (Canada Green Building Council, 2020), while certified green office buildings have grown from 6.4% of total office space in 2007 to 18.6% in 2018 across ten markets in Australia, Canada, Europe and the USA (Holtermans & Kok, 2018). The uptake of sustainable buildings is not fast enough to deal with the urgency of environmental degradation. Allen and Macomber (2020) describe healthy building growth through their 'Healthy Building Adoption Curve', adapted from the Rogers Adoption Curve. They suggest that market penetration of healthy buildings has passed through knowledge generators and early adopters, and in 2020, it was in the steep climb through leading markets, before it will arrive at developed markets and universal adoption. The key issue is the timescales over which these latter stages of adoption will be achieved. If progress can be gauged by sustainable building uptake, growth of healthy buildings may require regulation or financial incentives to significantly increase uptake. Residential buildings may lag behind commercial sectors due to the reduced financial incentives for developers, particularly for affordable and social housing tenures. It is a priority to identify healthy development models that are suited to diverse locations, markets and governance structures. Translating and adopting voluntary healthy built environment standards internationally will have benefits and drawbacks (Callway et al., 2020) and it is unlikely to be the only solution.

Achieving transformative change in cities requires interventions at multiple scales. Crane et al. (2021) analysed theories and levers for transformative urban change towards healthy and sustainable cities, drawing upon resilience theory and socio-technical studies (STS). They explain that transformational change requires 'integrated multi-scalar systems actors and actions operating across city sectors' (p. 6). From the STS literature, this means bottom-up and top-down actions across three interrelated scales:

> (i) the micro or niche level (localised innovations); (ii) the meso or 'regime' level, representing the institutional structure of the system; and (iii) the wider or macro socio-technical landscape, or exogenous environments. (ibid., p. 5)

The potential levers for transformative change include sustainable urban planning and infrastructure development, supported by technical and social innovations and behaviour change. An example of the types of changes needed can be found in the IPBES (2019) report on losses to biodiversity and ecosystem services which illustrate five 'levers' for transformative change to reverse environmental destruction:

> (1) developing incentives and widespread capacity for environmental responsibility and eliminating perverse incentives; (2) reforming sectoral and segmented decision-making to promote integration across sectors and jurisdictions; (3) taking pre-emptive and precautionary actions in regulatory and management institutions and businesses to avoid, mitigate and remedy the deterioration of nature, and monitoring their outcomes; (4) managing for resilient social and ecological systems in the face of uncertainty and complexity, to deliver decisions that are robust in a wide range of scenarios; and (5) strengthening environmental laws and policies and their implementation, and the rule of law more generally. (p. 9)

These actions can be mapped across the micro, meso and macro levels at which change is required for urban health and sustainability, but most require macro-scale action. All of these levers should be characterised by collaboration across sectors, disciplines and with the public.

Based on Crane et al.'s review, many of the case study examples referenced in this book could be seen as transformational change for urban health and sustainability because they were interventions at micro or meso scales that influenced wider system changes. The problem of how procedural or technological innovations can be scaled up beyond exemplar projects in a particular social, environmental and political context is a central issue for socio-technical studies. In Strategic Niche Management, the concept of the niche is used to describe 'spaces that allow nurturing and experimentation with the co-evolution of technology, user practices, and regulatory structures' that support regime change (Sengers et al., 2019, p. 155). For instance, the use of innovative technologies at the Bullitt Center helped to change Seattle's regulatory environment for new healthy and sustainable development. As described in Chap. 4, the Nightingale Housing model has been explored as an 'unprotected niche' (Doyon & Moore, 2019) that constitutes a rapidly growing alternative and more sustainable model to Australian housing. Another STS perspective to scalability stems from research on bounded socio-technical experiments in which participants would go through a process of social learning that they share with others to scale up the initiative (Sengers et al., 2019). Community-led co-housing projects could function through this mechanism and with adequate design expertise such projects can be healthy and sustainable. Theory from socio-technical studies shows how there is a need for bottom-up and top-down actions by diverse actors to support healthy urbanism. In the context of increasing deregulation in many settings, STS underscores the value of incremental small-scale action which may serve to shift wider structures and political systems, leading to large-scale transformational change. Communication of the benefits of small-scale actions (or exemplar projects) is a key part of how they could lead to systematic changes, therefore appropriately 'framing' these messages for the intended audience is crucial.

9.4 Framing Healthy Urbanism to Empower Practitioners

Learning how to positively 'frame' a complex policy problem is an increasingly important skill for built environment and public health professionals. The significant social and environmental challenges posed by environmental degradation and inequitable distribution of resources within cities can feel insurmountable, leading people to ignore the issue or become hopeless, both of which lead to inaction. Proponents of healthy urbanism need to find ways to frame urban health problems, including but not limited to climate change, in ways that acknowledge the reality and scale of the challenge whilst offering clear solutions. This book has tried to emulate this approach by describing the health impacts of environmental threats, but quickly moving to design and planning solutions, first through frameworks and principles and then through case study examples. This way of communicating is recommended by the FrameWorks Institute which specialises in 'framing' solutions for social issues, including health and climate change. They provide free briefing notes and toolkits based on research with varied professional groups and sectors about the messages that resonate to create understanding and motivate action. Useful strategies that can be adopted for healthy urbanism include the following: using metaphors for health, supporting messengers to champion the topic within their organisation or sector, describing collaboration as empowerment, communicating the 'value of investment', sharing success stories and using positive messages (Moyer et al., 2019). These strategies are considered, in turn, below using academic literature and findings from Moyer et al.'s research, forming a selection of wider strategies that could be adopted according to local political and cultural factors.

A lot has been written about the problem of communicating the meaning of the 'social determinants of health' to audiences without health backgrounds. Even the word 'health' is seen as problematic in particular places, exemplified by the WHO Healthy Cities movement finding more traction with 'environmental sustainability' as a policy goal than 'health' (de Leeuw, 2017b). The FrameWorks Institute found that 'social determinants of health' is an unclear term that 'without sufficient context and explanation, invites multiple problematic interpretations', such as blaming individuals for poor health or overemphasising government intervention (Moyer et al., 2019, p. 6). They recommend using more 'proactive, intersectional, and structural terms', specifically through the metaphor of the 'Foundation of Community Health' described as follows:

> The health of our community is like a building—it depends on a strong and stable foundation. Things like quality education, safe and affordable housing, access to healthcare, and employment opportunities structure positive health outcomes for everyone in important ways. As public health professionals, it's our mission to build thriving communities, so we work closely with many other sectors to assemble a solid foundation that supports long-lasting good health for us all. (ibid., p. 7)

This metaphor emphasises the active nature of creating health through collaboration across sectors. The authors also believe that this metaphor works well to communicate the need to work with local communities using an asset-based approach to create change. Rather than describing 'determinants' of health (of any kind, e.g. social, upstream, wider and environmental), those working on healthy urbanism could build on this 'foundations' metaphor.

Part of the reason why collaboration is important for urban health is that multiple sectors and disciplines are required to act, but they each have their own cultures, norms and languages that should be accounted for in any 'calls to action'. Moyer et al. recommend that a 'messenger' from another sector or discipline (e.g. housing or planning) who understands health can become an ally that is more effective at communicating to their own professional group. Their research found that other sectors may not be receptive to messages from public health professionals, viewing their arguments with 'a degree of uncertainty or skepticism' and harbouring 'fears about being "told what to do"', which can inhibit collaboration (ibid., 10–11). There are an increasing number of built environment professionals with training in healthy design and planning who could become these allies and collaborators with public health, but it could just as easily be an interested stakeholder with no specialist knowledge. Anybody who takes on this public health advocacy role should be aware of potential fears that non-health sectors may have about public health directing or derailing their work. This concern was identified by Pineo's (2019) study of the value of urban health indicators in urban planning, described in Chap. 8. Some public health interview participants who developed urban health indicators described cases where planners did not engage with them due to fear that the purpose of such engagement would be to obstruct new development or infrastructure to avoid potential negative health impacts. According to Moyer et al., these concerns need to be addressed proactively by 'affirming the priorities of other sectors, and acknowledging their valuable contributions' (2019, p. 11). They recommend using terms such as empower, fuel, energise and strengthen to describe public health's role in non-health sector activities. For instance, planners may collaborate with public health teams to identify the quantifiable health benefits of new active travel infrastructure as part of the justification for public funding.

The current urban governance context in most countries necessitates a business case to persuade public and private sector organisations to spend money on healthy and sustainable development, despite the difficulties with identifying returns for investors at the right timescale, as described in Chap. 8. Similarly, the FrameWorks Institute emphasises the 'major concern' of 'economic sustainability' for all sectors, noting that it can be a barrier to collaboration with public health due to 'small-picture thinking' and because 'it causes reticence about investing precious resources into new partnerships or unverified social endeavours, which can seem like irresponsible risks' (Moyer et al., 2019, p. 12). Given the risk averse nature of the construction industry, it is unsurprising that many developers would be reluctant to make commitments about developing a 'healthy' building or community, as expressed by international interview participants in research by Pineo and Moore (2021). There is a sense among some professionals that achieving 'health' through property development is less predictable and riskier than committing to reduce carbon emissions or achieve other sustainability outcomes. After all, future residents or those neighbouring new development could make claims about negative health impacts arising from many factors (e.g. the construction process or poor maintenance of mechanical systems) that would reflect negatively on the developer's reputation but may be outside of their control. To identify and persuade developers (or public sector commissioners) of the 'Value of Investment', it should be framed in relation to their 'own immediate financial objectives and decision-making to long-term social and economic goals' (Moyer et al., 2019, p. 12). There is also a need to reduce the actual or perceived risks of healthy development, which may be supported by standardisation and increasing learning within the sector. Close monitoring and evaluation of new buildings and infrastructure across multiple parameters (e.g. health benefits, reducing operational costs, increased productivity, etc.) will be required to enable learning and build many cases studies of increased value at different points in a project's lifecycle.

This book has gathered examples of equitable, sustainable and inclusive design and planning projects to enable readers to rethink what is possible and to challenge the status quo, in line with recommendations about framing social issues. Each of the examples has criticisable points, which could be viewed as an inevitable part of the compromises that characterise urban planning and development. But the examples do not have to be perfect to be effective at showing a different possible future. In its guidance for communicating about climate change during the difficult COVID-19 pandemic, the FrameWorks Institute described the importance of giving solutions, recommending that messages 'balance "we can see a problem ahead" with "there's a way to steer around it" ...[to] build realistic hope' (FrameWorks Institute, 2020). Case studies of innovative developments, technologies and policies provide clear evidence that change is possible. In researching case studies for this book, equity and inclusion were not as clearly driving decision-making and outcomes compared to environmental sustainability. At the time of writing, the Broadway Corridor project in Portland, Oregon, USA, was reported as a new example where racial equity was prioritised in the community benefit agreement (Walker, 2021). The case study describes years of strong relationship and trust building between a group called the Healthy Communities Coalition (an alliance of organisations representing communities of colour, people with disabilities, business, labour, environmental justice and others) and the city's economic and development agency, Prosper Portland, to negotiate the agreement. There is a need for more case studies about these complex projects that represent a significant and much needed shift in ways of working for built environment professionals. These examples are important because they subtly shift the way people think and talk about planning for equitable and inclusive places, in turn shifting action.

Using positive language is the final strategy for framing health issues described here. Numerous health and environmental reports have recently emphasised that 'doom and gloom' messages are not effective at mobilising action among the general public or built environment decision-makers. Moyer et al. (2019) describe how public health messages have fallen flat or even backfired in their aim to increase collaboration with non-health sectors:

> Public health professionals frequently want to sound the alarm for professionals in other sectors and make them see that working together is urgently needed in order to meet the challenges our communities face. Given other sectors' limited understandings of the field and how it relates to them, however, crisis messages not only fail to engage potential collaborators and inspire action, they trigger suspicion, degrade trust, and feed fatalism. (p. 17)

A similar effect from alarmist messaging was raised with regard to communicating about climate change, during the 'noisy, uncertain time' of the pandemic, where climate-related messages can 'add to people's sense of distress or fear' and that this can cause harm to the audience and the wider cause (FrameWorks Institute, 2020, p. 1). Built environment and health professionals need to find ways of communicating positively about the solutions to these significant health crises, and this can be done through the language they use. Although it may not match the way that professionals feel, Moyer et al. urge them to 'drop the crisis frame' and use a 'balanced or even optimistic tone' that emphasises solutions (2019, p. 18).

New guidance from the WHO about communicating the risks of air pollution adopts similar strategies to those described earlier. When communicating to the general public, the WHO recommends avoiding general statistics about the victims of air pollution and focusing instead on local environmental threats and clear personal actions that reflect 'the geography, economic conditions, culture, expectations and norms in each setting' (2020, p. 50). They also recommend highlighting policy solutions using positive language and imagery, such as photos of healthy active streets with greenspace and children playing, rather than images of polluted environments. Another example of building a positive story about environmental change is from a scientific review article in *Nature* where the authors state that rebuilding marine life could be achieved by 2050 (Duarte et al., 2020). In line with Moyer et al.'s advice, the article describes the problem but foregrounds the optimistic possibility of recovery. Duarte et al. detail how over-fishing and pollution have threatened marine life, but then they state: 'Nevertheless, biodiversity losses in the ocean are less pronounced than on land and many marine species are capable of recovery once pressures are reduced or removed' (2020, p. 39). This optimistic statement is followed by a list of potential interventions and case studies of successful rehabilitation efforts. This kind of framing should become a routine part of communicating healthy urbanism challenges and solutions.

9.5 Smart Cities and Health

The optimism surrounding new technology and smart systems to solve urban health challenges was not a core subject of this book, but one that will undoubtedly shape the future of healthy urbanism. In his book, *The Smart Enough City*, Ben Green (2019) contends that smart city technologies will drive urban transformation, but 'rather than by creating any sort of technological utopia', the core of this change will be by 'transforming the landscape of urban politics and power' (p. 12). One example germane to health is the autonomous vehicle, which is promoted with benefits for urban liveability such as removing the need for traffic lights and street parking, reducing traffic and increasing travel speeds. Green unpicks these claims arguing that AVs are a 'solution' from technology enthusiasts who 'ignore the multiplicity of needs in cities and the complexities of traffic, instead devising narrow solutions that revolve around technology' (p. 23). He suggests that through greater travel speeds and the promise of working or relaxing whilst travelling, AVs could contribute to urban sprawl because users could live further away from the urban core without the subsequent decline in quality of life usually caused by long-distance commuting. This 'AV-enabled sprawl' would, in turn, lead to disinvestment in city centres and increased greenhouse gas emissions, with clear negative health impacts from disconnected and unwalkable settlements. There are many other criticisms of AVs related to health, including their current limitations in recognising pedestrians, particularly those who are not of average size, such as children, and quickly moving street users, such as cyclists. Green suggests that such technological solutions distract attention from other approaches that are known to improve urban liveability, such as active travel infrastructure. City leaders and built environment professionals may not be able to stop AVs, but they can proactively consider how these technologies will serve their current goals and facilitate wide dialogue about whose problems they will solve or exacerbate.

Another smart city preoccupation with likely implications for health are the data platforms which promise to enable efficient management of urban systems. Data will increasingly be collected from sensors and electric devices in buildings and urban infrastructure that are analysed to inform everything from policing to maintenance. This book has echoed the calls of other urban health and sustainability transitions scholars to monitor and evaluate urban change processes and outcomes to ensure that the intended objectives are achieved. However, the collection and analysis of data from urban infrastructure and smartphones raises important issues of power and privacy, as evidenced by Google's abandoned development in Toronto via the company's Sidewalk Labs affiliate. An analysis by Goodman and Powles (2019) explains that the Sidewalk Toronto development on the city's waterfront was to be an exemplar of sustainable development with a digital layer running throughout the site and consisting of 'data and the things data touches, like sensors and cameras, data analytics and storage, wireless and wired infrastructure, and portals and devices' (p. 476). Their research focused on the governance challenges raised through this project, noting its significance for issues related to: 'innovation, privatization, privacy, surveillance, control, and the future of cities and urban life' (p. 459). A key contention among critics of the project was related to governance rather than privacy—in other words, who would be involved in decisions about the collection and use of data from residents and visitors.

Through the subject of city administration apps, Green (2019) characterises the tension between divergent perspectives of rational urban management versus the complex reality of urban governance. He argues that the 'core limitations on democratic decision-making and civic engagement—power, politics, public motivation and capacity' are misdiagnosed as 'problems of inefficiency and insufficient information' (p. 42). In other words, data platforms will not solve problems of urban governance. Healthy urbanism brings together the potential of public health—with its capabilities in data science—and the messy reality of achieving equitable, inclusive and sustainable outcomes through urban change. Data collected through smart urban systems could be linked to individual or neighbourhood-level health data to provide valuable insights about environments that support or hinder health-related behaviours (Sabel et al., 2021). This data could inform urban planning and decision-making, helping professionals to understand what works and to adapt ineffective design or policy measures. These analyses have already started, and some are governed by strict ethics and data protection procedures. However, as new data sources are generated and linked without public accountability, there is potential for organisations to undertake these analyses in ways that create adverse outcomes for privacy, which could compromise trust and ultimately undermine the potential for community benefit. This is an area for further research and civic debate, particularly because prior research has shown that existing data sources about urban health are not used in a rational and linear way for urban planning policy and decision-making, instead their value is more diffuse, resulting from long-term collaborations between diverse professionals and the community about what to measure and why (Pineo, 2019). The outputs of some data analyses may ironically be less important than the collaborative processes that determined their scope and use in urban governance.

9.6 Conclusion

The concepts behind healthy urbanism advocated in this book are not new, but through THRIVES, this book has reframed ideas of healthy place-making, shifting attention to the structural barriers to health that are driven by inequitable development and environmental degradation. In using the term 'healthy urbanism', a new so-called adjectival urbanism, the book does not intend to rival or displace other urbanisms (e.g. new urbanism and sustainable urbanism), but instead hopes to contribute to a healthy plurality of thought, practice and theory about the urban condition (Kim, 2014; Carmona, 2021). Healthy urbanism is not a prescriptive approach to design and planning but instead builds on existing theories to consider and describe the processes, principles and goals that can be applied to support health through urban planning and development. Through new understanding of the central principles of healthy urbanism—equity, inclusion and sustainability—practitioners can critically evaluate the merits of potential technologies and design solutions for health.

In the spirit of focusing on action and what can be achieved, this book has described the societal and environmental impacts of unhealthy planning, but more emphasis has been given to design and planning solutions, first through frameworks and principles and then through case study examples. The goal is for different audiences (planners, developers, urban designers, etc.) to find inspiration in the examples and the lessons they share. For planning policy-makers, the examples showcased the value of multi-sectoral collaboration, working closely with communities to understand diverse needs and solutions (including those which are community-led), the important role of data and monitoring, and the potentially transformative impact of demonstration and pilot projects. Developers may benefit from seeing examples, such as Nightingale Housing and Grow Community, that developed financially viable affordable housing using participatory processes and sustainable design. None of the case studies in this book are presented as perfect, instead they represent new ways of thinking and working that have achieved healthy and sustainable outcomes, partly through reflective learning and adjustment to their approach over time. Readers can extract the parts of these examples that best suit or can be adapted for their location and needs, over time transforming cities for health.

This book was written amid a period of great uncertainty, characterised by the COVID-19 pandemic, increasing climate change-induced disasters, and growing calls for social change in racial and gender equity. Healthy urbanism intersects through all of these issues and demands new skills and knowledge from professionals who aim to support health through the urban environment. In addition to their own disciplinary expertise (e.g. planning or architecture), professionals need general understanding of the epidemiological evidence base, knowledge of systems thinking and new skills in cultural competency and communication. Facing sustainability challenges, Green (2018) said, 'Professionals therefore need access to different bodies of knowledge or, phrased rather differently, professionals need to be committed to learning across artificially created professional boundaries' (p. 190). This imperative requires changes to higher education curriculum and professional bodies' competency frameworks (Pineo et al., 2021b). There are some undergraduate and postgraduate qualifications in public health and planning (particularly in the USA), but more of these programmes are needed, in addition to individual courses that are available to students and working professionals in a wide range of disciplines. Such healthy urbanism courses will enable participants to build shared understanding and strategies to work through any tensions that may arise between perspectives. Rather than expecting a single profession to have all of the required knowledge, healthy urbanism should be pursued in transdisciplinary teams who are not only diverse in disciplines and sectors but also in terms of their personal characteristics and life experiences.

References

Aldred, R., Verlinghieri, E., Sharkey, M., Itova, I., & Goodman, A. (2021). Equity in new active travel infrastructure: A spatial analysis of London's new Low Traffic Neighbourhoods. *Journal of Transport Geography, 96*, 103194.

Allen, J. G., & Macomber, J. D. (2020). *Healthy buildings: How indoor spaces can drive performance and productivity*. Harvard University Press.

Callway, R., Pineo, H., & Moore, G. (2020). Understanding the role of standards in the negotiation of a healthy built environment. *Sustainability, 12*, 9884.

Canada Green Building Council. (2020). Canada's green building engine: Market impact & opportunities in a critical decade.

Carmona, M. (2021). *Public places urban spaces: The dimensions of urban design* (3rd ed.). Routledge.

Corburn, J., & Sverdlik, A. (2017). Slum upgrading and health equity. *International Journal of Environmental Research and Public Health, 14*, 342.

Corburn, J., Vlahov, D., Mberu, B., Riley, L., Caiaffa, W. T., Rashid, S. F., Ko, A., Patel, S., Jukur, S., Martínez-Herrera, E., Jayasinghe, S., Agarwal, S., Nguendo-Yongsi, B., Weru, J., Ouma, S., Edmundo, K., Oni, T., & Ayad, H. (2020). Slum health: Arresting COVID-19 and improving well-being in urban informal settlements. *Journal of Urban Health, 97*, 348–357.

Crane, M., Lloyd, S., Haines, A., Ding, D., Hutchinson, E., Belesova, K., Davies, M., Osrin, D., Zimmermann, N., Capon, A., Wilkinson, P., & Turcu, C. (2021). Transforming cities for sustainability: A health perspective. *Environment International, 147*, 106366.

de Leeuw, E. (2017b). Engagement of sectors other than health in integrated health governance, policy, and action. *Annual Review of Public Health, 38*, 329–349.

Diez Roux, A. V., Barrientos-Gutierrez, T., Caiaffa, W. T., Miranda, J. J., Rodriguez, D., Sarmiento, O. L., Slesinski, S. C., & Vergara, A. V. (2020). Urban health and health equity in Latin American cities: What COVID-19 is teaching us. *Cities Health, 0*, 1–5.

Doyon, A., & Moore, T. (2019). The acceleration of an unprotected niche: The case of Nightingale Housing, Australia. *Cities, 92*, 18–26.

Duarte, C. M., Agusti, S., Barbier, E., Britten, G. L., Castilla, J. C., Gattuso, J.-P., Fulweiler, R. W., Hughes, T. P., Knowlton, N., Lovelock, C. E., Lotze, H. K., Predragovic, M., Poloczanska, E., Roberts, C., & Worm, B. (2020). Rebuilding marine life. *Nature, 580*, 39–51.

FrameWorks Institute. (2020). *Communicating about climate change in the time of COVID-19*. FrameWorks Institute.

Goodman, A., Laverty, A. A., Thomas, A., & Aldred, R. (2021). The impact of 2020 low traffic neighbourhoods on fire service emergency response times, in London, UK. *Findings* 23568.

Goodman, E. P., & Powles, J. (2019). Urbanism under Google: Lessons from Sidewalk Toronto Symposium: Rise of the machines: Artificial intelligence, robotics, and the reprogramming of law. *Fordham Law Review, 88*, 457–498.

Green, B. (2019). *The smart enough city: Putting technology in Its place to reclaim our urban future, strong ideas*. MIT Press.

Green, S. (2018). Sustainable construction. In *Sustainable futures in the built environment to 2050* (pp. 172–193). John Wiley & Sons, Ltd.

Holtermans, R., & Kok, N. (2018). *International green building adoption index*. Finance, GSBE Theme Sustainable Development.

IPBES. (2019). *Summary for policymakers of the global assessment report on biodiversity and ecosystem services of the Intergovernmental Science-Policy Platform on Biodiversity and Ecosystem Services*. IPBES Secretariat, Bonn, Germany.

Kim, D. (2014). Learning from adjectival urbanisms: The pluralistic urbanism. In *102nd ACSA Annual Meeting Proceedings Globalizing Architecture/Flows and Disruptions* (pp. 439–447), 2.

Laverty, A. A., Aldred, R., & Goodman, A. (2021). The impact of introducing low traffic neighbourhoods on road traffic injuries. *Findings* 18330.

Lewer, D., Braithwaite, I., Bullock, M., Eyre, M. T., White, P. J., Aldridge, R. W., Story, A., & Hayward, A. C. (2020). COVID-19 among people experiencing homelessness in England: A modelling study. *The Lancet Respiratory Medicine, 8*, 1181–1191.

Mayor of London. (2020b). Mayor of London leads the way in making homes fit for the future. Retrieved July 29, 2021, from https://www.london.gov.uk//press-releases/mayoral/mayor-introduces-programme-to-update-london-homes

Morawska, L., Allen, J., Bahnfleth, W., Bluyssen, P. M., Boerstra, A., Buonanno, G., Cao, J., Dancer, S. J., Floto, A., Franchimon, F., Greenhalgh, T., Haworth, C., Hogeling, J., Isaxon, C., Jimenez, J. L., Kurnitski, J., Li, Y., Loomans, M., Marks, G., … Yao, M. (2021). A paradigm shift to combat indoor respiratory infection. *Science, 372*, 689–691.

Moyer, J., L'Hôte, E., & Levay, K. (2019). *Public health reaching across sectors: Strategies for communicating effectively about public health and cross-sector collaboration with professionals from other sectors*. FrameWorks Institute, Washington, DC.

National Engineering Policy Centre. (2021). *Infection resilient environments: Buildings that keep us healthy and safe*, Initial Report. Royal Academy of Engineering, London, UK.

Pineo, H. (2019). *The value and use of urban health indicator tools in the complex urban planning policy and decision-making context* (Doctoral Thesis). University College London, London.

Pineo, H., & Moore, G. (2021). Built environment stakeholders' experiences of implementing healthy urban development: An exploratory study. *Cities Health, 0*, 1–15. https://doi.org/10.1080/23748834.2021.1876376

Pineo, H., Turnbull, E. R., Davies, M., Rowson, M., Hayward, A. C., Hart, G., Johnson, A. M., & Aldridge, R. W. (2021b). A new transdisciplinary research model to investigate and improve the health of the public. *Health Promotion International, 36*, 481–492.

Sabel, C., Amegbor, P. M., Zhang, Z., Chen, T.-H. K., Poulsen, M., Hertel, O., Sigsgaard, T., Horsdal, H. T., Pedersen, C. B., & Khan, J. (2021). Urban health and wellbeing. In W. Shi, M. F. Goodchild, M. Batty, M.-P. Kwan, & A. Zhang (Eds.), *Urban informatics* (pp. 259–280). Springer Nature.

Sengers, F., Wieczorek, A. J., & Raven, R. (2019). Experimenting for sustainability transitions: A systematic literature review. *Technological Forecasting and Social Change, 145*, 153–164.

Walker, M. (2021). How a diverse coalition in Portland, Ore. is centering racial equity in a large-scale development project. *Brookings*. Retrieved July 28, 2021, from https://www.brookings.edu/blog/the-avenue/2021/07/12/how-a-diverse-coalition-in-portland-ore-is-centering-racial-equity-in-a-large-scale-development-project/

WHO. (2020). *Personal interventions and risk communication on air pollution*. WHO, Geneva, Switzerland.

Wong, K. V., Paddon, A., & Jimenez, A. (2013). Review of world urban heat islands: Many linked to increased mortality. *Journal of Energy Resources Technology, 135*, 11.

References

Abubakar, I., Aldridge, R. W., Devakumar, D., Orcutt, M., Burns, R., Barreto, M. L., Dhavan, P., Fouad, F. M., Groce, N., Guo, Y., Hargreaves, S., Michael Knipper, J., Miranda, J., Madise, N., Kumar, B., Mosca, D., McGovern, T., Rubenstein, L., Sammonds, P., … Zhou, S. (2018). The UCL–Lancet Commission on migration and health: The health of a world on the move. *The Lancet, 392*, 2606–2654.

Acevedo-Garcia, D., Osypuk, T. L., Werbel, R. E., Meara, E. R., Cutler, D. M., & Berkman, L. F. (2004). Does housing mobility policy improve health? *Housing Policy Debate, 15*, 49–98.

Agyeman, J., Bullard, R. D., & Evans, B. (Eds.). (2003). *Just sustainabilities: Development in an unequal world*. Earthscan.

Agyeman, J., & Erickson, J. S. (2012). Culture, recognition, and the negotiation of difference: Some thoughts on cultural competency in planning education. *Journal of Planning Education and Research, 32*, 358–366. https://doi.org/10.1177/0739456X12441213

Agyeman, J. (2013). *Introducing just sustainabilities: Policy, planning and practice, Just sustainabilities*. Zed Books.

Aldred, R., Verlinghieri, E., Sharkey, M., Itova, I., & Goodman, A. (2021). Equity in new active travel infrastructure: A spatial analysis of London's new Low Traffic Neighbourhoods. *Journal of Transport Geography, 96*, 103194.

Aldridge, R. W., Lewer, D., Katikireddi, S. V., Mathur, R., Pathak, N., Burns, R., Fragaszy, E. B., Johnson, A. M., Devakumar, D., Abubakar, I., & Hayward, A. (2020). Black, Asian and Minority Ethnic groups in England are at increased risk of death from COVID-19: Indirect standardisation of NHS mortality data. *Wellcome Open Res., 5*, 88.

Aldridge, R.W., Pineo, H., Fragaszy, E., Eyre, M.T., Kovar, J., Nguyen, V., Beale, S., Byrne, T., Aryee, A., Smith, C., Devakumar, D., Taylor, J., Katikireddi, S.V., Fong, W.L.E., Geismar, C., Patel, P., Shrotri, M., Braithwaite, I., Patni, N., Navaratnam, A.M.D., Johnson, A., Hayward, A. (2021). Household overcrowding and risk of SARS-CoV-2: analysis of the Virus Watch prospective community cohort study in England and Wales [version 1; peer review: awaiting peer review]. Wellcome Open Res 2021, 6:347 (https://doi.org/10.12688/wellcomeopenres.17308.1)

Aldridge, R. W., Zenner, D., White, P. J., Williamson, E. J., Muzyamba, M. C., Dhavan, P., Davide Mosca, H., Thomas, L., Lalor, M. K., Abubakar, I., & Hayward, A. C. (2016). Tuberculosis in migrants moving from high-incidence to low-incidence countries: A population-based cohort study of 519 955 migrants screened before entry to England, Wales, and Northern Ireland. *The Lancet, 388*, 2510–2518.

Aletta, F., Oberman, T., & Kang, J. (2018). Associations between positive health-related effects and soundscapes perceptual constructs: A systematic review. *International Journal of Environmental Research and Public Health, 15*, 2392.

H. Pineo, *Healthy Urbanism*, Planning, Environment, Cities,
https://doi.org/10.1007/978-981-16-9647-3

Alidoust, S., & Bosman, C. (2015). Planning for an ageing population: Links between social health, neighbourhood environment and the elderly. *Australian Planner, 52*, 177–186.

Alirol, E., Getaz, L., Stoll, B., Chappuis, F., & Loutan, L. (2011). Urbanisation and infectious diseases in a globalised world. *The Lancet Infectious Diseases, 11*, 131–141.

Allen, J., & Balfour, R. (2014). *Natural solutions for tackling health inequalities*. UCL Institute of Health Equity.

Allen, J. G., MacNaughton, P., Laurent, J. G. C., Flanigan, S. S., Eitland, E. S., & Spengler, J. D. (2015). Green buildings and health. *Current Environmental Health Reports, 2*, 250–258.

Allen, J. G., & Macomber, J. D. (2020). *Healthy buildings: How indoor spaces can drive performance and productivity*. Harvard University Press.

Allen, L. N., & Feigl, A. B. (2017). What's in a name? A call to reframe non-communicable diseases. *The Lancet Global Health, 5*, e129–e130.

Andersson, A., Deng, J., Du, K., Zheng, M., Yan, C., Sköld, M., & Gustafsson, Ö. (2015). Regionally-varying combustion sources of the January 2013 severe haze events over Eastern China. *Environmental Science & Technology, 49*, 2038–2043.

Andres, L., Bryson, J. R., & Moawad, P. (2021). Temporary urbanisms as policy alternatives to enhance health and well-being in the post-pandemic city. *Current Environmental Health Reports, 8*(2), 167–176.

Angel, S., & Gregory, J. (2021). Does housing tenure matter? Owner-occupation and wellbeing in Britain and Austria. *Housing Studies, 0*, 1–21.

Angelakis, A. N., Antoniou, G. P., Yapijakis, C., & Tchobanoglous, G. (2020). History of hygiene focusing on the crucial role of water in the Hellenic Asclepieia (i.e., Ancient Hospitals). *Water, 12*, 754.

Anguelovski, I., Triguero-Mas, M., Connolly, J. J., Kotsila, P., Shokry, G., Pérez Del Pulgar, C., Garcia-Lamarca, M., Argüelles, L., Mangione, J., Dietz, K., & Cole, H. (2019). Gentrification and health in two global cities: A call to identify impacts for socially-vulnerable residents. *Cities Health, 4*, 1–10.

Annerstedt van den Bosch, M., & Depledge, M. H. (2015). Healthy people with nature in mind. *BMC Public Health, 15*, 1232.

Arao, B., & Clemens, K. (2013). From Safe Spaces to Brave Spaces, a new way to frame dialogue and diversity and social justice. In L. M. Landreman (Ed.), *The art of effective facilitation; Reflections from social justice educators* (pp. 135–150). Stylus Publishing.

Arnstein, S. R. (1969). A ladder of citizen participation. *Journal of the American Planning Association, 85*, 24–34.

Arup. (2017). *Oxford City Council, Barton NHS Healthy New Town and Underhill Circus redevelopment, health impact assessment*. Ove Arup & Partners Ltd.

Asian Development Bank, Revitalising Informal Settlements and their Environments (RISE). (2021). *Water-sensitive informal settlement upgrading: Overall principles and approach.*

Aubry, T., Bloch, G., Brcic, V., Saad, A., Magwood, O., Abdalla, T., Alkhateeb, Q., Xie, E., Mathew, C., Hannigan, T., Costello, C., Thavorn, K., Stergiopoulos, V., Tugwell, P., & Pottie, K. (2020). Effectiveness of permanent supportive housing and income assistance interventions for homeless individuals in high-income countries: A systematic review. *The Lancet Public Health, 5*, e342–e360.

Audrey, S., & Batista-Ferrer, H. (2015). Healthy urban environments for children and young people: A systematic review of intervention studies. *Health & Place, 36*, 97–117.

Babisch, W. (2006). Transportation noise and cardiovascular risk: Updated review and synthesis of epidemiological studies indicate that the evidence has increased. *Noise & Health, 8*, 1.

Bai, X., Dawson, R. J., Ürge-Vorsatz, D., Delgado, G. C., Barau, A. S., Dhakal, S., Dodman, D., Leonardsen, L., Masson-Delmotte, V., Roberts, D. C., & Schultz, S. (2018). Six research priorities for cities and climate change. *Nature, 555*, 23–25.

Bailey, Z. D., Krieger, N., Agénor, M., Graves, J., Linos, N., & Bassett, M. T. (2017). Structural racism and health inequities in the USA: Evidence and interventions. *The Lancet, 389*, 1453–1463.

Baker, E., Pham, N. T. A., Daniel, L., & Bentley, R. (2020). New evidence on mental health and housing affordability in cities: A quantile regression approach. *Cities, 96*, 102455.

Barlow, P. (2021). Regulation 28: Report to prevent future deaths. Ref: 2021-0113. Retrieved April 23, 2021 from https://www.judiciary.uk/publications/ella-kissi-debrah/

Barnes, M., Butt, S., & Tomaszewski, W. (2011). The duration of bad housing and children's well-being in Britain. *Housing Studies, 26*, 155–176.

Barnes, P. W., Williamson, C. E., Lucas, R. M., Robinson, S. A., Madronich, S., Paul, N. D., Bornman, J. F., Bais, A. F., Sulzberger, B., Wilson, S. R., Andrady, A. L., McKenzie, R. L., Neale, P. J., Austin, A. T., Bernhard, G. H., Solomon, K. R., Neale, R. E., Young, P. J., Norval, M., ... Zepp, R. G. (2019). Ozone depletion, ultraviolet radiation, climate change and prospects for a sustainable future. *Nature Sustainability, 2*, 569–579.

Barnett, C., & Parnell, S. (2016). Ideas, implementation and indicators: Epistemologies of the post-2015 urban agenda. *Environment and Urbanization, 28*, 87–98.

Barnett, D. W., Barnett, A., Nathan, A., Van Cauwenberg, J., Cerin, E., & on behalf of the Council on Environment and Physical Activity (CEPA)—Older Adults working group. (2017). Built environmental correlates of older adults' total physical activity and walking: A systematic review and meta-analysis. *International Journal of Behavioral Nutrition and Physical Activity, 14*, 103.

Barton, H. (2005). A health map for urban planners: Towards a conceptual model for healthy, sustainable settlements. *Built Environment, 31*, 339–355.

Barton, H. (2017). *City of well-being: A radical guide to planning*. Routledge; Taylor & Francis Group.

Barton, H., & Grant, M. (2006). A health map for the local human habitat. *The Journal of the Royal Society for the Promotion of Health, 126*, 252–253.

Barton, H., Grant, M., & Guise, R. (2003). *Shaping neighbourhoods: A guide for health, sustainability and vitality*. Spon.

Barton, H., Grant, M., & Guise, R. (2021). *Shaping neighbourhoods: For local health and global sustainability* (3rd ed.). Routledge, Taylor & Francis Group.

Bassett, E. (2018). Designing for health: fostering social capital formation through public space. In T. Beatley, C. Jones, & R. M. Rainey (Eds.), *Healthy environments, healing spaces: Practices and directions in health, planning, and design*. University of Virginia Press.

Beaglehole, R. (1993). *Basic epidemiology*. World Health Organization.

Beatley, T. (2016). *Handbook of biophilic city planning and design*. Island Press.

Beatley, T., Jones, C., & Rainey, R. M. (Eds.). (2018). *Healthy environments, healing spaces: Practices and directions in health, planning, and design*. University of Virginia Press.

Beato, C., & Velásquez, C. (2021). Participatory Slum upgrading and urban peacebuilding challenges in Favela settlements: The Vila Viva Program at Aglomerado da Serra (Belo Horizonte, Brazil). *The Journal of Illicit Economies and Development, 2*, 155–170.

Beaulac, J., Kristjansson, E., & Cummins, S. (2009). A systematic review of food deserts, 1966–2007. *Preventing Chronic Disease, 6*, A105.

Becerra, J. M., Reis, R. S., Frank, L. D., Ramirez-Marrero, F. A., Welle, B., Arriaga Cordero, E., Paz, F. M., Crespo, C., Dujon, V., Jacoby, E., Dill, J., Weigand, L., & Padin, C. M. (2013). Transport and health: A look at three Latin American cities. *Cadernos de Saúde Pública, 29*, 654–666.

Behbehani, F., Dombrowski, E., & Black, M. (2019). Systematic review of early child care centers in low- and middle-income countries and health, growth, and development among children aged 0–3 years (P11-052-19). *Current Developments in Nutrition, 3*.

Bell, S. (2015). Renegotiating urban water. *Progress in Planning, 96*, 1–28.

Ben-David, T., Rackes, A., & Waring, M. S. (2017). Alternative ventilation strategies in U.S. offices: Saving energy while enhancing work performance, reducing absenteeism, and considering outdoor pollutant exposure tradeoffs. *Building and Environment, 116*, 140–157.

Bentley, R. J., Pevalin, D., Baker, E., Mason, K., Reeves, A., & Beer, A. (2016). Housing affordability, tenure and mental health in Australia and the United Kingdom: A comparative panel analysis. *Housing Studies, 31*, 208–222.

Berke, E. M., & Vernez-Moudon, A. (2014). Built environment change: A framework to support health-enhancing behaviour through environmental policy and health research. *Journal of Epidemiology and Community Health, 68*, 586–590.

Berkowitz, R. L., Gao, X., Michaels, E. K., & Mujahid, M. S. (2020). Structurally vulnerable neighbourhood environments and racial/ethnic COVID-19 inequities. *Cities Health, 0*, 1–4. https://doi.org/10.1080/23748834.2020.1792069

Bicquelet-Lock, A., Divine, J., & Crabb, B. (2020). *Women and planning: An analysis of gender related barriers to professional advancement*. Royal Town Planning Institute.

Bircher, J., & Kuruvilla, S. (2014). Defining health by addressing individual, social, and environmental determinants: New opportunities for health care and public health. *Journal of Public Health Policy Basingstoke, 35*, 363–386.

Birkeland, J. (2008). *Positive development: From vicious circles to virtuous cycles through built environment design*. Earthscan.

Birkeland, J. (2017). Net-positive design and development. *Landscape Review, 17*, 83–87.

Birkeland, J. (2018). Challenging policy barriers in sustainable urban design. *Bulletin of Geography: Socio-Economic Series, 40*, 41–56.

Birkeland, J., & Knight-Lenihan, S. (2016). Biodiversity offsetting and net positive design. *Journal of Urban Design, 21*, 50–66.

Black, C., Moon, G., & Baird, J. (2014). Dietary inequalities: What is the evidence for the effect of the neighbourhood food environment? *Health & Place, 27*, 229–242.

Black, P., & Sonbli, T. E. (2019). *The urban design process*. Lund Humphries.

Bluyssen, P. M. (2010). Towards new methods and ways to create healthy and comfortable buildings. *Building and Environment, 45*, 808–818.

Bolden, K. (2020). Sustainable, healthy buildings and real estate. *Ernst Young*. Retrieved January 02, 2011, from https://www.ey.com/en_us/real-estate-hospitality-construction/sustainable-healthy-buildings-meeting-real-estate-expectations

Bonita, R., Beaglehole, R., & Kjellström, T. (2006). *Basic epidemiology* (2nd ed.). World Health Organization.

Brainard, J., Cooke, R., Lane, K., & Salter, C. (2019). Age, sex and other correlates with active travel walking and cycling in England: Analysis of responses to the Active Lives Survey 2016/17. *Preventive Medicine, 123*, 225–231.

Braithwaite, I., Zhang, S., Kirkbride, J. B., Osborn, D. P. J., & Hayes, J. F. (2019). Air pollution (particulate matter) exposure and associations with depression, anxiety, bipolar, psychosis and suicide risk: A systematic review and meta-analysis. *Environmental Health Perspectives, 127*, 126002.

Braubach, M., Jacobs, D., & Ormandy, D. (2011). *Environmental burden of disease associated with inadequate housing: A method guide to the quantification of health effects of selected housing risks in the WHO European region*. World Health Organization, Regional Office for Europe, Copenhagen.

Braveman, P. (2014). What are health disparities and health equity? We need to be clear. *Public Health Report, 129*(Suppl 2), 5–8.

BRE. (2016). Masthusen, Malmö, Sweden. *BREEAM*. Retrieved March 19, 2020, from https://www.breeam.com/case-studies/communities/masthusen-malmo-sweden/

Breed, M. F., Cross, A. T., Wallace, K., Bradby, K., Flies, E., Goodwin, N., Jones, M., Orlando, L., Skelly, C., Weinstein, P., & Aronson, J. (2020). Ecosystem restoration: A public health intervention. *EcoHealth, 18*, 269–271.

Breuer, D. (1998). *City health profiles: A review of progress*. World Health Organization, Regional Office for Europe.

Breysse, J., Jacobs, D. E., Weber, W., Dixon, S., Kawecki, C., Aceti, S., & Lopez, J. (2011). Health outcomes and green renovation of affordable housing. *Public Health Reports, 126*, 64–75.

Bristol City Council. (n.d.). The quality of life in Bristol. Retrieved October 30, 2016, from https://www.bristol.gov.uk/statistics-census-information/the-quality-of-life-in-bristol

Bristol, K. G. (1991). The Pruitt-Igoe myth. *Journal of Architectural Education, 44*, 163–171.

British Council for Offices. (2018). *Wellness matters: Health and wellbeing in offices and what to do about it.*

Britton, E., Kindermann, G., Domegan, C., & Carlin, C. (2020). Blue care: A systematic review of blue space interventions for health and wellbeing. *Health Promotion International, 35*, 50–69.

Brussoni, M., Gibbons, R., Gray, C., Ishikawa, T., Sandseter, E., Bienenstock, A., Chabot, G., Fuselli, P., Herrington, S., Janssen, I., Pickett, W., Power, M., Stanger, N., Sampson, M., & Tremblay, M. (2015). What is the relationship between risky outdoor play and health in children? A systematic review. *International Journal of Environmental Research and Public Health, 12*, 6423–6454.

Bull, J. W., Gordon, A., Watson, J. E. M., & Maron, M. (2016). Seeking convergence on the key concepts in 'no net loss' policy. *Journal of Applied Ecology, 53*, 1686–1693.

Bullard, R. D. (2005). *The quest for environmental justice: Human rights and the politics of pollution*. Sierra Club Books.

Bullard, R. D. (2007). Equity, unnatural man-made disasters, and race: Why environmental justice matters. In C. Wilkinson & R., R. Freudenburg, W. (Eds.), *Equity and the environment, research in social problems and public policy* (pp. 51–85). Emerald Group Publishing Limited.

Bullitt Center. (2021). The Bullitt Center composting toilet system a white paper on lessons learned. Retrieved November 26, 2021, from https://bullittcenter.org/wp-content/uploads/2021/03/The-Bullitt-Center-Composting-Toilet-System-FINAL.pdf

Bullitt Center. (n.d.). Greywater system. Retrieved April 14, 2021, from https://bullittcenter.org/building/building-features/wastewater-use/

Bullitt Foundation. (n.d.-a). History. Retrieved March 07, 2021, from https://www.bullitt.org/about/history/

Bullitt Foundation. (n.d.-b). Deep green buildings. Retrieved March 07, 2021, from https://www.bullitt.org/programs/deep-green-buildings/

Bullitt Foundation. (n.d.-c). About. Bullitt Found. Retrieved March 07, 2021, from https://www.bullitt.org/about/

Burbeck, A. (2020). *Chicana/o murals, placemaking, and the threat of street art in Denver's La Alma-Lincoln Park* (M.A.). University of Colorado at Boulder, United States, Colorado.

Burman, M., Brown, J., Tisdall, K., & Batchelor, S. (2000). *View from the girls: Exploring violence and violent behaviour*. ESRC End of Award Report. Economic and Social Research Council, Swindon.

Burpee, H., Beck, D. A. C., & Meschke, J. S. (2014). *Health impacts of green buildings*. American Institute of Architects.

Burpee, H., Gilbride, M., Douglas, K., Beck, D., & Meschke, J. S. (2015). *Health impacts of a living building*. University of Washington.

Burton, E. J., Mitchell, L., & Stride, C. B. (2011). Good places for ageing in place: Development of objective built environment measures for investigating links with older people's wellbeing. *BMC Public Health, 11*, 839.

Buse, C. G., Oestreicher, J. S., Ellis, N. R., Patrick, R., Brisbois, B., Jenkins, A. P., McKellar, K., Kingsley, J., Gislason, M., Galway, L., McFarlane, A., Walker, J., Frumkin, H., & Parkes, M. (2018). Public health guide to field developments linking ecosystems, environments and health in the Anthropocene. *Journal of Epidemiology and Community Health, 72*, 420–425.

Butler, C. D. (2016). Sounding the alarm: Health in the anthropocene. *International Journal of Environmental Research and Public Health, 13*, 665.

Buttorff, C., Ruder, T., & Bauman, M. (2017). *Multiple chronic conditions in the United States*. RAND Corporation.

Callway, R., Pineo, H., & Moore, G. (2020). Understanding the role of standards in the negotiation of a healthy built environment. *Sustainability, 12*, 9884.

Campbell, M., Escobar, O., Fenton, C., & Craig, P. (2018). The impact of participatory budgeting on health and wellbeing: A scoping review of evaluations. *BMC Public Health, 18*, 822.

Canada Green Building Council. (2020). Canada's green building engine: Market impact & opportunities in a critical decade.

Cancer Research UK. (2019). Obese people outnumber smokers two to one. Retrieved April 19, 2021, from https://www.cancerresearchuk.org/about-us/cancer-news/press-release/2019-07-03-obese-people-outnumber-smokers-two-to-one

Cappuccio, F., Miller, M. A., Lockley, S. L., & Rajaratnam, S. M. W. (2018). Sleep, health and society. In F. Cappuccio, M. A. Miller, S. W. Lockley, & S. M. W. Rajaratnam (Eds.), *Sleep, health, and society: From aetiology to public health* (pp. 1–9). Oxford Scholarship Online. Oxford University Press.

Capstick, S., Whitmarsh, L., Poortinga, W., Pidgeon, N., & Upham, P. (2015). International trends in public perceptions of climate change over the past quarter century. *WIREs Climate Change, 6*, 35–61.

Carey, G., Crammond, B., & De Leeuw, E. (2015). Towards health equity: A framework for the application of proportionate universalism. *International Journal for Equity in Health, 14*, 81.

Carmichael, L., Barton, H., Gray, S., Lease, H., & Pilkington, P. (2012). Integration of health into urban spatial planning through impact assessment: Identifying governance and policy barriers and facilitators. *Environmental Impact Assessment Review, 32*, 187–194.

Carmichael, L., Townshend, T. G., Fischer, T. B., Lock, K., Petrokofsky, C., Sheppard, A., Sweeting, D., & Ogilvie, F. (2019). Urban planning as an enabler of urban health: Challenges and good practice in England following the 2012 planning and public health reforms. *Land Use Policy, 84*, 154–162.

Carmona, M. (2019). Place value: Place quality and its impact on health, social, economic and environmental outcomes. *Journal of Urban Design, 24*, 1–48.

Carmona, M. (2021). *Public places urban spaces: The dimensions of urban design* (3rd ed.). Routledge.

Carrere, J., Reyes, A., Oliveras, L., Fernández, A., Peralta, A., Novoa, A. M., Pérez, K., & Borrell, C. (2020). The effects of cohousing model on people's health and wellbeing: A scoping review. *Public Health Reviews, 41*, 22.

Carrus, G., Scopelliti, M., Lafortezza, R., Colangelo, G., Ferrini, F., Salbitano, F., Agrimi, M., Portoghesi, L., Semenzato, P., & Sanesi, G. (2015). Go greener, feel better? The positive effects of biodiversity on the well-being of individuals visiting urban and peri-urban green areas. *Landscape and Urban Planning, 134*, 221–228.

Carvalho, A. M., Rezende, B. de M., Santos, D. G. O., Miranda, I. G., Merladet, F. A. D., Coelho, L. X. P., de Oliveira, R. A. P., & Isaías, T. L. S. (2012). Vila Viva, a project of urban, social and political organization of Aglomerado da Serra: Analysis of effects.

Carver, A., Timperio, A., & Crawford, D. (2008). Playing it safe: The influence of neighbourhood safety on children's physical activity—A review. *Health & Place, 14*, 217–227.

Caspi, C. E., Sorensen, G., Subramanian, S. V., & Kawachi, I. (2012). The local food environment and diet: A systematic review. *Health & Place, 18*, 1172–1187.

Cave, B. (2015). Assessing the potential health effects of policies, plans, programmes and projects. In H. Barton, S. Thompson, M. Grant, & S. Burgess (Eds.), *The Routledge handbook of planning for health and well-being: Shaping a sustainable and healthy future* (pp. 371–385). Taylor and Francis.

Cedeño-Laurent, J. G., Allen, J., & Spengler, J. D. (2018a). The built environment and sleep. In F. Cappuccio, M. A. Miller, S. W. Lockley, & S. M. W. Rajaratnam (Eds.), *Sleep, health, and society: From aetiology to public health, Oxford Scholarship Online*. Oxford University Press.

Cedeño-Laurent, J. G., Williams, A., MacNaughton, P., Cao, X., Eitland, E., & Spengler, J. D. (2018b). Building evidence for health: Green buildings, current science, and future challenges. *Annual Review of Public Health, 39*, 291–308.

Center for Active Design, United Nations Environment Programme Finance Initiative, BentallGreenOak. (2021). *A new investor consensus: The rising demand for healthy buildings*.

Central Committee of Chinese Communist Party, State Council. (2016). *The plan for healthy China 2030. Government of the People's Republic of China*, Beijing.

Centric Lab. (2020). *Air pollution and health in Southall*. A report for C.A.S.H.

Cerchiai, L. (2017). Urban civilization. In A. Naso (Ed.), *Etruscology*. De Gruyter, Berlin.

Chadwick, E. (1842). *Report to Her Majesty's principal secretary of state for the home department, from the poor law commissioners, on an inquiry into the sanitary condition of the labouring population of Great Britain* (House of Commons Sessional Paper). W. Clowes and Sons, London.

Chanan, G., Morton, K., & Harris, K. (2019). *Barton healthy new town concluding evaluation: Three key activities*. Health Empowerment Leverage Project.

Chang, M. (2018). *Securing constructive collaboration and consensus for planning healthy developments: A report from the Developers and Wellbeing project*. Town and Country Planning Association.

Chastin, S. F. M., Abaraogu, U., Bourgois, J. G., Dall, P. M., Darnborough, J., Duncan, E., Dumortier, J., Pavón, D. J., McParland, J., Roberts, N. J., & Hamer, M. (2021). Effects of regular physical activity on the immune system, vaccination and risk of community-acquired infectious disease in the general population: Systematic review and meta-analysis. *Sports Medicine, 51*, 1673–1686.

Chatterjee, H. J., Camic, P. M., Lockyer, B., & Thomson, L. J. M. (2018). Non-clinical community interventions: A systematised review of social prescribing schemes. *Arts Health, 10*, 97–123.

Chen, J., & Krieger, N. (2020). *Revealing the unequal burden of COVID-19 by income, race/ethnicity, and household crowding: US county vs. ZIP code analyses* (HCPDS Working Paper, Volume 19, Number 1). Harvard T.H. Chan School of Public Health, Boston, MA.

Choi, J.-H., Beltran, L. O., & Kim, H.-S. (2012). Impacts of indoor daylight environments on patient average length of stay (ALOS) in a healthcare facility. *Building and Environment, 50*, 65–75.

Christian, H., Knuiman, M., Divitini, M., Foster, S., Hooper, P., Boruff, B., Bull, F., & Giles-Corti, B. (2017). A longitudinal analysis of the influence of the neighborhood environment on recreational walking within the neighborhood: Results from RESIDE. *Environmental Health Perspectives, 125*, 077009.

Chung, C. K. L., Zhang, F., & Wu, F. (2018). Negotiating green space with landed interests: The urban political ecology of Greenway in the Pearl River Delta, China. *Antipode, 50*, 891–909.

City of New York. (2010). *Active design guidelines: Promoting physical activity and health in design*.

City of Seattle. (2015). Seattle launches Vision Zero plan to end traffic deaths and injuries by 2030 [WWW Document]. *Seattle.gov*. Retrieved February 08, 2017, from http://murray.seattle.gov/seattle-launches-vision-zero-plan-to-end-traffic-deaths-and-injuries-by-2030/

City of Seattle. (n.d.). Vision Zero—transportation. Retrieved April 16, 2021, from https://www.seattle.gov/transportation/projects-and-programs/safety-first/vision-zero

City of Vienna. (2013). *Gender mainstreaming in urban planning and urban development*. MA 18—Urban Development and Planning, Vienna.

Clements, R. (2004). An investigation of the status of outdoor play. *Contemporary Issues in Early Childhood, 5*, 68–80.

Clements-Croome, D., Turner, B., & Pallaris, K. (2019). Flourishing workplaces: A multisensory approach to design and POE. *Intelligent Buildings International, 0*, 1–14.

Clifford, B., Canelas, P., Ferm, J., & Livingstone, N. (2020). *Research into the quality standard of homes delivered through change of use permitted development rights*. Ministry of Housing, Communities and Local Government.

Clifford, B., Ferm, J., Livingstone, N., & Canelas, P. (2018). *Assessing the impacts of extending permitted development rights to office-to-residential change of use in England*. Royal Institution of Chartered Surveyors.

Clifford, M. (2018). Investors see returns in healthy buildings. *Jones Lang LaSalle*. Retrieved January 02, 2021, from https://www.jll.co.uk/en/trends-and-insights/investor/investors-see-returns-in-healthy-buildings

Cole, H. V. S., Anguelovski, I., Connolly, J. J. T., García-Lamarca, M., Perez-del-Pulgar, C., Shokry, G., & Triguero-Mas, M. (2021). Adapting the environmental risk transition theory for urban health inequities: An observational study examining complex environmental riskscapes in seven neighborhoods in Global North cities. *Social Science & Medicine, 277*, 113907.

Cole, H. V. S., Lamarca, M. G., Connolly, J. J. T., & Anguelovski, I. (2017). Are green cities healthy and equitable? Unpacking the relationship between health, green space and gentrification. *Journal of Epidemiology and Community Health, 71*, 1118–1121.

Cole, M. A., Ozgen, C., & Strobl, E. (2020). Air pollution exposure and Covid-19 in Dutch municipalities. *Environmental and Resource Economics, 76*, 581–610.

Coleman, S., Touchie, M. F., Robinson, J. B., & Peters, T. (2018). Rethinking performance gaps: A Regenerative sustainability approach to built environment performance assessment. *Sustainability, 10*, 4829.

Colton, M. D., Laurent, J. G. C., MacNaughton, P., Kane, J., Bennett-Fripp, M., Spengler, J., & Adamkiewicz, G. (2015). Health benefits of green public housing: Associations with asthma morbidity and building-related symptoms. *American Journal of Public Health, 105*, 2482–2489.

CSDH. (2007). *A conceptual framework for action on the social determinants of health.* Discussion paper for the Commission on Social Determinants of Health. World Health Organization, Geneva.

Conticini, E., Frediani, B., & Caro, D. (2020). Can atmospheric pollution be considered a co-factor in extremely high level of SARS-CoV-2 lethality in Northern Italy? *Environmental Pollution, 261*, 114465.

Convention on Biological Diversity. (2006). Article 2: Use of terms. Retrieved March 11, 2021, from https://www.cbd.int/convention/articles/?a=cbd-02

Corburn, J. (2013). *Healthy city planning: From neighbourhood to national health equity* (Planning, History and Environment Series). Routledge.

Corburn, J. (2015). Urban inequities, population health and spatial planning. In H. Barton, S. Thompson, M. Grant, & S. Burgess (Eds.), *The Routledge handbook of planning for health and well-being: Shaping a sustainable and healthy future* (pp. 37–47). Taylor and Francis.

Corburn, J. (2017). Equitable and healthy city planning: Towards healthy urban governance in the century of the city. In E. De Leeuw & J. Simos (Eds.), *Healthy cities: The theory, policy, and practice of value-based urban planning* (pp. 31–41). Springer.

Corburn, J., & Cohen, A. K. (2012). Why we need urban health equity indicators: Integrating science, policy, and community. *PLoS Medicine, 9*, e1001285.

Corburn, J., & Sverdlik, A. (2017). Slum upgrading and health equity. *International Journal of Environmental Research and Public Health, 14*, 342.

Corburn, J., Vlahov, D., Mberu, B., Riley, L., Caiaffa, W. T., Rashid, S. F., Ko, A., Patel, S., Jukur, S., Martínez-Herrera, E., Jayasinghe, S., Agarwal, S., Nguendo-Yongsi, B., Weru, J., Ouma, S., Edmundo, K., Oni, T., & Ayad, H. (2020). Slum health: Arresting COVID-19 and improving well-being in urban informal settlements. *Journal of Urban Health, 97*, 348–357.

Corkery, L. (2015). Urban greenspaces and human well-being. In H. Barton, S. Thompson, M. Grant, & S. Burgess (Eds.), *The Routledge Handbook of planning for health and well-being: Shaping a Sustainable and healthy future* (pp. 239–253). Taylor and Francis.

Cortright, J. (2009). *Walking the walk: How walkability raises home values in US cities.* CEOs for Cities.

Costa, S., Benjamin-Neelon, S. E., Winpenny, E., Phillips, V., & Adams, J. (2019). Relationship between early childhood non-parental childcare and diet, physical activity, sedentary behaviour, and sleep: A systematic review of longitudinal studies. *International Journal of Environmental Research and Public Health, 16*, 4652.

Costello, A., Abbas, M., Allen, A., Ball, S., Bell, S., Bellamy, R., Friel, S., Groce, N., Johnson, A., Kett, M., Lee, M., Levy, C., Maslin, M., McCoy, D., McGuire, B., Montgomery, H., Napier, D., Pagel, C., Patel, J., … Patterson, C. (2009). Managing the health effects of climate change. *The Lancet, 373*, 1693–1733.

Cousins, E. H. (2020). *The sanctuary at bath in the Roman Empire, Cambridge classical studies*. Cambridge University Press.
Cowan, S., Davies, B., Diaz, D., Enelow, N., & Halsey, K. (2014). *Optimizing urban ecosystem services: The Bullitt center case study*. Ecotrust.
Crane, M., Lloyd, S., Haines, A., Ding, D., Hutchinson, E., Belesova, K., Davies, M., Osrin, D., Zimmermann, N., Capon, A., Wilkinson, P., & Turcu, C. (2021). Transforming cities for sustainability: A health perspective. *Environment International, 147*, 106366.
Criado-Perez, C., Collins, C. G., Jackson, C. J., Oldfield, P., Pollard, B., & Sanders, K. (2020). Beyond an 'informed opinion': Evidence-based practice in the built environment. *Architectural Engineering and Design Management, 16*, 23–40.
CSDH. (2008). *Closing the gap in a generation: Health equity through action on the social determinants of health*. Final Report of the Commission on Social Determinants of Health. World Health Organization, Geneva.
Curtis, V. A. (2007). Dirt, disgust and disease: A natural history of hygiene. *Journal of Epidemiology and Community Health, 61*, 660–664.
Dahlgren, G., & Whitehead, M. (1991). *Policies and strategies to promote social equity in health*. Institute for Futures Studies
Dahlgren, G., & Whitehead, M. (2006). *European strategies for tackling social inequities in health: Levelling up Part 2*. WHO Collaborating Centre for Policy Research on Social Determinants University of Liverpool of Health, Liverpool.
Dakubo, C. Y. (2011). *Ecosystems and human health: A critical approach to ecohealth research and practice*. Springer.
Dallat, M. A. T., Soerjomataram, I., Hunter, R. F., Tully, M. A., Cairns, K. J., & Kee, F. (2014). Urban greenways have the potential to increase physical activity levels cost-effectively. *European Journal of Public Health, 24*, 190–195.
de Coninck, H., Revi, A., Babiker, M., Bertoldi, P., Buckeridge, M., Cartwright, A., … Waterfield, T. (Eds.). (2018). Global warming of 1.5°C. An IPCC Special Report on the impacts of global warming of 1.5°C above pre-industrial levels and related global greenhouse gas emission pathways, in the context of strengthening the global response to the threat of climate change, sustainable development, and efforts to eradicate poverty (p. 132). Intergovernmental Panel on Climate Change.
de Dear, R., Kim, J., Candido, C., & Deuble, M. (2015). Adaptive thermal comfort in Australian school classrooms. *Building Research and Information, 43*, 383–398.
de Jong, M., Yu, C., Joss, S., Wennersten, R., Yu, L., Zhang, X., & Ma, X. (2016). Eco city development in China: Addressing the policy implementation challenge. *Journal of Cleaner Production*, Special Volume: Transitions to Sustainable Consumption and Production in Cities, *134*, 31–41.
de Leeuw, E. (2017a). Cities and health from the neolithic to the anthropocene. In E. De Leeuw & J. Simos (Eds.), *Healthy cities: The theory, policy, and practice of value-based urban planning* (pp. 3–30). Springer.
de Leeuw, E. (2017b). Engagement of sectors other than health in integrated health governance, policy, and action. *Annual Review of Public Health, 38*, 329–349.
de Leeuw, E., & Simos, J. (Eds.). (2017). *Healthy cities: The theory, policy, and practice of value-based urban planning*. Springer.
de Leeuw, E., & Skovgaard, T. (2005). Utility-driven evidence for healthy cities: Problems with evidence generation and application. *Social Science & Medicine, 61*, 1331–1341.
de Leeuw, E., Tsouros, A. D., Dyakova, M., & Green, G. (Eds.) (2014). *Healthy cities, promoting health and equity, evidence for local policy and practice: Summary evaluation of Phase V of the WHO European Healthy Cities Network*. World Health Organization Regional Office for Europe, Copenhagen, Denmark.
Dennis, S. F., Gaulocher, S., Carpiano, R. M., & Brown, D. (2009). Participatory photo mapping (PPM): Exploring an integrated method for health and place research with young people. *Health & Place, 15*, 466–473.

Department for Communities and Local Government. (2011). *Cost of building to the code for sustainable homes: Updated cost review*. Department for Communities and Local Government, London.
Department for Communities and Local Government. (2015). *Technical housing standards: Nationally described space standard*. Department for Communities and Local Government, London.
Department for Transport. (2020). Travel to school. Retrieved February 25, 2021, from https://www.ethnicity-facts-figures.service.gov.uk/culture-and-community/transport/travel-to-school/latest
Després, C. (1991). The meaning of home: Literature review and directions for future research and theoretical development. *Journal of Architectural and Planning Research, 8*, 96–115.
Deutch, J. (2020). Is net zero carbon 2050 possible? *Joule, 4*, 2237–2240.
Dias, M. A. d. S., Friche, A. A. d. L., de Oliveira, V. B., & Caiaffa, W. T. (2015). The Belo Horizonte Observatory for Urban Health: Its history and current challenges. *Cadernos de Saúde Pública, 31*, 277–285.
Dias, M. A. d. S., Friche, A. A. d. L., Mingoti, S. A., Costa, D. A. d. S., Andrade, A. C. d. S., Freire, F. M., de Oliveira, V. B., & Caiaffa, W. T. (2019). Mortality from homicides in slums in the city of Belo Horizonte, Brazil: An evaluation of the impact of a re-urbanization project. *International Journal of Environmental Research and Public Health, 16*, 154.
Dicker, D., Nguyen, G., Abate, D., Abate, K. H., Abay, S. M., Abbafati, C. M., Abbasi, N., Abbastabar, H., Abd-Allah, F., Abdela, J., Abdelalim, A., Abdel-Rahman, O., Abdi, A., Abdollahpour, I., Abdulkader, R. S., Abdurahman, A. A., Abebe, H. T., Abebe, M., … Murray, C. J. L. (2018). Global, regional, and national age-sex-specific mortality and life expectancy, 1950–2017: A systematic analysis for the Global Burden of Disease Study 2017. *The Lancet, 392*, 1684–1735.
Diez Roux, A. V., Barrientos-Gutierrez, T., Caiaffa, W. T., Miranda, J. J., Rodriguez, D., Sarmiento, O. L., Slesinski, S. C., & Vergara, A. V. (2020). Urban health and health equity in Latin American cities: What COVID-19 is teaching us. *Cities Health, 0*, 1–5.
Dixon, B. N., Ugwoaba, U. A., Brockmann, A. N., & Ross, K. M. (2020). Associations between the built environment and dietary intake, physical activity, and obesity: A scoping review of reviews. *Obesity Reviews, 22*, 13171.
Djongyang, N., Tchinda, R., & Njomo, D. (2010). Thermal comfort: A review paper. *Renewable and Sustainable Energy Reviews, 14*, 2626–2640.
Dodge, R., Daly, A. P., Huyton, J., & Sanders, L. D. (2012). The challenge of defining wellbeing. *International Journal of Wellbeing, 2*, 222–235.
Dora, C., Haines, A., Balbus, J., Fletcher, E., Adair-Rohani, H., Alabaster, G., Hossain, R., de Onis, M., Branca, F., & Neira, M. (2015). Indicators linking health and sustainability in the post-2015 development agenda. *The Lancet, 385*, 380–391.
Dovjak, M., & Kukec, A. (2019). *Creating healthy and sustainable buildings: An assessment of health risk factors*. Springer International Publishing.
Downing, J. (2016). The health effects of the foreclosure crisis and unaffordable housing: A systematic review and explanation of evidence. *Social Science & Medicine, 162*, 88–96.
Doyon, A., & Moore, T. (2019). The acceleration of an unprotected niche: The case of Nightingale Housing, Australia. *Cities, 92*, 18–26.
Droste, C. (2015). German co-housing: An opportunity for municipalities to foster socially inclusive urban development? *Urban Research and Practice, 8*, 79–92.
Duarte, C. M., Agusti, S., Barbier, E., Britten, G. L., Castilla, J. C., Gattuso, J.-P., Fulweiler, R. W., Hughes, T. P., Knowlton, N., Lovelock, C. E., Lotze, H. K., Predragovic, M., Poloczanska, E., Roberts, C., & Worm, B. (2020). Rebuilding marine life. *Nature, 580*, 39–51.
Dunn, J. (2002). Housing and inequalities in health: A study of socioeconomic dimensions of housing and self reported health from a survey of Vancouver residents. *Journal of Epidemiology and Community Health, 56*, 671–681.
Dunn, J. R. (2020). Housing and healthy child development: Known and potential impacts of interventions. *Annual Review of Public Health, 41*, 381–396.

Dyer, C. (2020). Air pollution from road traffic contributed to girl's death from asthma, coroner concludes. *BMJ, 371*, m4902.
ECDG. (2015). Towards an EU research and innovation policy agenda for nature-based solutions & re-naturing cities: Final report of the Horizon 2020 expert group on 'Nature based solutions and re naturing cities': (full version). Publications Office, Brussels.
EcoDistricts. (2018). *EcoDistricts protocol: The standard for urban and community development*, version 1.3. EcoDistricts, Portland, Oregon.
Edwards, P., & Tsouros, A. D. (2008). *A healthy city is an active city: A physical activity planning guide*. WHO Regional Office for Europe, Denmark.
Egan, M., Kearns, A., Katikireddi, S. V., Curl, A., Lawson, K., & Tannahill, C. (2016). Proportionate universalism in practice? A quasi-experimental study (GoWell) of a UK neighbourhood renewal programme's impact on health inequalities. *Social Science & Medicine, 152*, 41–49.
EIDD. (2009). EIDD Stockholm declaration 2004.
Eime, R. M., Young, J. A., Harvey, J. T., Charity, M. J., & Payne, W. R. (2013). A systematic review of the psychological and social benefits of participation in sport for adults: Informing development of a conceptual model of health through sport. *International Journal of Behavioral Nutrition and Physical Activity, 10*, 135.
Ellen MacArthur Foundation. (2015). *Growth within: A circular economy vision for a competitive Europe*. Ellen MacArthur Foundation.
Elwood, M. (2017). Critical appraisal of epidemiological studies and clinical trials. In *The diagnosis of causation* (4th ed.). Oxford University Press.
Engel, G. L. (1977). The need for a new medical model: A challenge for biomedicine. *Science, 196*, 129–136.
Esteban-Cornejo, I., Carlson, J. A., Conway, T. L., Cain, K. L., Saelens, B. E., Frank, L. D., Glanz, K., Roman, C. G., & Sallis, J. F. (2016). Parental and adolescent perceptions of neighborhood safety related to adolescents' physical activity in their neighborhood. *Research Quarterly for Exercise and Sport, 0*, 1–9.
European Environment Agency. (2020). Environmental noise in Europe. Publications Office, Luxembourg.
Executive Management Committee. (2018). *Metro board report: Metro equity platform framework* (Board Report). Los Angeles County Metropolitan Transportation Authority, Los Angeles, CA.
Farre, A., & Rapley, T. (2017). The new old (and old new) medical model: Four decades navigating the biomedical and psychosocial understandings of health and illness. *Healthcare, 5*, 1–9.
Farrell, K. (2017). The rapid urban growth triad: A new conceptual framework for examining the urban transition in developing countries. *Sustainability, 9*, 1407.
Fazli, G. S., Creatore, M. I., Matheson, F. I., Guilcher, S., Kaufman-Shriqui, V., Manson, H., Johns, A., & Booth, G. L. (2017). Identifying mechanisms for facilitating knowledge to action strategies targeting the built environment. *BMC Public Health, 17*, 1.
Fiscella, N. A., Case, L. K., Jung, J., & Yun, J. (2021). Influence of neighborhood environment on physical activity participation among children with autism spectrum disorder. *Autism Research, 14*, 560–570.
Floater, G., & Rode, P. (2014). *Cities and the new climate economy: The transformative role of global urban growth, cities and the new climate economy*. London School of Economics and Political Science, LSE Cities.
Flynn, C., Yamasumi, E., Fisher, S., Snow, D., Grant, Z., Kirby, M., Browning, P., Rommerskirchen, M., & Russell, I. (2021). *People's climate vote results*. United Nations Development Programme and University of Oxford.
Foley, R. (2014). The Roman–Irish Bath: Medical/health history as therapeutic assemblage. *Social Science & Medicine, 106*, 10–19.
Fontaine, K. R. (2000). Physical activity improves mental health. *The Physician and Sportsmedicine, 28*, 83–84.
FrameWorks Institute. (2020). *Communicating about climate change in the time of COVID-19*. FrameWorks Institute.

Frank, L. D., Iroz-Elardo, N., MacLeod, K. E., & Hong, A. (2019). Pathways from built environment to health: A conceptual framework linking behavior and exposure-based impacts. *Journal of Transport and Health, 12*, 319–335.

Friche, A. A. d. L., Dias, M. A. d. S., Reis, P. B. d., Dias, C. S., & Caiaffa, W. T. (2015). Urban upgrading and its impact on health: A "quasi-experimental" mixed-methods study protocol for the BH-Viva Project. *Cadernos de Saúde Pública, 31*, 51–64.

Frumkin, H. (2004). *Urban sprawl and public health: Designing, planning, and building for healthy communities*. Island Press.

Fuller, R. A., Irvine, K. N., Devine-Wright, P., Warren, P. H., & Gaston, K. J. (2007). Psychological benefits of greenspace increase with biodiversity. *Biology Letters, 3*, 390–394.

Fusco Girard, L., & Nocca, F. (2019). Moving towards the circular economy/city model: Which tools for operationalizing this model? *Sustainability, 11*, 6253.

Galea, S., & Vlahov, D. (2005). Urban health: Evidence, challenges, and directions. *Annual Review of Public Health, 26*, 341–365.

Gallagher, J., Baldauf, R., Fuller, C. H., Kumar, P., Gill, L. W., & McNabola, A. (2015). Passive methods for improving air quality in the built environment: A review of porous and solid barriers. *Atmospheric Environment, 120*, 61–70.

Gallent, N., Morphet, J., Chiu, R. L. H., Filion, P., Fischer, K. F., Gurran, N., Li, P., Li, P., Schwartzf, A., & Stead, D. (2020). International experience of public infrastructure delivery in support of housing growth. *Cities, 107*, 102920.

Gardner, T. A., Von Hase, A., Brownlie, S., Ekstrom, J. M. M., Pilgrim, J. D., Savy, C. E., Theo Stephens, R. T., Treweek, J., Ussher, G. T., Ward, G., & Kate, K. T. (2013). Biodiversity offsets and the challenge of achieving no net loss. *Conservation Biology, 27*, 1254–1264.

Garrard, G. E., Williams, N. S. G., Mata, L., Thomas, J., & Bekessy, S. A. (2018). Biodiversity sensitive urban design. *Conservation Letters, 11*, e12411.

Gascon, M., Zijlema, W., Vert, C., White, M. P., & Nieuwenhuijsen, M. J. (2017). Outdoor blue spaces, human health and well-being: A systematic review of quantitative studies. *International Journal of Hygiene and Environmental Health, 220*, 1207–1221.

Gaster, S. (1991). Urban children's access to their neighborhood: Changes over three generations. *Environment and Behavior, 23*, 70–85.

Gatzweiler, F. W., Reis, S., Zhang, Y., & Jayasinghe, S. (2018). Lessons from complexity science for urban health and well-being. *Cities Health, 1*, 210–223.

Gatzweiler, F. W., Zhu, Y.-G., Roux, A. V. D., Capon, A., Donnelly, C., Salem, G., Ayad, H. M., Speizer, I., Nath, I., Boufford, J. I., Hanaki, K., Rietveld, L. C., Ritchie, P., Jayasinghe, S., Parnell, S., & Zhang, Y. (2017). *Advancing health and wellbeing in the changing urban environment: Implementing a systems approach, urban health and wellbeing*. Springer.

Gehl Institute. (2018). *Inclusive healthy places: A guide to inclusion & health in public space: Learning globally to transform locally*.

Gehl, J. (1986). "Soft edges" in residential streets. *Scandinavian Housing & Planning Research, 3*, 89–102.

Gelormino, E., Melis, G., Marietta, C., & Costa, G. (2015). From built environment to health inequalities: An explanatory framework based on evidence. *Preventive Medical Reports, 2*, 737–745.

Gharaveis, A., & Kazem-Zadeh, M. (2018). The role of environmental design in cancer prevention, diagnosis, treatment, and survivorship: A systematic literature review. *HERD Health Environments Research & Design Journal., 11*, 18–32.

Ghiassee, C., Urquhart, G., Duarte-Davidson, R., Wilding, J., Landeg-Cox, C., & Gittins, A. (2014). Key concepts and framework for investigation. In N. Bradley, H. Harrison, G. Hodgson, R. Kamanyire, A. Kibble, & V. Murray (Eds.), *Essentials of environmental public health science* (pp. 20–56). Oxford University Press.

Gibson, J. J. (1979). *The ecological approach to visual perception*. Houghton Mifflin.

Gibson, M., Petticrew, M., Bambra, C., Sowden, A. J., Wright, K. E., & Whitehead, M. (2011). Housing and health inequalities: A synthesis of systematic reviews of interventions aimed at different pathways linking housing and health. *Health Place*, Health Geographies of Voluntarism, *17*, 175–184.

Gilbride, M., Loveland, J., Burpee, H., Kriegh, J., & Meek, C. (2016). Occupant-behavior-driven energy savings at the Bullitt Center in Seattle, Washington. In *From components to systems, from buildings to communities* (p. 15). Presented at the ACEEE Summer Study on Energy Efficiency in Buildings, American Council for an Energy Efficient Economy, Pacific Grove, California.

Giles-Corti, B., Bull, F., Knuiman, M., McCormack, G., Van Niel, K., Timperio, A., Christian, H., Foster, S., Divitini, M., Middleton, N., & Boruff, B. (2013). The influence of urban design on neighbourhood walking following residential relocation: Longitudinal results from the RESIDE study. *Social Science & Medicine, 77*, 20–30.

Giles-Corti, B., Vernez-Moudon, A., Reis, R., Turrell, G., Dannenberg, A. L., Badland, H., Foster, S., Lowe, M., Sallis, J. F., Stevenson, M., & Owen, N. (2016). City planning and population health: A global challenge. *The Lancet, 388*, 2912–2924.

Gill, T. (2021). *Urban playground: How child-friendly planning and design can save cities*. RIBA Publishing.

Glouberman, S., Gemar, M., Campsie, P., Miller, G., Armstrong, J., Newman, C., Siotis, A., & Groff, P. (2006). A framework for improving health in cities: A discussion paper. *Journal of Urban Health, 83*, 325–338.

Goodman, A., Laverty, A. A., Thomas, A., & Aldred, R. (2021). The impact of 2020 low traffic neighbourhoods on fire service emergency response times, in London, UK. *Findings* 23568.

Goodman, E. P., & Powles, J. (2019). Urbanism under Google: Lessons from Sidewalk Toronto Symposium: Rise of the machines: Artificial intelligence, robotics, and the reprogramming of law. *Fordham Law Review, 88*, 457–498.

Götschi, T., Kahlmeier, S., Castro, A., Brand, C., Cavill, N., Kelly, P., Lieb, C., Rojas-Rueda, D., Woodcock, J., & Racioppi, F. (2020). Integrated impact assessment of active travel: Expanding the scope of the health economic assessment tool (HEAT) for walking and cycling. *International Journal of Environmental Research and Public Health, 17*, 7361.

Green, B. (2019). *The smart enough city: Putting technology in Its place to reclaim our urban future, strong ideas*. MIT Press.

Green, S. (2018). Sustainable construction. In *Sustainable futures in the built environment to 2050* (pp. 172–193). John Wiley & Sons, Ltd.

Greenhalgh, T. (2019). *How to read a paper: The basics of evidence-based medicine and healthcare* (6th ed.). John Wiley & Sons Ltd.

Griffin, J. (2019). Families hit out at London gasworks redevelopment. *The Observer*. May 5, 2019. Retrieved from https://www.theguardian.com/environment/2019/may/04/brownfield-site-new-homes-building-wrecking-health-southall

Griffin, J. (2020a). Residents demand new clean air rules for former gasworks sites in England. *The Guardian*. September 22, 2020. Retrieved from https://www.theguardian.com/environment/2020/sep/22/residents-demand-new-clean-air-rules-for-former-gasworks-sites-in-england

Griffin, J. (2020b). Londoners claim toxic air from gasworks damaging their health. *The Guardian*. August 27, 2020. Retrieved from https://www.theguardian.com/environment/2020/aug/27/londoners-claim-toxic-air-from-gasworks-damaging-their-health

Groshong, L., Wilhelm Stanis, S. A., Kaczynski, A. T., & Hipp, J. A. (2020). Attitudes about perceived park safety among residents in low-income and high minority Kansas City, Missouri, Neighborhoods. *Environment and Behavior, 52*, 639–665.

Hahad, O., Kröller-Schön, S., Daiber, A., & Münzel, T. (2019). The cardiovascular effects of noise. *Deutsches Ärzteblatt International, 116*, 245–250.

Hahn, R. A., & Truman, B. I. (2015). Education improves public health and promotes health equity. *International Journal of Health Services, 45*, 657–678.

Haines, T. (2020). Micro-housing in Seattle update: Combating "Seattle-ization". *Seattle University School of Law Review Supra 43*, 6.

Halawa, F., Madathil, S. C., Gittler, A., & Khasawneh, M. T. (2020). Advancing evidence-based healthcare facility design: A systematic literature review. *Health Care Management Science, 23*, 453–480.

Hamiduddin, I. (2015). Social sustainability, residential design and demographic balance: Neighbourhood planning strategies in Freiburg, Germany. *Town Planning Review, 86*, 29–52.

Hamiduddin, I., & Gallent, N. (2016). Self-build communities: The rationale and experiences of group-build (Baugruppen) housing development in Germany. *Housing Studies, 31*, 365–383.

Hancock, T. (1985). The mandala of health: A model of the human ecosystem. *Family & Community Health, 8*, 1–10.

Hancock, T. (1993). The evolution, impact and significance of the health cities/healthy communities movement. *Journal of Public Health Policy, 14*, 5–18.

Hancock, T. (2011). The Ottawa charter at 25. *Canadian Journal of Public Health*. Can. Santee Publique, *102*, 404–406.

Hancock, T., & Duhl, L. J. (1986). *Healthy cities: Promoting health in the urban context* (No. WHO Healthy Cities Paper #1). World Health Organization Regional Office for Europe.

Harris, P., Sainsbury, P., & Kemp, L. (2014). The fit between health impact assessment and public policy: Practice meets theory. *Social Science & Medicine, 108*, 46–53.

Harrison, A. P., Cattani, I., & Turfa, J. M. (2010). Metallurgy, environmental pollution and the decline of Etruscan civilisation. *Environmental Science and Pollution Research, 17*, 165–180.

Hedman, E. (2008). *A history of the Swedish system of non-profit municipal housing*. The Swedish Board of Housing, Building and Planning.

Heise, T. L., Romppel, M., Molnar, S., Buchberger, B., Berg, A. v. d., Gartlehner, G., & Lhachimi, S. K. (2017). Community gardening, community farming and other local community-based gardening interventions to prevent overweight and obesity in high-income and middle-income countries: Protocol for a systematic review. *BMJ Open, 7*, e016237.

Henneberry, J., Lange, E., Moore, S., Morgan, E., & Zhao, N. (2011). Physical-financial modelling as an aid to developers' decision-making. In S. Tiesdell & D. Adams (Eds.), *Urban Design in the real estate development process* (pp. 219–235). John Wiley & Sons, Ltd.

Heylighen, A., Van der Linden, V., & Van Steenwinkel, I. (2017). Ten questions concerning inclusive design of the built environment. *Building and Environment, 114*, 507–517.

Heymann, D. L., & Shindo, N. (2020). COVID-19: What is next for public health? *The Lancet, 395*, 542–545.

Hidalgo, D., Pereira, L., Estupiñán, N., & Jiménez, P. L. (2013). TransMilenio BRT system in Bogota, high performance and positive impact—Main results of an ex-post evaluation. *Research in Transportation Economics*, THREDBO 12: Recent developments in the reform of land passenger transport, *39*, 133–138.

Hippensteel, C. L., Sadler, R. C., Milam, A. J., Nelson, V., & Debra Furr-Holden, C. (2019). Using zoning as a public health tool to reduce oversaturation of alcohol outlets: An examination of the effects of the new "300 Foot Rule" on packaged goods stores in a Mid-Atlantic City. *Prevention Science, 20*, 833–843.

Hoffman, J. S., Shandas, V., & Pendleton, N. (2020). The effects of historical housing policies on resident exposure to intra-urban heat: A study of 108 US urban areas. *Climate, 8*, 12.

Holden, M. (2007). Revisiting the local impact of community indicators projects: Sustainable Seattle as prophet in its own land. *Applied Research in Quality of Life, 1*, 253–277.

Holden, M. (2008). Social learning in planning: Seattle's sustainable development codebooks. *Progress in Planning, 69*, 1–40.

Holtermans, R., & Kok, N. (2018). *International green building adoption index*. Finance, GSBE Theme Sustainable Development.

Hoornweg, D., & Bhada-Tata, P. (2012). *What a waste: A global review of waste management*. Urban Development Series; Knowledge Papers no. 15. World Bank, Washington, DC.

Howard, E. (1902). *Garden cities of to-morrow* (New rev edn/with an introduction by Ray Thomas). New towns bibliography by Stephen Potter. ed. Attic, Eastbourne.

Howden-Chapman, P., Pierse, N., Nicholls, S., Gillespie-Bennett, J., Viggers, H., Cunningham, M., Phipps, R., Boulic, M., Fjällström, P., Free, S., Chapman, R., Lloyd, B., Wickens, K., Shields, D., Baker, M., Cunningham, C., Woodward, A., Bullen, C., & Crane, J. (2008). Effects of improved home heating on asthma in community dwelling children: Randomised controlled trial. *BMJ, 337*, a1411–a1411.

Hsu, A., Sheriff, G., Chakraborty, T., & Manya, D. (2021). Disproportionate exposure to urban heat island intensity across major US cities. *Nature Communications, 12*, 2721.

Huang, G., Zhou, W., Qian, Y., & Fisher, B. (2019). Breathing the same air? Socioeconomic disparities in PM2.5 exposure and the potential benefits from air filtration. *Science of the Total Environment, 657*, 619–626.

Huber, M., Knottnerus, J. A., Green, L., Horst, H. van der, Jadad, A. R., Kromhout, D., Leonard, B., Lorig, K., Loureiro, M. I., van der Meer, Jos W. M, Schnabel, P., Smith, R., van Weel, C., & Smid, H. (2011). How should we define health? *BMJ* Online London, *343*, d4163.

Hunter, R. F., Christian, H., Veitch, J., Astell-Burt, T., Hipp, J. A., & Schipperijn, J. (2015). The impact of interventions to promote physical activity in urban green space: A systematic review and recommendations for future research. *Social Science & Medicine, 124*, 246–256.

Ige, J., Pilkington, P., Orme, J., Williams, B., Prestwood, E., Black, D., Carmichael, L., & Scally, G. (2018). The relationship between buildings and health: A systematic review. *Journal of Public Health, 41*(2), e121–e132.

Ige-Elegbede, J., Pilkington, P., Bird, E. L., Gray, S., Mindell, J. S., Chang, M., Stimpson, A., Gallagher, D., & Petrokofsky, C. (2020). Exploring the views of planners and public health practitioners on integrating health evidence into spatial planning in England: A mixed-methods study. *Journal of Public Health, 43*, 664–672.

IHME. (2018). *Financing global health 2017: Funding universal health coverage and the unfinished HIV/AIDS agenda*. IHME, Seattle, WA.

IHME. (2021). COVID-19 results briefing: Global. IHME, Seattle, WA.

Innes, J. E. (1998). Information in Communicative Planning. *Journal of the American Planning Association, 64*, 52–63.

Innes, J. E., & Booher, D. E. (2010). *Planning with complexity: An introduction to collaborative rationality for public policy*. Routledge.

Institute of Health Equity. (2014). *Local action on health inequalities: Improving access to green spaces*. Health Equity Briefing 8. Public Health England, London.

Institution of Mechanical Engineers. (2020). The energy hierarchy: Supporting policy making for "Net Zero." Institute of Mechanical Engineers.

International Organization for Standardization. (2014). *ISO 12913-1:2014 Acoustics—soundscape—Part 1: Definition and conceptual framework*. ISO, Geneva, Switzerland.

IPBES. (2019). *Summary for policymakers of the global assessment report on biodiversity and ecosystem services of the Intergovernmental Science-Policy Platform on Biodiversity and Ecosystem Services*. IPBES Secretariat, Bonn, Germany.

Jedwab, R., Christiaensen, L., & Gindelsky, M. (2017). Demography, urbanization and development: Rural push, urban pull and … urban push? *Journal of Urban Economics, 98*, 6–16.

Joseph, M. (2019). *Mixed-income communities as a strategic lever to impact health equity: Lessons from the field and implications for strategy and investment*. CSSP.

Joss, S. (2015). *Sustainable cities: Governing for urban innovation, Planning, environment, cities*. Palgrave Macmillan.

Kahlmeier, S., & World Health Organization, Regional Office for Europe. (2011). *Health economic assessment tools (HEAT) for walking and for cycling: Methodology and user guide: Economic assessment of transport infrastructure and policies*. World Health Organisation, Regional Office for Europe, Copenhagen.

Kaplan, R., & Kaplan, S. (1989). *The experience of nature: A psychological perspective*. Cambridge University Press.

Kaplan, S. (1995). The restorative benefits of nature: Toward an integrative framework. *Journal of Environmental Psychology, 15*, 169–182.

Karvonen, A. (2010). Metronatural™: Inventing and reworking urban nature in Seattle. Prog. Plan., Metronatural™: Inventing and Reworking Urban Nature in Seattle. *Progress in Planning, 74*, 153–202.

Keeble, M., Burgoine, T., White, M., Summerbell, C., Cummins, S., & Adams, J. (2019). How does local government use the planning system to regulate hot food takeaway outlets? A census of current practice in England using document review. *Health & Place, 57*, 171–178.

Keis, O., Helbig, H., Streb, J., & Hille, K. (2014). Influence of blue-enriched classroom lighting on students' cognitive performance. *Trends Neurosci. Educ., 3*, 86–92.

Keller, J. (2012). Mapping Hurricane Sandy's Deadly Toll. *N. Y. Times.*

Kelly, F. J., & Fussell, J. C. (2019). Improving indoor air quality, health and performance within environments where people live, travel, learn and work. *Atmospheric Environment, 200*, 90–109.

Kelly, M. P., & Russo, F. (2018). Causal narratives in public health: The difference between mechanisms of aetiology and mechanisms of prevention in non-communicable diseases. *Sociology of Health & Illness, 40*, 82–99.

Kent, J., & Thompson, S. (2019). Planning Australia's healthy built environments. In *Routledge research in planning and urban design*. Routledge.

Kenworthy, J. R. (2006). The eco-city: Ten key transport and planning dimensions for sustainable city development. *Environment and Urbanization, 18*, 67–85.

Kern, L. (2020). *Feminist city: Claiming space in a man-made world*. Verso.

Kickbusch, I., & Gleicher, D. (2012). *Governance for health in the 21st century*. World Health Organization, Regional Office for Europe, Copenhagen.

Kim, D. (2014). Learning from adjectival urbanisms: The pluralistic urbanism. In *102nd ACSA Annual Meeting Proceedings Globalizing Architecture/Flows and Disruptions* (pp. 439–447), 2.

Kisacky, J. S. (2017). *Rise of the modern hospital: An architectural history of health and healing, 1870–1940*. University of Pittsburgh Press.

Kondo, M. C., South, E. C., & Branas, C. C. (2015). Nature-based strategies for improving urban health and safety. *Journal of Urban Health, 92*, 800–814.

Kraft, A. N., Thatcher, E. J., & Zenk, S. N. (2020). Neighborhood food environment and health outcomes in U.S. Low-socioeconomic status, racial/ethnic minority, and rural populations: A systematic review. *Journal of Health Care for the Poor and Underserved, 31*, 1078–1114.

Kramer, A., Lassar, T. J., Federman, M., & Hammerschmidt, S. (2014). *Building for wellness: The business case*. Urban Land Institute.

Krausmann, F., Wiedenhofer, D., Lauk, C., Haas, W., Tanikawa, H., Fishman, T., Miatto, A., Schandl, H., & Haberl, H. (2017). Global socioeconomic material stocks rise 23-fold over the 20th century and require half of annual resource use. *Proceedings of the National Academy of Sciences, 114*(8), 1880–1885.

Kremen, C. (2005). Managing ecosystem services: What do we need to know about their ecology? *Ecology Letters, 8*, 468–479.

Krieger, J., & Higgins, D. L. (2002). Housing and health: Time again for public health action. *American Journal of Public Health, 92*, 758–768.

Krieger, N. (1994). Epidemiology and the web of causation: Has anyone seen the spider? *Social Science & Medicine, 39*, 887–903.

Krieger, N. (2001). Theories for social epidemiology in the 21st century: An ecosocial perspective. *International Journal of Epidemiology, 30*, 668–677.

Krieger, N. (2008). Proximal, distal, and the politics of causation: What's level got to do with it? *American Journal of Public Health, 98*, 221–230.

Lalonde, M. (1974). *A new perspective on the health of Canadians: A working document*. Minister of National Health and Welfare.

Lam, T. M., Vaartjes, I., Grobbee, D. E., Karssenberg, D., & Lakerveld, J. (2021). Associations between the built environment and obesity: An umbrella review. *International Journal of Health Geographics, 20*, 7.

Lambert, A., Vlaar, J., Herrington, S., & Brussoni, M. (2019). What is the relationship between the neighbourhood built environment and time spent in outdoor play? A systematic review. *International Journal of Environmental Research and Public Health, 16*, 3840.
Lang, T., & Rayner, G. (2012). Ecological public health: The 21st century's big idea? An essay by Tim Lang and Geof Rayner. *BMJ, 345*, e5466.
Laverty, A. A., Aldred, R., & Goodman, A. (2021). The impact of introducing low traffic neighbourhoods on road traffic injuries. *Findings* 18330.
Lawrance, E., Thompson, R., Fontana, G., & Jennings, N. (2021). *The impact of climate change on mental health and emotional wellbeing: Current evidence and implications for policy and practice*. Grantham Institute.
Lee, I.-M., Shiroma, E. J., Lobelo, F., Puska, P., Blair, S. N., & Katzmarzyk, P. T. (2012). Effect of physical inactivity on major non-communicable diseases worldwide: An analysis of burden of disease and life expectancy. *The Lancet, 380*, 219–229.
Lee, J. Y., & Van Zandt, S. (2019). Housing tenure and social vulnerability to disasters: A Review of the evidence. *Journal of Planning Literature, 34*, 156–170.
Lehmann, S. (2010). *The principles of green urbanism: Transforming the city for sustainability*. Earthscan
Lehmann, S. (2014). Introduction Low carbon cities: More than just buildings. In S. Lehmann (Ed.), *Low carbon cities: Transforming urban systems* (pp. 1–55). Routledge.
Leising, E., Quist, J., & Bocken, N. (2018). Circular economy in the building sector: Three cases and a collaboration tool. *Journal of Cleaner Production, 176*, 976–989.
Lekaviciute, J., Kephalopoulos, S., Clark, C., & Stansfeld, S. (2013). *Institute for health and consumer protection, 2013*. Final Report: ENNAH European Network on Noise and Health. Publications Office, Luxembourg.
Lennon, M., Douglas, O., & Scott, M. (2017). Urban green space for health and well-being: Developing an 'affordances' framework for planning and design. *Journal of Urban Design, 22*, 778–795.
Lewer, D., Braithwaite, I., Bullock, M., Eyre, M. T., White, P. J., Aldridge, R. W., Story, A., & Hayward, A. C. (2020). COVID-19 among people experiencing homelessness in England: A modelling study. *The Lancet Respiratory Medicine, 8*, 1181–1191.
Liu, D., & Wang, S. (2019). The global issue of foreign waste. *Lancet Planetary Health, 3*, e120.
Lledó, R. (2019). Human centric lighting, a new reality in healthcare environments. In T. P. Cotrim, F. Serranheira, P. Sousa, S. Hignett, S. Albolino, & R. Tartaglia (Eds.), *Health and social care systems of the future: Demographic changes, digital age and human factors, advances in intelligent systems and computing* (pp. 23–26). Springer International Publishing.
London Legacy Development Corporation. (2013). Legacy communities scheme: Biodiversity action plan 2014–2019. Retrieved March 5, 2021 from https://www.queenelizabetholympicpark.co.uk/-/media/lldc/sustainability-and-biodiversity/legacy-communities-scheme-biodiversity-action-plan-2014-2019.ashx?la=en
Lopez, E., Carvalho, P., & Pedrosa Nahasunder, M. I. (2010). Belo Horizonte, Brazil, Villa Viva Programme—Aglomerado da Serra. UCLG Committee on Social Inclusion, Participatory Democracy and Human Rights.
Los Angeles County Metropolitan Transportation Authority. (2018). Metro equity platform framework.
Loukaitou-Sideris, A., & Fink, C. (2008). Addressing women's fear of victimization in transportation settings: A survey of U.S. transit agencies. *Urban Affairs Review, 44*, 554–587.
Lowe, M., Whitzman, C., & Giles-Corti, B. (2018). Health-promoting spatial planning: Approaches for strengthening urban policy integration. *Planning Theory and Practice, 19*, 180–197.
Luke, D. A., & Stamatakis, K. A. (2012). Systems science methods in public health: Dynamics, networks, and agents. *Annual Review of Public Health, 33*, 357–376.
Lundberg, O. (2020). Next steps in the development of the social determinants of health approach: The need for a new narrative. *Scandinavian Journal of Public Health*. https://doi.org/10.1177/1403494819894789.

Lusk, A. C., Willett, W. C., Morris, V., Byner, C., & Li, Y. (2019). Bicycle facilities safest from crime and crashes: Perceptions of residents familiar with higher crime/lower income neighborhoods in Boston. *International Journal of Environmental Research and Public Health, 16*, 484.

Lynch, K. (1960). *The image of the city*, Publication of the Joint Center for Urban Studies. MIT Press.

Lynch, K. (1981). *A theory of good city form*. MIT Press.

MacArthur, D. E. (2015). *The surprising thing I learned sailing solo around the world*. TED. Retrieved March 18, 2021 from https://www.ted.com/talks/dame_ellen_macarthur_the_surprising_thing_i_learned_sailing_solo_around_the_world

Mackenbach, J. D., Rutter, H., Compernolle, S., Glonti, K., Oppert, J.-M., Charreire, H., De Bourdeaudhuij, I., Brug, J., Nijpels, G., & Lakerveld, J. (2014). Obesogenic environments: A systematic review of the association between the physical environment and adult weight status, the SPOTLIGHT project. *BMC Public Health, 14*, 1.

MacNaughton, P., Satish, U., Laurent, J. G. C., Flanigan, S., Vallarino, J., Coull, B., Spengler, J. D., & Allen, J. G. (2017). The impact of working in a green certified building on cognitive function and health. *Building and Environment, 114*, 178–186.

Madanipour, A. (Ed.). (2010). *Whose public space? International case studies in urban design and development*. Routledge.

Madeddu, M., Gallent, N., & Mace, A. (2015). Space in new homes: Delivering functionality and liveability through regulation or design innovation? *The Town Planning Review, 86*, 73–95.

Marmot, A. F. (1981). The legacy of Le Corbusier and high-rise housing. *Built Environment, 7*, 82–95.

Marmot, M., Allen, J., Boyce, T., Goldblatt, P., & Morrison, J. (2020). *Health equity in England: The Marmot Review 10 years on*. Institute of Health Equity, London.

Marmot, M., Allen, J., Goldblatt, P., Boyce, T., McNeish, D., Grady, M., & Geddes, I. (2010). Fair society, healthy lives: The Marmot review; strategic review of health inequalities in England post-2010. *Marmot Review*, London.

Marsh, H. W., Huppert, F. A., Donald, J. N., Horwood, M. S., & Sahdra, B. K. (2020). The well-being profile (WB-Pro): Creating a theoretically based multidimensional measure of well-being to advance theory, research, policy, and practice. *Psychological Assessment, 32*, 294–313.

Mavrogianni, A., Taylor, J., Symonds, P., Oikonomou, E., Pineo, H., Zimmermann, N., & Davies, M. (2022). Cool cities by design: Shaping a healthy and equitable London in a warming climate. In G. McGregor & C. Ren (Eds.), *Urban climate science for planning healthy cities*. Springer.

Mayor of London. (2016). The London plan, Chapter 3. *London's People*.

Mayor of London. (2020a). Energy assessment guidance: Greater London Authority guidance on preparing energy assessments as part of planning applications. Greater London Authority.

Mayor of London. (2020b). Mayor of London leads the way in making homes fit for the future. Retrieved July 29, 2021, from https://www.london.gov.uk//press-releases/mayoral/mayor-introduces-programme-to-update-london-homes

McArthur, J. J., & Powell, C. (2020). Health and wellness in commercial buildings: Systematic review of sustainable building rating systems and alignment with contemporary research. *Building and Environment, 171*, 106635.

McCaughey Centre, VicHealth Centre for the Promotion of Mental Health and Community Wellbeing, University of Melbourne. (n.d.). Community indicators Victoria: Data framework. Retrieved February 17, 2016, from http://www.communityindicators.net.au/metadata_items

McDonald, R. I., Weber, K., Padowski, J., Flörke, M., Schneider, C., Green, P. A., Gleeson, T., Eckman, S., Lehner, B., Balk, D., Boucher, T., Gril, G., & Montgomery, M. (2014). Water on an urban planet: Urbanization and the reach of urban water infrastructure. *Global Environmental Change, 27*, 96–105.

McKinnon, G., Pineo, H., Chang, M., Taylor-Green, L., Johns, A., & Toms, R. (2020). Strengthening the links between planning and health in England. *BMJ, 369*. https://doi.org/10.1136/bmj.m795

Meadows, D. H. (2008). *Thinking in systems: A primer*. Chelsea Green Pub.

Medved, P. (2018). Exploring the 'Just City principles' within two European sustainable neighbourhoods. *Journal of Urban Design, 23*, 414–431.

Mell, I. (2019). The impact of austerity on funding green infrastructure: A DPSIR evaluation of the Liverpool Green & Open Space Review (LG&OSR), UK. *Land Use Policy, 91*, 104284.

Mendenhall, E., & Singer, M. (2019). The global syndemic of obesity, undernutrition, and climate change. *The Lancet, 393*, 741.

Millennium Ecosystem Assessment. (2005). *Ecosystems and human well-being: Synthesis*. Island Press, Washington, DC.

Mindell, J. S., & Karlsen, S. (2012). Community severance and health: What do we actually know? *Journal of Urban Health, 89*, 232–246.

Ministry of Housing, Communities and Local Government. (2021). National planning policy framework.

Mitchell, L., Burton, E., Raman, S., Blackman, T., Jenks, M., & Williams, K. (2003). Making the outside world dementia-friendly: Design issues and considerations. *Environment and Planning B: Planning and Design, 30*, 605–632.

Mithun. (2010). *Executive summary*, South Lincoln Redevelopment Master Plan Final Report. Mithun, Seattle, WA.

Moore, J. (2015). Ecological footprints and lifestyle archetypes: Exploring dimensions of consumption and the transformation needed to achieve urban sustainability. *Sustainability, 7*, 4747–4763.

Moore, T., & Doyon, A. (2018). The uncommon nightingale: Sustainable housing innovation in Australia. *Sustainability, 10*, 3469.

Moore, T., & Higgins, D. (2016). Influencing urban development through government demonstration projects. *Cities, 56*, 9–15.

Moore, T., Horne, R., & Morrissey, J. (2014). Zero emission housing: Policy development in Australia and comparisons with the EU, UK, USA and California. *Environmental Innovation and Societal Transitions, 11*, 25–45.

Morawska, L., Allen, J., Bahnfleth, W., Bluyssen, P. M., Boerstra, A., Buonanno, G., Cao, J., Dancer, S. J., Floto, A., Franchimon, F., Greenhalgh, T., Haworth, C., Hogeling, J., Isaxon, C., Jimenez, J. L., Kurnitski, J., Li, Y., Loomans, M., Marks, G., … Yao, M. (2021). A paradigm shift to combat indoor respiratory infection. *Science, 372*, 689–691.

Moyer, J., L'Hôte, E., & Levay, K. (2019). *Public health reaching across sectors: Strategies for communicating effectively about public health and cross-sector collaboration with professionals from other sectors*. FrameWorks Institute, Washington, DC.

Mumovic, D., & Santamouris, M. (2019). *A handbook of sustainable building design and engineering: An integrated approach to energy, health and operational performance* (2nd ed.). Routledge.

Næss, P. (2016). Built environment, causality and urban planning. *Planning Theory & Practice, 17*, 52–71.

Næss, P., Nicolaisen, M. S., & Strand, A. (2012). Traffic forecasts ignoring induced demand: A Shaky fundament for cost-benefit analyses. *European Journal of Transport and Infrastructure Research, 12*, 291–309.

Napier, D., Depledge, M., Knipper, M., Lovell, R., Ponarin, E., Sanabria, E., & Thomas, F. (2017). *Culture matters: Using a cultural contexts of health approach to enhance policy-making*. World Health Organization.

Nassauer, J. I. (1995). Messy ecosystems, orderly frames. *Landscape Journal, 14*, 161–170.

National Center for Chronic Disease Prevention and Health Promotion. (2020). Health and economic costs of chronic disease. Retrieved February 04, 2020, from https://www.cdc.gov/chronicdisease/about/costs/index.htm

National Center for Healthy Housing, American Public Health Association. (2014). National Healthy Housing Standard.

National Engineering Policy Centre. (2021). *Infection resilient environments: Buildings that keep us healthy and safe*, Initial Report. Royal Academy of Engineering, London, UK.

National Health Commission. (2019). Healthy China action (2019–2030). Retrieved October 25, 2020, from http://www.gov.cn/xinwen/2019-07/15/content_5409694.htm

Natural England. (2009). *Childhood and nature: A survey on changing relationships with nature across generations*. Natural England, Warboys, England.

Nayar, J. (2015). The Bullitt center: Raising the bar with the living building challenge [WWW Document]. Interface Human Spaces. Retrieved March 19, 2021, from https://blog.interface.com/raising-the-bar-bullitt-center/

Neiderud, C.-J. (2015). How urbanization affects the epidemiology of emerging infectious diseases. *Infection Ecology & Epidemiology, 5*, 27060.

Newlin, K., Dyess, S. M., Allard, E., Chase, S., & Melkus, G. D. (2012). A methodological review of faith-based health promotion literature: Advancing the science to expand delivery of diabetes education to Black Americans. *Journal of Religion and Health, 51*, 1075–1097.

NHS England. (2019a). *Putting health into place: Executive summary*. NHS England, London.

NHS England. (2019b). *Putting health into place: Develop and provide healthcare services*. NHS England, London.

NHS London Healthy Urban Development Unit. (2019). *HUDU rapid health impact assessment tool*.

Nicol, S., Roys, M., Ormandy, D., & Ezratty, V. (2017). *The cost of poor housing in the European Union*. BRE Press Briefing Paper.

Nieuwenhuijsen, M., & Khreis, H. (Eds.). (2019). *Integrating human health into urban and transport planning: A framework*. Springer International Publishing.

Nieuwenhuijsen, M. J. (2016). Urban and transport planning, environmental exposures and health-new concepts, methods and tools to improve health in cities. *Environmental Health, 15*, S38.

Nightingale, C. M., Limb, E. S., Ram, B., Shankar, A., Clary, C., Lewis, D., Cummins, S., Ellaway, A., Giles-Corti, B., Whincup, P. H., Rudnicka, A. R., Cook, D. G., & Owen, C. G. (2019). The effect of moving to East Village, the former London 2012 Olympic and Paralympic Games Athletes' Village, on physical activity and adiposity (ENABLE London): A cohort study. *The Lancet Public Health, 4*, e421–e430.

Noonan, R. J. (2021). Family income matters! Tracking of habitual car use for school journeys and associations with overweight/obesity in UK youth. *Journal of Transport and Health, 20*, 100979.

Northridge, D. M. E., Sclar, D. E. D., & Biswas, M. P. (2003). Sorting out the connections between the built environment and health: A conceptual framework for navigating pathways and planning healthy cities. *Journal of Urban Health, 80*, 556–568.

Northridge, M. E., & Freeman, L. (2011). Urban planning and health equity. *Journal of Urban Health, 88*, 582–597.

OECD. (2015). *Ageing in cities*. OECD Publishing.

Office of the United Nations High Commissioner for Human Rights, UN Habitat. (2009). *The right to adequate housing*. Fact Sheet No. 21/Rev.1. Office of the United Nations High Commissioner for Human Rights, Geneva, Switzerland.

Ohly, H., Gentry, S., Wigglesworth, R., Bethel, A., Lovell, R., & Garside, R. (2016). A systematic review of the health and well-being impacts of school gardening: Synthesis of quantitative and qualitative evidence. *BMC Public Health, 16*, 286.

Oja, P., Titze, S., Kokko, S., Kujala, U. M., Heinonen, A., Kelly, P., Koski, P., & Foster, C. (2015). Health benefits of different sport disciplines for adults: Systematic review of observational and intervention studies with meta-analysis. *British Journal of Sports Medicine, 49*, 434–440.

Oliveira, A., Monteiro, Â., Jácome, C., Afreixo, V., & Marques, A. (2017). Effects of group sports on health-related physical fitness of overweight youth: A systematic review and meta-analysis. *Scandinavian Journal of Medicine & Science in Sports, 27*, 604–611.

Omran, A. R. (1971). The epidemiologic transition: A theory of the epidemiology of population change. *Milbank Memorial Fund Quarterly, 49*, 509–538.

Osibona, O., Solomon, B. D., & Fecht, D. (2021). Lighting in the home and health: A systematic review. *International Journal of Environmental Research and Public Health, 18*, 609.
Parker, G., & Street, E. (2021). Conceptualising the contemporary planning profession. In G. Parker & E. Street (Eds.), *Contemporary planning practice: Skills, specialisms and knowledges* (pp. 1–11). Red Globe Press.
Parnell, S. (2016). Defining a global urban development agenda. *World Development, 78*, 529–540.
Pauleit, S., Zölch, T., Hansen, R., Randrup, T. B., & Konijnendijk van den Bosch, C. (2017). Nature-Based Solutions and Climate Change—Four Shades of Green. In N. Kabisch, H. Korn, J. Stadler, & A. Bonn (Eds.), *Nature-based solutions to climate change adaptation in urban areas: Linkages between science, policy and practice, theory and practice of urban sustainability transitions* (pp. 29–49). Springer International Publishing.
Peinhardt, K., & Storring, N. (2019). *A playbook for inclusive placemaking: Community process*. Project for Public Spaces. Retrieved May 19, 2021, from https://www.pps.org/article/a-playbook-for-inclusive-placemaking-community-process
Peña, R., Meek, C., & Davis, D. (2017). The Bullitt center: A comparative analysis between simulated and operational performance. *Technology|Architecture + Design, 1*, 163–173.
Peña, R. B. (2014). *Living proof: The Bullitt Center*. High Performance Building Case Study. University of Washington Center for Integrated Design.
Perry, C. A. (1929). City planning for neighborhood life. *Social Forces, 8*, 98–100.
Peters, R., Peters, J., Booth, A., & Mudway, I. (2015). Is air pollution associated with increased risk of cognitive decline? A systematic review. *Age and Ageing, 44*, 755–760.
Petrokofsky, C., & Davis, A. (2016). *Working together to promote active travel: A briefing for local authorities*. Public Health England.
Pickett, K. E., & Wilkinson, R. G. (2015). Income inequality and health: A causal review. *Social Science & Medicine, 128*, 316–326.
Pineo, H. (2015). Housing for inclusive communities—an experimental model in Gothenburg. *Town & Country Planning, 84*, 566–571.
Pineo, H. (2016). The value of healthy places—for developers, occupants and society. *Town and Country Planning*, Securing Outcomes from United Action, *85*, 477–480.
Pineo, H. (2019). *The value and use of urban health indicator tools in the complex urban planning policy and decision-making context* (Doctoral Thesis). University College London, London.
Pineo, H. (2020). Towards healthy urbanism: Inclusive, equitable and sustainable (THRIVES)—an urban design and planning framework from theory to praxis. *Cities Health, 0*, 1–19. https://doi.org/10.1080/23748834.2020.1769527
Pineo, H., Glonti, K., Rutter, H., Zimmermann, N., Wilkinson, P., & Davies, M. (2018a). Urban health indicator tools of the physical environment: A systematic review. *Journal of Urban Health, 95*, 613–646.
Pineo, H., Glonti, K., Rutter, H., Zimmermann, N., Wilkinson, P., & Davies, M. (2020a). Use of urban health indicator tools by built environment policy- and decision-makers: A systematic review and narrative synthesis. *Journal of Urban Health, 97*, 418–435.
Pineo, H., & Moore, G. (2021). Built environment stakeholders' experiences of implementing healthy urban development: An exploratory study. *Cities Health, 0*, 1–15. https://doi.org/10.1080/23748834.2021.1876376
Pineo, H., Moore, G., & Braithwaite, I. (2020b). Incorporating practitioner knowledge to test and improve a new conceptual framework for healthy urban design and planning. *Cities Health, 0*, 1–16. https://doi.org/10.1080/23748834.2020.1773035
Pineo, H., & Rydin, Y. (2018). *Cities, health and well-being*. Royal Institution of Chartered Surveyors.
Pineo, H., Turnbull, E. R., Davies, M., Rowson, M., Hayward, A. C., Hart, G., Johnson, A. M., & Aldridge, R. W. (2021b). A new transdisciplinary research model to investigate and improve the health of the public. *Health Promotion International, 36*, 481–492.
Pineo, H., Zhou, K., Niu, Y., Hale, J., Willan, C., Crane, M., Zimmermann, N., Michie, S., Liu, Q., & Davies, M. (2021a). Evidence-informed urban health and sustainability governance in two Chinese cities. *Buildings and Cities, 2*, 550–567. https://doi.org/10.5334/bc.90

Pineo, H., Zimmermann, N., Cosgrave, E., Aldridge, R. W., Acuto, M., & Rutter, H. (2018b). Promoting a healthy cities agenda through indicators: Development of a global urban environment and health index. *Cities Health, 2*, 27–45.

Pineo, H., Zimmermann, N., & Davies, M. (2019). Urban planning: Leveraging the urban planning system to shape healthy cities. In S. Galea, C. K. Ettman, & D. Vlahov (Eds.), *Urban health* (pp. 198–206). Oxford University Press.

Pineo, H., Zimmermann, N., & Davies, M. (2020c). Integrating health into the complex urban planning policy and decision-making context: A systems thinking analysis. *Palgrave Communications, 6*, 1–14.

Pizarro, R. (2009). Urban Form and Climate Change: Towards appropriate development patterns to mitigate and adapt to climate change. In S. Davoudi, J. Crawford, & A. Mehmood (Eds.), *Planning for climate change: Strategies for mitigation and adaptation for spatial planners* (pp. 33–45). Earthscan.

Planning Out. (2019). LGBT+ Placemaking Toolkit. Retrieved December 10, 2021, from https://becg.com/lgbt-placemaking-toolkit-2/

Platt, H. L. (2007). From Hygeia to the garden city: Bodies, houses, and the rediscovery of the slum in Manchester, 1875–1910. *Journal of Urban History, 33*, 756–772.

Pollard, E. L., & Lee, P. D. (2003). Child well-being: A systematic review of the literature. *Social Indicators Research, 61*, 59–78.

Porta, M. (2014). *A dictionary of epidemiology*. Oxford University Press.

Powell-Wiley, T. M., Ayers, C. R., Lemos, J. A. d., Lakoski, S. G., Vega, G. L., Grundy, S., Das, S. R., Banks-Richard, K., & Albert, M. A. (2013). Relationship between perceptions about neighborhood environment and prevalent obesity: Data from the Dallas heart study. *Obesity, 21*, E14–E21.

Pozoukidou, G., & Chatziyiannaki, Z. (2021). 15-Minute city: Decomposing the new urban planning Eutopia. *Sustainability, 13*, 928.

Pretty, J., & Barton, J. (2020). Nature-based interventions and mind–body interventions: Saving public health costs whilst increasing life satisfaction and happiness. *International Journal of Environmental Research and Public Health, 17*, 7769.

Prüss-Üstün, A., Wolf, J., Corvalan, C., Bos, R., & Neira, M. (2016). *Preventing disease through healthy environments: A global assessment of the burden of disease from environmental risks*. World Health Organization.

Public Health England. (n.d.). *Health matters: Air pollution* [WWW Document]. GOV.UK. Retrieved April 13, 2021, from https://www.gov.uk/government/publications/health-matters-air-pollution/health-matters-air-pollution

Qin, B., & Zhang, Y. (2014). Note on urbanization in China: Urban definitions and census data. *China Economic Review, 30*, 495–502. https://doi.org/10.1016/j.chieco.2014.07.008

Radicchi, A., Henckel, D., & Memmel, M. (2016). Citizens as smart, active sensors for a quiet and just city. The case of the "open source soundscapes" approach to identify, assess and plan "everyday quiet areas" in cities. *Noise Mapping, 5*, 1–20.

Radicchi, A., Yelmi, P. C., Chung, A., Jordan, P., Stewart, S., Tsaligopoulos, A., McCunn, L., & Grant, M. (2021). Sound and the healthy city. *Cities Health, 5*, 1–13.

Rapkins, C. (2017). *Barton healthy new town*. Oxford City Council, Final Report.

Raworth, K. (2017). *Doughnut economics: Seven ways to think like a 21st-century economist*. Random House Business Books.

Rayner, G., & Lang, T. (2012). *Ecological public health: Reshaping the conditions for good health*. Routledge.

Reed, R., Bilos, A., Wilkinson, S., & Schulte, K.-W. (2009). International comparison of sustainable rating tools. *Journal of Sustainable Real Estate, 1*, 1–22.

Register, R. (1987). *Ecocity Berkeley: Building cities for a healthy future*. North Atlantic Books.

Retzlaff, R. C. (2009). The use of LEED in planning and development regulation: An exploratory analysis. *Journal of Planning Education and Research, 29*, 67–77.

RIBA, Hay, R., Bradbury, S., Dixon, D., Martindale, K., Samuel, F., & Tait, A. (2017). *Building knowledge: Pathways to post occupancy evaluation: Value of architects*. University of Reading.
Rice, J. L. (2010). Climate, carbon, and territory: Greenhouse gas mitigation in Seattle, Washington. *Annals of the Association of American Geographers, 100*, 929–937.
Richardson, B. W. (1876). *Hygeia, a city of health*. Macmillan.
Richardson, M., Maspero, M., Golightly, D., Sheffield, D., Staples, V., & Lumber, R. (2017). Nature: A new paradigm for well-being and ergonomics. *Ergonomics, 60*, 292–305.
Ritchie, H., & Roser, M. (2017). Air pollution [WWW Document]. *Our World Data*. Retrieved April 13, 2021, from https://ourworldindata.org/air-pollution
Ritchie, H., & Roser, M. (n.d.). Emissions by sector [WWW Document]. *Our World Data*. Retrieved April 16, 2021, from https://ourworldindata.org/emissions-by-sector
Robinson, J., & Cole, R. J. (2015). Theoretical underpinnings of regenerative sustainability. *Building Research and Information, 43*, 133–143.
Rockström, J., Steffen, W., Noone, K., Persson, Å., Chapin, F. S. I., Lambin, E., Lenton, T. M., Scheffer, M., Folke, C., Schellnhuber, H., Nykvist, B., De Wit, C. A., Hughes, T., van der Leeuw, S., Rodhe, H., Sörlin, S., Snyder, P. K., Costanza, R., Svedin, U., … Foley, J. (2009). Planetary boundaries: Exploring the safe operating space for humanity. *Ecology and Society, 14*, 32.
Romanelli, C., Cooper, H. D., Campbell-Lendrum, D., Maiero, M., Karesh, W. B., Hunter, D., & Golden, C. D. (2015). *Connecting global priorities: Biodiversity and human health, a state of knowledge review*. World Health Organization and Secretariat for the Convention on Biological Diversity.
Rosberg, G. (n.d.). Western Harbour—A new sustainable city district in Malmö.
Ross, S. E., Flynn, J. I., & Pate, R. R. (2016). What is really causing the obesity epidemic? A review of reviews in children and adults. *Journal of Sports Sciences, 34*, 1148–1153.
Roth, G. A., Abate, D., Abate, K. H., Abay, S. M., Abbafati, C., Abbasi, N., Abbastabar, H., Abd-Allah, L., Abdela, J., Abdelalim, A., Abdollahpour, I., Abdulkader, R. S., Abebe, H. T., Abebe, M., Abebe, Z., Abejie, A. N., Abera, S. F., Abil, O. Z., Abraha, H. N., & Murray, C. J. L. (2018). Global, regional, and national age-sex-specific mortality for 282 causes of death in 195 countries and territories, 1980–2017: A systematic analysis for the Global Burden of Disease Study 2017. *The Lancet, 392*, 1736–1788.
Rothman, L., Howard, A., Buliung, R., Macarthur, C., Richmond, S. A., & Macpherson, A. (2017). School environments and social risk factors for child pedestrian-motor vehicle collisions: A case-control study. *Accident; Analysis and Prevention, 98*, 252–258.
Rudge, G. M., Suglani, N., Saunders, P., & Middleton, J. (2013). OP24 are fast food outlets concentrated in more deprived areas? A geo-statistical analysis of an urban area in Central England. *Journal of Epidemiology and Community Health, 67*(A14), 1–A14.
Rutter, H. (2018). The complex systems challenge of obesity. *Clinical Chemistry, 64*, 44–46.
Rydin, Y. (2010). *Governing for sustainable urban development* (1st ed.). Earthscan.
Rydin, Y. (2013). *The future of planning: Beyond growth dependence*. Policy Press.
Rydin, Y., Bleahu, A., Davies, M., Dávila, J. D., Friel, S., De Grandis, G., Groce, N., Hallal, P. C., Hamilton, I., Howden Chapman, P., Ka-Man Lai, C. J., Lim, J. M., Osrin, D., Ridley, I., Scott, I., Taylor, M., Wilkinson, P., & Wilson, J. (2012). Shaping cities for health: Complexity and the planning of urban environments in the 21st century. *The Lancet, 379*, 2079–2108.
Sabel, C., Amegbor, P. M., Zhang, Z., Chen, T.-H. K., Poulsen, M., Hertel, O., Sigsgaard, T., Horsdal, H. T., Pedersen, C. B., & Khan, J. (2021). Urban health and wellbeing. In W. Shi, M. F. Goodchild, M. Batty, M.-P. Kwan, & A. Zhang (Eds.), *Urban informatics* (pp. 259–280). Springer Nature.
Saelens, B. E., Sallis, J. F., & Frank, L. D. (2016). Environmental correlates of walking and cycling: Findings from the transportation, urban design, and planning literatures. *Annals of Behavioral Medicine, 25*, 80–91.
Saelens, B. E., Sallis, J. F., Frank, L. D., Couch, S. C., Zhou, C., Colburn, T., Cain, K. L., Chapman, J., & Glanz, K. (2012). Obesogenic neighborhood environments, child and parent obesity. *American Journal of Preventive Medicine, 42*, e57–e64.

Sahlin, E., Ahlborg, G., Tenenbaum, A., & Grahn, P. (2015). Using nature-based rehabilitation to restart a stalled process of rehabilitation in individuals with stress-related mental illness. *International Journal of Environmental Research and Public Health, 12*, 1928–1951.

Sallis, J. F., Cerin, E., Conway, T. L., Adams, M. A., Frank, L. D., Pratt, M., Salvo, D., Schipperijn, J., Smith, G., Cain, K. L., Davey, R., Kerr, J., Lai, P.-C., Mitáš, J., Reis, R., Sarmiento, O. L., Schofield, G., Troelsen, J., Van Dyck, D., … Owen, N. (2016). Physical activity in relation to urban environments in 14 cities worldwide: A cross-sectional study. *The Lancet, 387*, 2207–2217.

Sanbonmatsu, L., Marvakov, J., Potter, N. A., Yang, F., Adam, E. K., Congdon, W. J., Duncan, G. J., Gennetian, L. A., Katz, L. F., Kling, J. R., Kessler, R. C., Lindau, S. T., Ludwig, J., & McDade, T. W. (2012). The long-term effects of moving to opportunity on adult health and economic self-sufficiency. *City, 14*, 109–136.

Sanchez-Guevara, C., Núñez Peiró, M., Taylor, J., Mavrogianni, A., & Neila González, J. (2019). Assessing population vulnerability towards summer energy poverty: Case studies of Madrid and London. *Energy and Buildings, 190*, 132–143.

Santosa, A., Wall, S., Fottrell, E., Högberg, U., & Byass, P. (2014). The development and experience of epidemiological transition theory over four decades: A systematic review. *Global Health Action, 7*, 23574.

Satterthwaite, D., Archer, D., Colenbrander, S., Dodman, D., Hardoy, J., & Patel, S. (2018). *Responding to climate change in cities and in their informal settlements and economies*. International Institute for Environment and Development.

Saunders, P., Saunders, A., & Middleton, J. (2015). Living in a 'fat swamp': Exposure to multiple sources of accessible, cheap, energy-dense fast foods in a deprived community. *The British Journal of Nutrition, 113*, 1828–1834.

Savaget, P., Geissdoerfer, M., Kharrazi, A., & Evans, S. (2019). The theoretical foundations of sociotechnical systems change for sustainability: A systematic literature review. *Journal of Cleaner Production, 206*, 878–892.

Savoie-Roskos, M. R., Wengreen, H., & Durward, C. (2017). Increasing fruit and vegetable intake among children and youth through gardening-based interventions: A systematic review. *Journal of the Academy of Nutrition and Dietetics, 117*, 240–250.

Schnake-Mahl, A. S., Jahn, J. L., Subramanian, S. V., Waters, M. C., & Arcaya, M. (2020). Gentrification, neighborhood change, and population health: A systematic review. *Journal of Urban Health, 97*, 1–25.

Schug, G. R., Blevins, K. E., Cox, B., Gray, K., & Mushrif-Tripathy, V. (2013). Infection, disease, and biosocial processes at the end of the Indus civilization. *PLoS One San Franc., 8*, e84814.

Schultz, J. M. (2019). Disasters. In S. Galea, C. K. Ettman, & D. Vlahov (Eds.), *Urban health* (pp. 156–165). Oxford University Press.

Schulz, A., & Northridge, M. E. (2004). Social determinants of health: Implications for environmental health promotion. *Health Education & Behavior, 31*, 455–471.

Seaman, R., Mitchell, R., Dundas, R., Leyland, A. H., & Popham, F. (2015). How much of the difference in life expectancy between Scottish cities does deprivation explain? *BMC Public Health Lond., 15*, 1057.

Seattle Department of Construction & Inspections. (n.d.). Living building & 2030 challenge pilots. Retrieved March 22, 2021, from https://www.seattle.gov/sdci/permits/green-building/living-building-and-2030-challenge-pilots

Sellers, S., Ebi, K. L., & Hess, J. (2019). Climate change, human health, and social stability: Addressing interlinkages. *Environmental Health Perspectives, 127*, 045002.

Sen, A. (1999). *Development as freedom*. Oxford University Press.

Sengers, F., Wieczorek, A. J., & Raven, R. (2019). Experimenting for sustainability transitions: A systematic literature review. *Technological Forecasting and Social Change, 145*, 153–164.

Seto, K. C., Dhakal, S., Bigio, A., Blanco, H., Delgado, G. C., Dewar, D., … Zwickel, T. (Eds.). (2014). Climate change 2014: Mitigation of climate change. In *Contribution of Working*

Group III to the Fifth Assessment Report of the Intergovernmental Panel on Climate Change. Cambridge University Press, Cambridge, United Kingdom and New York, NY, USA.

Shackelford, B. B., Cronk, R., Behnke, N., Cooper, B., Tu, R., D'Souza, M., Bartram, J., Schweitzer, R., & Jaff, D. (2020). Environmental health in forced displacement: A systematic scoping review of the emergency phase. *Science of the Total Environment, 714*, 136553.

Shanahan, D. F., Fuller, R. A., Bush, R., Lin, B. B., & Gaston, K. J. (2015). The health benefits of urban nature: How much do we need? *Bioscience, 65*, 476–485.

Shannon, H., Allen, C., Clarke, M., Dávila, D., Fletcher-Wood, L., Gupta, S., Keck, K., Lang, S., & Allen Kahangire, D. (2018). Web Annex A: Report of the systematic review on the effect of household crowding on health, In *WHO housing and health guidelines*. World Health Organization, Geneva, Switzerland.

Shaw, M. (2004). Housing and public health. *Annual Review of Public Health, 25*, 397–418.

Shrubsole, C., Macmillan, A., Davies, M., & May, N. (2014). 100 Unintended consequences of policies to improve the energy efficiency of the UK housing stock. *Indoor and Built Environment, 23*, 340–352.

Silveira, D. C., Carmo, R. F., & Luz, Z. M. P. (2019). Planning in four areas of the Vila Viva Program in the city of Belo Horizonte, Brazil: A documentary analysis. *Ciência & Saúde Coletiva, 24*, 1165–1174.

Singer, M., & Clair, S. (2003). Syndemics and public health: Reconceptualizing disease in bio-social context. *Medical Anthropology Quarterly, 17*, 423–441.

Singh, A., Daniel, L., Baker, E., & Bentley, R. (2019). Housing disadvantage and poor mental health: A systematic review. *American Journal of Preventive Medicine, 57*, 262–272.

Sisk, A., MacLeish-White, O., Gavin, V., Butler, T., Ogbu, L., Davis, V. O., Chaudhury, N., Hamdi, H., Worden, K., Kabane, N., Poticha, S., & Pathuis, H. (2020). Confronting power and privilege for inclusive, equitable, and healthy communities. *The BMJ*. Retrieved June 23, 2021, from https://blogs.bmj.com/bmj/2020/04/16/confronting-power-and-privilege-for-inclusive-equitable-and-healthy-communities/

Sletten, T. L., Ftouni, S., Nicholas, C. L., Magee, M., Grunstein, R. R., Ferguson, S., Kennaway, D. J., O'Brien, D., Lockley, S. W., & Rajaratnam, S. M. W. (2017). Randomised controlled trial of the efficacy of a blue-enriched light intervention to improve alertness and performance in night shift workers. *Occupational and Environmental Medicine, 74*, 792–801.

Smith, M. (2019). International poll: Most expect to feel impact of climate change, many think it will make us extinct. *YouGov.co.uk*.

Smith, M. E., & Hein, C. (2017). The ancient past in the urban present. In C. Hein (Ed.), *The Routledge handbook of planning history* (1st ed., pp. 109–120). Routledge.

Snow, J. (1854). The Cholera near golden square, and at Deptford, letter to the editor. *Medical times and Gazette, 9*, 321–322.

Stanaway, J. D., Afshin, A., Gakidou, E., Lim, S. S., Abate, D., Abate, K. H., Abbafati, C., Abbasi, N., Abbastabar, H., Abd-Allah, F., Abdela, J., Abdelalim, A., Abdollahpour, I., Abdulkader, R. S., Abebe, M., Abebe, Z., Abera, S. F., Abil, O. Z., Abraha, H. N., … Murray, C. J. L. (2018). Global, regional, and national comparative risk assessment of 84 behavioural, environmental and occupational, and metabolic risks or clusters of risks for 195 countries and territories, 1990–2017: A systematic analysis for the Global Burden of Disease Study 2017. *The Lancet, 392*, 1923–1994.

Steg, L., Bolderdijk, J. W., Keizer, K., & Perlaviciute, G. (2014). An integrated framework for encouraging pro-environmental behaviour: The role of values, situational factors and goals. *Journal of Environmental Psychology, 38*, 104–115.

Sterman, J. D. (2000). *Business dynamics: Systems thinking and modeling for a complex world*. Irwin/McGraw-Hill.

Sterman, J. D. (2006). Learning from evidence in a complex world. *American Journal of Public Health, 96*, 505–514.

Stiglitz, J. E., Sen, A., & Fitoussi, J. -P. (2017). *Report by the commission on the measurement of economic performance and social progress*. French National Institute of Statistics and Economic Studies.

Stigsdotter, U. K., Palsdottir, A. M., Burls, A., Chermaz, A., Ferrini, F., & Grahn, P. (2011). Nature-based therapeutic interventions. In K. Nilsson, M. Sangster, C. Gallis, T. Hartig, S. de Vries, K. Seeland, & J. Schipperijn (Eds.), *Forests, trees and human health* (pp. 309–342). Springer.

Stone, S. B., Myers, S. S., & Golden, C. D. (2018). Cross-cutting principles for planetary health education. *Lancet Planet Health, 2*, e192–e193.

Sundell, J. (2017). Reflections on the history of indoor air science, focusing on the last 50 years. *Indoor Air, 27*, 708–724.

Talen, E. (2013). *Charter of the new urbanism: Congress for the new urbanism* (2nd ed.). McGraw-Hill Education.

Tan, D. T., Siri, J. G., Gong, Y., Ong, B., Lim, S. C., MacGillivray, B. H., & Marsden, T. (2019). Systems approaches for localising the SDGs: Co-production of place-based case studies. *Globalization and Health, 15*, 85.

Tang, Y.-T., Chan, F. K. S., O'Donnell, E. C., Griffiths, J., Lau, L., Higgitt, D. L., & Thorne, C. R. (2018). Aligning ancient and modern approaches to sustainable urban water management in China: Ningbo as a "Blue-Green City" in the "Sponge City" campaign. *Journal of Flood Risk Management, 11*, e12451.

Thompson, C., Cummins, S., Brown, T., & Kyle, R. (2013). Understanding interactions with the food environment: An exploration of supermarket food shopping routines in deprived neighbourhoods. *Health & Place, 19*, 116–123.

Thomson, H., Atkinson, R., Petticrew, M., & Kearns, A. (2006). Do urban regeneration programmes improve public health and reduce health inequalities? A synthesis of the evidence from UK policy and practice (1980–2004). *Journal of Epidemiology and Community Health, 60*, 108–115.

Thomson, H., & Thomas, S. (2015). Developing empirically supported theories of change for housing investment and health. *Social Science & Medicine, 124*, 205–214.

Tiesdell, S., & Oc, T. (1998). Beyond 'Fortress' and 'Panoptic' Cities—Towards a Safer Urban Public Realm. *Environment and Planning B: Planning and Design, 25*, 639–655.

Tonne, C., Beevers, S., Armstrong, B., Kelly, F., & Wilkinson, P. (2008). Air pollution and mortality benefits of the London Congestion Charge: Spatial and socioeconomic inequalities. *Occupational and Environmental Medicine, 65*, 620–627.

Townshend, T., & Lake, A. (2017). Obesogenic environments: Current evidence of the built and food environments. *Perspectives in Public Health, 137*, 38–44.

Transport for London. (2017). FOI request detail. Retrieved April 16, 2021, from https://www.tfl.gov.uk/corporate/transparency/freedom-of-information/foi-request-detail

Trasande, L., Zoeller, R. T., Hass, U., Kortenkamp, A., Grandjean, P., Myers, J. P., DiGangi, J., Bellanger, M., Hauser, R., Legler, J., Skakkebaek, N. E., & Heindel, J. J. (2015). Estimating Burden and Disease Costs of Exposure to Endocrine-Disrupting Chemicals in the European Union. *The Journal of Clinical Endocrinology and Metabolism, 100*, 1245–1255.

Tulier, M. E., Reid, C., Mujahid, M. S., & Allen, A. M. (2019). "Clear action requires clear thinking": A systematic review of gentrification and health research in the United States. *Health & Place, 59*, 102173.

Turner, M., Kooshian, C., & Winkelman, S. (2012). *Colombia's Bus Rapid Transit (BRT) development and expansion*. Report produced for Mitigation Action Implementation Network (MAIN) by CCAP.

Twohig-Bennett, C., & Jones, A. (2018). The health benefits of the great outdoors: A systematic review and meta-analysis of greenspace exposure and health outcomes. *Environmental Research, 166*, 628–637.

Ulrich, R. S., Zimring, C., Zhu, X., DuBose, J., Seo, H.-B., Choi, Y.-S., Quan, X., & Joseph, A. (2008). A review of the research literature on evidence-based healthcare design. *HERD Health Environ. Res. Des. J., 1*, 61–125.

UNEP, International Energy Agency. (2017). *Towards a zero-emission, efficient, and resilient buildings and construction sector*. Global Status Report 2017. UNEP.
UNESCO. (2001). *UNESCO universal declaration on cultural diversity*.
UNFCCC. (2010). *CDM project co-benefits in Bogotá*. Rapid and reliable bus transport for urban communities, Colombia.
UN-Habitat, World Health Organization. (2020). *Integrating health in urban and territorial planning: A sourcebook*. UN-Habitat and World Health Organization, Geneva, Switzerland.
United Nations. (2015). *The millennium development goals report*. United Nations, New York.
United Nations, Department of Economic and Social Affairs, Population Division. (2019a). World urbanization prospects: The 2018 revision, File 3: Urban Population at Mid-Year by Region, Subregion, Country and Area, 1950–2050 (thousands).
United Nations, Department of Economic and Social Affairs, Population Division. (2019b). World population prospects 2019: Volume II: Demographic profiles.
United Nations, Department of Economic and Social Affairs, Population Division. (2019c). World population prospects 2019: Highlights (ST/ESA/SER.A/423).
United Nations Environment Programme. (2021a). *Making peace with nature: A scientific blueprint to tackle the climate, biodiversity and pollution emergencies*. UNEP, Nairobi, Kenya.
United Nations Environment Programme. (2021b). *Beating the heat: A sustainable cooling handbook for cities*. UNEP, Nairobi, Kenya.
United Nations General Assembly. (2015). Resolution adopted by the general assembly on 25 September 2015: Transforming our world: The 2030 Agenda for Sustainable Development. United Nations.
Urban Land Institute. (2015). *ULI case studies*. Bullitt Center. Retrieved September 07, 2020 from https://casestudies.uli.org/wp-content/uploads/2015/02/TheBullittCenter1.pdf
Utzet, M., Valero, E., Mosquera, I., & Martin, U. (2020). Employment precariousness and mental health, understanding a complex reality: A systematic review. *International Journal of Occupational Medicine and Environmental Health, 33*, 569–598.
Van Buren, N., Demmers, M., Van der Heijden, R., & Witlox, F. (2016). Towards a circular economy: The role of Dutch logistics industries and governments. *Sustainability, 8*, 647.
van den Bosch, M., & Ode Sang, Å. (2017). Urban natural environments as nature-based solutions for improved public health—A systematic review of reviews. *Environmental Research, 158*, 373–384.
van der Noordt, M., IJzelenberg, H., Droomers, M., & Proper, K. I. (2014). Health effects of employment: A systematic review of prospective studies. *Occupational and Environmental Medicine, 71*, 730–736.
van Hoof, J., Kazak, J. K., Perek-Białas, J. M., & Peek, S. T. M. (2018). The challenges of urban ageing: Making cities age-friendly in Europe. *International Journal of Environmental Research and Public Health, 15*, 2473.
van Kamp, I., Leidelmeijer, K., Marsman, G., & de Hollander, A. (2003). Urban environmental quality and human well-being: Towards a conceptual framework and demarcation of concepts; a literature study. *Landscape and Urban Planning*, Urban environmental quality and human wellbeing, *65*, 5–18.
Van Renterghem, T., & Botteldooren, D. (2016). View on outdoor vegetation reduces noise annoyance for dwellers near busy roads. *Landscape and Urban Planning, 148*, 203–215.
Vásquez-Vera, H., Palència, L., Magna, I., Mena, C., Neira, J., & Borrell, C. (2017). The threat of home eviction and its effects on health through the equity lens: A systematic review. *Social Science & Medicine, 175*, 199–208.
Vaughan, A. (2020). Landmark ruling says air pollution contributed to death of 9-year-old. *New Scientist*. December 16, 2020. Retrieved from https://www.newscientist.com/article/2263165-landmark-ruling-says-air-pollution-contributed-to-death-of-9-year-old/
Vincent, J. M. (2014). Joint use of public schools: A framework for promoting healthy communities. *Journal of Planning Education and Research, 34*, 153–168.

Viola, A. U., James, L. M., Schlangen, L. J., & Dijk, D.-J. (2008). Blue-enriched white light in the workplace improves self-reported alertness, performance and sleep quality. *Scandinavian Journal of Work, Environment & Health, 34*, 297–306.

Vlachokostas, C., Banias, G., Athanasiadis, A., Achillas, C., Akylas, V., & Moussiopoulos, N. (2014). Cense: A tool to assess combined exposure to environmental health stressors in urban areas. *Environment International, 63*, 1–10.

Vlahov, D., Galea, S., & Freudenberg, N. (2005). The urban health "Advantage". *Journal of Urban Health, 82*, 1–4.

Walch, J. M., Rabin, B. S., Day, R., Williams, J. N., Choi, K., & Kang, J. D. (2005). The effect of sunlight on postoperative analgesic medication use: A prospective study of patients undergoing spinal surgery. *Psychosomatic Medicine, 67*, 156–163.

Walker, M. (2021). How a diverse coalition in Portland, Ore. is centering racial equity in a large-scale development project. *Brookings*. Retrieved July 28, 2021, from https://www.brookings.edu/blog/the-avenue/2021/07/12/how-a-diverse-coalition-in-portland-ore-is-centering-racial-equity-in-a-large-scale-development-project/

Wang, L., Zhang, F., Pilot, E., Yu, J., Nie, C., Holdaway, J., Yang, L., Li, Y., Wang, W., Vardoulakis, S., & Krafft, T. (2018). Taking action on air pollution control in the Beijing-Tianjin-Hebei (BTH) region: Progress, challenges and opportunities. *International Journal of Environmental Research and Public Health, 15*, 306.

Wang, Q., Deng, Y., Li, G., Meng, C., Xie, L., Liu, M., & Zeng, L. (2020). The current situation and trends of healthy building development in China. *Chinese Science Bulletin, 65*, 246–255.

Wang, Q., Meng, C., & Li, G. (2017). Development demands and prospect of healthy buildings. *Heating, Ventilation and Air-Conditioning, 47*, 32–35.

Watts, N., Amann, M., Arnell, N., Ayeb-Karlsson, S., Beagley, J., Belesova, K., Boykoff, M., Byass, P., Cai, W., Campbell-Lendrum, D., Capstick, S., Chambers, J., Coleman, S., Dalin, C., Daly, M., Dasandi, N., Dasgupta, S., Davies, M., Di Napoli, C., … Costello, A. (2021). The 2020 report of The Lancet Countdown on health and climate change: Responding to converging crises. *The Lancet, 397*, 129–170.

Weller, S. J., Crosby, L. J., Turnbull, E. R., Burns, R., Miller, A., Jones, L., & Aldridge, R. W. (2019). The negative health effects of hostile environment policies on migrants: A cross-sectional service evaluation of humanitarian healthcare provision in the UK. *Wellcome Open Res., 4*, 109.

Whitehead, S. J., & Ali, S. (2010). Health outcomes in economic evaluation: The QALY and utilities. *British Medical Bulletin, 96*, 5–21.

Whitmee, S., Haines, A., Beyrer, C., Boltz, F., Capon, A. G., de Souza Dias, B. F., Ezeh, A., Frumkin, H., Gong, P., Head, P., Horton, R., Mace, G. M., Marten, R., Myers, S. S., Nishtar, S., Osofsky, S. A., Pattanayak, S. K., Pongsiri, M. J., Romanelli, C., … Yach, D. (2015). Safeguarding human health in the Anthropocene epoch: Report of The Rockefeller Foundation–Lancet Commission on planetary health. *The Lancet, 386*, 1973–2028.

WHO. (1946). Preamble to the Constitution of the World Health Organization as adopted by the International Health Conference, 19–22 June 1946; and entered into force on 7 April 1948. WHO, New York.

WHO. (1986). *Ottawa charter for health promotion*. World Health Organization, Geneva, Switzerland.

WHO. (2007). *Global age-friendly cities: A guide*. WHO, Geneva, Switzerland.

WHO. (2012). *Addressing the social determinants of health: The urban dimension and the role of local government*. WHO, Copenhagen, Denmark.

WHO. (2013). *Review of evidence on health aspects of air pollution—REVIHAAP project: Technical report*. WHO Regional Office for Europe.

WHO. (2014). *Global status report on noncommunicable diseases 2014: Attaining the nine global noncommunicable diseases targets; a shared responsibility*. WHO, Geneva, Switzerland.

WHO. (2016a). *Global report on urban health: Equitable, healthier cities for sustainable development*. WHO, Geneva, Switzerland.

WHO. (2016b). *Waste and human health: Evidence and needs*. WHO Meeting Report, 5–6 November 2015, Bonn, Germany. WHO Regional Office for Europe, Copenhagen, Denmark.

WHO. (2018a). *Global status report on road safety*. WHO, Geneva, Switzerland.

WHO. (2018b). *WHO housing and health guidelines*. WHO, Geneva, Switzerland.

WHO. (2018c). *Air pollution and child health: Prescribing clean air*. WHO, Geneva, Switzerland.

WHO. (2018d). *Global status report on road safety*. WHO, Geneva, Switzerland.

WHO. (2018e). *WHO housing and health guidelines*. WHO, Geneva, Switzerland.

WHO. (2018f). *Air pollution and child health: Prescribing clean air*. WHO, Geneva, Switzerland.

WHO. (2018g). *Global action plan on physical activity 2018–2030: More active people for a healthier world*. WHO, Geneva, Switzerland.

WHO. (2018h). *The global network for age-friendly cities and communities: Looking back over the last decade, looking forward to the next*. WHO, Geneva, Switzerland.

WHO. (2018i). *Environmental noise guidelines for the European Region*. WHO Regional Office for Europe, Copenhagen, Denmark.

WHO. (2020). *Personal interventions and risk communication on air pollution*. WHO, Geneva, Switzerland.

WHO. (2021a). *WHO global air quality guidelines: Particulate matter (PM2.5 and PM10), ozone, nitrogen dioxide, sulfur dioxide and carbon monoxide*. WHO, Geneva, Switzerland.

WHO. (2021b). *Policies, regulations and legislation promoting healthy housing: A review*. WHO, Geneva, Switzerland.

WHO, United Nations Children's Fund. (2017). *Progress on drinking water, sanitation and hygiene. 2017 update and SDG baseline*. WHO and United Nations Children's Fund, Geneva, Switzerland.

WHO, United Nations Human Settlements Programme (Eds.). (2010). *Hidden cities: Unmasking and overcoming health inequities in urban settings*. WHO; UN-HABITAT, Kobe, Japan.

Wilkie, S., Townshend, T., Thompson, E., & Ling, J. (2018). Restructuring the built environment to change adult health behaviors: A scoping review integrated with behavior change frameworks. *Cities Health, 2*, 198–211.

Wilkinson, S., Remøy, H. T., & Langston, C. (2014). *Sustainable building adaptation: Innovations in decision-making, Innovation in the built environment*. Wiley-Blackwell.

Williams, J. (2016). *Circular cities: Strategies, challenges and knowledge gaps*. Circular Cities Hub.

Williams, J. (2019). Circular cities: Challenges to implementing looping actions. *Sustainability, 11*, 423.

Won, J., Lee, C., Forjuoh, S. N., & Ory, M. G. (2016). Neighborhood safety factors associated with older adults' health-related outcomes: A systematic literature review. *Social Science & Medicine, 165*, 177–186.

Wong, K. V., Paddon, A., & Jimenez, A. (2013). Review of world urban heat islands: Many linked to increased mortality. *Journal of Energy Resources Technology, 135*, 11.

World Bank. (2020). *Handbook for gender-inclusive urban planning and design*. World Bank Group.

World GBC. (2014). *Health, wellbeing and productivity in offices—The next chapter for green building*.

World GBC. (2016). *Building the business case: Health, wellbeing and productivity in Green Offices*. World Green Building Council, Toronto.

World GBC. (2020). *Health & wellbeing framework: Six principles for a healthy, sustainable built environment*. Executive Report.

Wu, X., Nethery, R. C., Sabath, M. B., Braun, D., & Dominici, F. (2020). Air pollution and COVID-19 mortality in the United States: Strengths and limitations of an ecological regression analysis. *Science Advances, 6*, eabd4049.

Yang, J., Siri, J. G., Remais, J. V., Cheng, Q., Zhang, H., Chan, K. K. Y., Sun, Z., Zhao, Y., Cong, N., Li, X., Zhang, W., Bai, Y., Bi, J., Cai, W., Chan, E. Y. Y., Chen, W., Fan, W., Hua, F.,

He, J., … Gong, P. (2018). The Tsinghua—Lancet commission on healthy cities in China: Unlocking the power of cities for a healthy China. *The Lancet, 391*, 2140–2184.

Yang, K., & Victor, C. (2011). Age and loneliness in 25 European nations. *Ageing and Society, 31*, 1368–1388.

Yolande, M., Kerry, F., Paul, H., Louise, U., Sian, M., & George, K. (2014). Contaminated land and public health. In N. Bradley, H. Harrison, G. Hodgson, R. Kamanyire, A. Kibble, & V. Murray (Eds.), *Essentials of environmental public health science* (pp. 115–145). Oxford University Press.

Younger, M., Morrow-Almeida, H. R., Vindigni, S. M., & Dannenberg, A. L. (2008). The built environment, climate change, and health. *American Journal of Preventive Medicine, 35*, 517–526.

Zhu, Y., Xie, J., Huang, F., & Cao, L. (2020). Association between short-term exposure to air pollution and COVID-19 infection: Evidence from China. *Science of the Total Environment, 727*, 138704.

Index

H. Pineo, *Healthy Urbanism*, Planning, Environment, Cities,
https://doi.org/10.1007/978-981-16-9647-3

Printed by Printforce, United Kingdom